# *Salmonella* spp. - A Global Challenge

*Edited by Alexandre Lamas, Patricia Regal and Carlos Manuel Franco*

Published in London, United Kingdom

IntechOpen

*Supporting open minds since 2005*

*Salmonella* spp. - A Global Challenge
http://dx.doi.org/10.5772/intechopen.91517
Edited by Alexandre Lamas, Patricia Regal and Carlos Manuel Franco

Contributors
Sarika kombade, Navneet Kaur, Hidajah Rachmawati, Raditya Weka Nugraheni, Firasti Agung Nugrahening Sumadi, Mamta Pandey, Emmagouni Sharath Kumar Goud, Weam S. Al-Hamadany, Abdulmaleek Idanesimhe Sado, Aduke Sado, Delia Lacasta, Inés Rubira, Luis Pedro Figueras, José Calasanz Jiménez, Marta Ruiz de Arcaute, Héctor Ruíz, José Antonio Ventura, Noor S.K Salman Al-Khafaji, Hussein O. M. Al-Dahmoshi, Ali M.K. Al-Bayati, Chandrajit Lahiri, Rohit Mishra, Yong Chiang Tan, Amr Adel Ahmed Abd El-Aal, Julia D. Toscano-Garibay, Nancy Jannet Ruiz-Pérez, Jaime Sánchez-Navarrete, Dele Ogunremi, Ruimin Gao, Rosemary Slowey, Shu Chen, Olga Andrievskaia, Sadjia Bekal, Lawrence Goodridge, Roger C. Levesque, Martha R.J. Clokie, Anisha M. Thanki, Steve Hooton, Robert J. Atterbury, Adriano M. Gigante, Alexandre Lamas, Patricia Regal, Carlos Franco, Aberto Cepeda, Aroa López-Santamarina, Laura Sanjulián



First published in London, United Kingdom, 2021 by IntechOpen
IntechOpen is the global imprint of INTECHOPEN LIMITED, registered in England and Wales, registration number: 11086078, 5 Princes Gate Court, London, SW7 2QJ, United Kingdom
Printed in Croatia

British Library Cataloguing-in-Publication Data
A catalogue record for this book is available from the British Library

Additional hard and PDF copies can be obtained from orders@intechopen.com

*Salmonella* spp. - A Global Challenge
Edited by Alexandre Lamas, Patricia Regal and Carlos Manuel Franco
p. cm.
Print ISBN 978-1-83969-017-4
Online ISBN 978-1-83969-018-1
eBook (PDF) ISBN 978-1-83969-019-8

# Meet the editors

Alexandre Lamas is a postdoctoral fellow in the Department of Analytical Chemistry, Nutrition and Bromatology, University of Santiago de Compostela, Spain. He obtained a Ph.D. in Innovation in Food Safety and Technologies from the same university in 2019. He is the author of more than thirty-five scientific articles in international peer-reviewed journals as well as four book chapters and two chapters in encyclopedias. He is an active reviewer for different journals of food science and microbiology and a member of the editorial board of two international journals. His research focuses on the study of foodborne pathogens, especially Salmonella, new control methods, the study of food microbiota, and more recently the study of foodborne pathogens bacteriophages.

Carlos Manuel Franco is a full professor . He obtained a Ph.D. in Nutrition and Bromatology from the same university in 1994. He is the author of more than 120 international publications in scientific journals and several book chapters. He is currently a member of the scientific committee of the Spanish Security Agency Food and Nutrition (AESAN). He is the editor-in-chief and editorial board member of diverse food science and microbiology journals. His research is focused on new methods of food control, detection of pathogenic bacteria in food and other matrices, and the effect of various substances on pathogenic bacteria.

Patricia Regal is a full-time assistant professor in the Department of Analytical Chemistry, Nutrition and Bromatology, University of Santiago de Compostela, Spain. She graduated in Veterinary Science from the Faculty of Veterinary Science of Lugo, in 2007, and obtained her Ph.D. (cum laude) in 2012. Dr. Regal has published more than fifty scientific articles in prestigious international journals. She has co-authored four book chapters, various open-access articles, national contributions, and two chapters in encyclopedias. She has wide demonstrable experience in the field of food safety, food composition, omic technologies, biostatistics, biomarkers discovery, and chromatographic techniques and mass spectrometry applied to biological matrices.

# Contents

# Preface

*Salmonella* spp. is a global pathogen responsible for millions of deaths each year. *Salmonella* is a complex genus comprising two species, S. enterica and S. bongori, and more than 2600 different serotypes. S. enterica is composed of six different subspecies: enterica (I), salamae (II), arizonae (IIIa), diarizonae (IIIb), houtenae (IV), and indica (VI). Serotypes of S. enterica subsp. enterica are responsible for more than 99% of human infections. Another way to classify *Salmonella* is according to the type of disease it causes. In general terms, there is typhoid fever caused by S. Typhi and infections caused by non-typhoidal *Salmonella*.

Typhoid fever is caused by the ingestion of water or food contaminated with fecal material carrying S. Typhi. As described by Al-Khafaji et al. in Chapter 1, a tool of mechanisms is responsible for the virulence of S. Typhi. The Vi capsule antigen can inhibit phagocytosis and complement C3. This capsule decreases the recognition of somatic antigens by antibodies. Flagella also contribute to virulence by interacting with host epithelial cells, macrophages, and immune evasion. S. Typhi also encodes in its genome some pathogenic island with the genes necessary to invade the host and effectively evade the immune system. Although typhoid fever is distributed worldwide, improvements in water supply and sewerage systems have resulted in a decreased incidence.

In low- and middle-income countries where sanitary conditions continue to be a problem, typhoid fever is still highly prevalent. In developed countries, typhoid fever is a travel-related disease. It is estimated that there are 26.9 million S. Typhi infections annually. However, these data are misleading. The fact that this disease presents non-specific symptoms along with the lack of diagnostic tests and under-reporting of cases in some regions of the world suggests that the real prevalence of this disease is much greater. In Chapter 2, Sado and Sado describe the importance of enteric fever diagnosis in primary care. The development of nonspecific symptoms complicates disease diagnosis. It is therefore important to follow a series of criteria established by health authorities to facilitate diagnosis.

Prevention is undoubtedly the key and vaccines can certainly play a key role in this regard. Although there are some vaccines commercially available, their efficacy can still be improved. In Chapter 3, Rachmawati et al. provide an overview of the in silico approach for the S. Typhi epitope vaccine. There are three types of S. Typhi vaccines: live-attenuated, inactivated, and sub-unit vaccines. The authors further elaborate on the steps to be followed for developing a new type of vaccine using only the part of the subunit that is recognized by B and T cells of the immune system, the epitope area. The development of bioinformatic tools, omic technologies, and recombinant DNA technologies give these types of vaccines enormous potential. In Chapter 4, Mishra et al. describe how computational tools can help in the development of effective vaccines against multidrug-resistant S. enterica strains.

Nontyphoidal *Salmonella* primarily causes gastroenteritis, bacteremia, and focal infection, mainly related to the consumption of food contaminated with this pathogen. Livestock, especially poultry, can carry *Salmonella* in their gut without symptoms. Therefore, *Salmonella* can contaminate products produced from these animals, move through the food chain, and reach the consumer. In Chapter 5, Pandey and Goud highlight the importance of control measures to prevent the spread of this pathogen. Another cause for concern is the increase in antibiotic resistance. The Exponential increase in antibiotic resistance in recent years has resulted in increased hospitalization and deaths. In Chapter 6, Al-Hamadany describes how *Salmonella* must evade the host immune system to survive. The main serotypes involved in human cases of salmonellosis are S. Typhimurium and S. Enteritidis. The virulence and invasiveness of these and other serotypes are mainly related to the presence of *Salmonella* Pathogenicity Islands (SPIs), as discussed in Chapter 7 by Sarika. In these SPIs are located the main genes involved in invasion, survival, and extra-intestinal spread. These SPIs are also of great interest from an evolutionary point of view as they can be acquired by horizontal transmission. SPI1 and SPI2 are the main types of pathogenicity islands, and they encode the main virulent genes and Type III secretor systems necessary for host cell invasion. However, in *Salmonella* there are more than 10 different SPIs, some of which are specific to certain subspecies.

S. Enteritidis is one of the main serotypes causing infections in humans. As reviewed in Chapter 8 by Ogunremi et al., it is important to characterize the S. Enteritidis strain implicated in food salmonellosis. Although pulsed-field gel electrophoresis (PFGE) is a widely used typing tool, its discriminatory capacity in S. Enteritidis is low due to the clonality of this serotype. In the genomic era, whole genome sequencing is an attractive tool for a full and comprehensive characterization of the genetic attributes of bacteria. Despite massive information obtained with this technology, this approach is not useful for generating a useful nomenclature-based description of S. Enteritidis subtypes. In this sense, the characterization of 60 polymorphic loci by a single nucleotide-based genotypic polymerase chain reaction assay (SNP-PCR) allowed to define 25 circulating clades of S. Enteritidis. This approach is an ideal subtyping test, being highly discriminatory, low cost, rapid, and reproducible. It is useful to identify the subtype designation of an isolate for outbreak surveillance.

Non-enterica subspecies of S. enterica are mainly related to cold-blooded animals. However, it has been observed that these subspecies can also colonize warm-blooded animals. For example, in Chapter 9, Rubira et al. describe the adaption of S. enterica subsp. diarizonae serotype 61: k: 1,5, (7) to sheep. This serotype has become highly prevalent in sheep herds in Sweden, Norway, Switzerland, and the United Kingdom. In this sense, most animals are asymptomatic carriers and rarely show clinical symptoms. Studies carried out in slaughterhouses have isolated this serotype from the intestinal content and respiratory track of healthy animals. Rarely these S. enterica subsp. diarizonae serotype 61: k: 1,5, (7) cause health disorders such as chronic proliferative abortions, testicular lesions in rams, or tract disorders in young animals. This serotype is endemic in sheep herds in Sweden, with 40% of large herds being positive for this bacterium. In addition, 1.8% of sheep carcasses of the largest slaughterhouse in Sweden were positive for this serotype. This could suggest that there is a direct link between the consumption of sheep products and cases of human salmonellosis caused by S. enterica subsp. diarizonae serotype 61: k: 1,5, (7). However, the number of reported cases due to this serotype is residual. These data highlight the low pathogenicity of this serotype. Consequently, an exception for this serotype was made in the Swedish *Salmonella* control program. However, the pathogenicity of this serotype could change over time and an increase of human salmonellosis cases due to S. enterica subsp. diarizonae serotype 61: k: 1,5, (7) should result in the revision of this exception. A similar situation is observed with serotype S. enterica subsp. salamae 4, [5], 12:b:- (commonly known as S. Sofia) in Australia broiler flocks. This serotype has also become highly prevalent, but no direct relationship can be established with an increase in cases of human salmonellosis due to this bacterium. This study reflects the adaption of non-enterica serotypes of S. enterica to livestock. Comparative genomic studies are necessary to fully understand the potential pathogenicity of these serotypes and if they can be considered as commensal microorganisms.

Antimicrobial resistance is one of the main challenges of global public health. The indiscriminatory use of antimicrobials in both humans and animals has increased the number of resistant bacteria exponentially. For that reason, it is necessary to make rational use of antibiotics and to find new alternatives to them. The food production chain is one of the points where it is important to reduce the use of antibiotics. The transmission of multidrug-resistant foodborne pathogens from food to consumers is one of the main areas of concern. In the last years, a great effort has been made to introduce new control measures from farm to fork.

Bacteriophages (phages) are prokaryotic viruses that can infect and kill bacterial pathogens. Phages were discovered in the beginning of the 20th century, but due to the discovery of antibiotics they were relegated to the background during the subsequent decade. However, the need for antibiotic alternatives has resulted in the rediscovery of phages. In Chapter 10, Thanki et al. describe the application of phages in different points of the poultry and swine production chain. They can be applied to farm animals through feed or water, in slaughterhouses to reduce *Salmonella* load in chicken and swine skin, or in food packaging to inhibit the growth of this pathogen during storage. There are some commercially available *Salmonella* phage cocktails approved by the US Food and Drug Administration to be used in the food chain. However, phage therapy needs to overcome some challenges such as host range, resistance development, and phage delivery. Tools like genetic engineering and phage encapsulation could help to solve the actual limitations of this antibiotic alternative.

In Chapter 11, Ruiz-Pérez et al. describe the potential of natural products to control the growth of *Salmonella* in the food production chain. Different bacteria-, fungus-, animal-, and plant-derived products have been tested against different foodborne pathogens such as *Salmonella*. Some of them have shown promising results, but commercial production and application of these products is still a limiting factor. Biofilm formation is another problem associated with *Salmonella* in the food production chain. In Chapter 12, Lamas et al. explain that bacteriocins and phages can be applied to combat biofilms. However, they also have some limitations to kill biofilm cells, mainly due to protection offered by the characteristic extracellular substances produced in this type of bacterial life.

This book offers a global vision of the *Salmonella* genus and its implications for human health, from host-specific S. Typhi to serotypes transmitted through the food chain. The guest editors would like to thank the editorial team for their invaluable assistance.

**Alexandre Lamas, Patricia Regal and Carlos Manuel Franco**
Department of Analytical Chemistry,
Nutrition and Bromatology,
Universidade de Santiago de Compostela,
Lugo, Spain

Section 1

# An Overview of *Salmonella Typhi* and Vaccine Development

Chapter 1

# Virulence Factors of *Salmonella Typhi*

*Noor S.K. Al-Khafaji, Ali M.K. Al-Bayati and Hussein O.M. Al-Dahmoshi*

## Abstract

S. Typhi is an enteric bacillus which belongs,to the genus *Salmonella* in the family Enterobacteriacaea and it is a multi–organs pathogen which inhibits the lymphatic tissues of the small intestine, liver, spleen, and blood stream of infected humans. S. Typhi has a mixture of features that make it an efficient pathogen. This species contains an endotoxin that is characteristic of Gram-negative organisms, as well as the virulence-enhancing Vi antigen. Many of the S. Typhi virulence factors are clustered in some areas of the chromosome known as *Salmonella* pathogenicity islands (SPI), such as adhesion, invasion, and toxin genes. A protein known as invasin that permits non-phagocytic cells is also produced and excreted by the bacterium., Where it is capable of intracellular living. The oxidative burst of leukocytes may also be inhibited, making innate immune reaction ineffective.

**Keywords:** S. Typhi-virulence factors -endotoxin, enterotoxin, cytotoxin

## 1. Introduction

It was quite a long time before typhoid fever was differentiated from other febrile disorders. Pierre Louis was the first who used the word " typhoid " and give the classical picture of typhoid in 1829 and described in detail post-mortem finding, especially the enlargement and ulceration of peyer's patches. However, he did not clearly differentiate between typhoid and typhus. In 1837, Gerhard was the first who clearly differentiate typhoid from typhus fever and William Budd described the contagious nature of the disease and Incriminated transmission of facially polluted water supplies in 1873 [1].

In 1873, William Budd, a physician in Bristol who was interested in cholera and intestinal fever, showed that typhoid fever could be transmitted by a particular toxin found in the excrement and that this propagation was responsible for the contamination of water by the feces of patients. Each case was linked to another anterior case, according to Budd. A significant number of doctors and scientists have attempted to discover the nature of the disease-causing microorganism and have experienced considerable difficulty in isolating the bacillus. It was Karl Joseph Eberth, Rudolf Virchow's doctor and pupil, who discovered the bacillus in the abdominal lymph nodes and the spleen in 1879. In 1880 and 1881, he reported his findings. The genus '*Salmonella*' was named after Daniel Elmer Salmon, an American veterinary pathologist, who was the administrator of the USDA (United States Department of Agriculture) research program. His discovery was then tested and confirmed by German and English bacteriologists, including Robert Koch., Thus, despite the fact that a number of scientists had contributed to the quest [2, 3], the organism was named after him. *Salmonella* has thus become new scientific knowledge and thus the mechanisms of infection and the presence of healthy carriers have been relatively nascent [4]. Recent reports suggest that there are approximately 20 million cases of typhoid each year, resulting in deaths of 100,000-200,000 [5]. Karl J. Eberth, who isolated the bacterium from spleen parts and lymph nodes from a patient who died of typhoid fever and discovered the typhoid agglutinins and their diagnostic application, first isolated S. Typhi in 1880. In 1881, Robert Koch succeeded in cultivating the bacterium. However the isolation of typhoid bacillus from other enteric bacteria was unclear due to the lack of differential characters [6, 7].

*Salmonella* is a genus of rod shaped (bacillus) gram negative bacteria related to family Enterobacteriaceae. They have two species which are *Salmonella enterica* and *Salmonella bongori*. *S. enterica* is the kind species and is further divided into sex subspecies [8]. that contain over 2,600 serotypes [9]. *Salmonella* species are non spore forming, predominantly motile enterobacteria for cell diameters between on 0.7 and 1.5 µm, lengths for 2 to 5 µm, and peritrichous flagella (all concerning the cell body [10]. exceptions *S. Gallinarum* and *S. Pullorum* [11, 12]. The bacterial strain was named after the American pathologist, Dr. Daniel Elmer Salmon, who collaborated with Smith. The *Salmonella* nomenclature is controversial and still changing. The Centers for Disease Control and Prevention (CDC) is currently using the *Salmonella* nomenclature system suggested by the World Health Organization (WHO) Collaborating Centre as a nomenclature system.: Species: *Salmonella enterica* serotype Typhi. [13].

## 2. Virulence factors of *Salmonella Typhi*

Virulence factors in *Salmonella Typhi* are involved in the various stages of infection, namely: the production of toxins (LPS) endotoxin, enterotoxin, cytotoxin), colonization, adhesion and invasion, as well as survival inside the host cells [14] (**Figure 1**).

### 2.1 Vi antigen

The capsular Vi antigen is a linear homopolymer of alpha 1–4 linked to galactose aminouronic acid which is variably acetylated at the C3 position. This antigen is

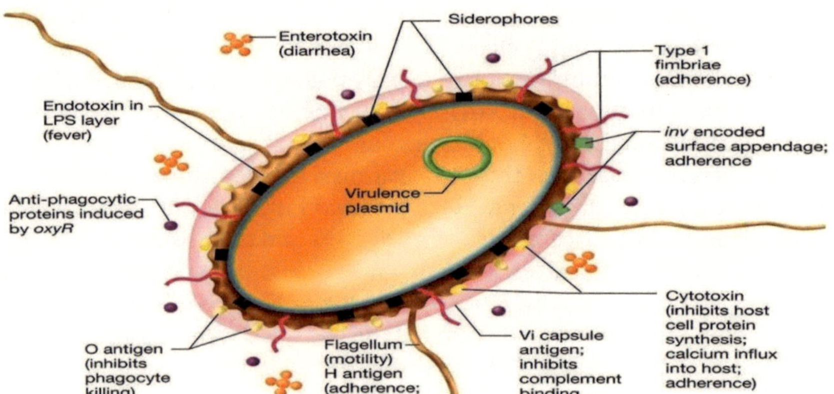


**Figure 1.**
Salmonella *virulence factors [14].*

believed to inhibit phagocytosis and complement C3 activation thus inhibiting non-specific opsonization, "[15] One of the main characteristics that distinguishes S. Typhi from (non typhoid *Salmonella*) NTS is the production of a polysaccharide capsule named the Vi antigen. The Vi capsule inhibits phagocytosis and confers serum resistance [16, 17], likely by shielding the O-antigen from antibodies [16]. The genes encoding the Vi capsule comprise the viaB locus within *Salmonella* pathogenicity island (SPI)-7, which also encodes the type III secretion system (T3SS) effector SopE and a type IVB pilus [18].

### 2.2 The SPI-1, SPI-2 and type III secretion systems

Common to both typhoidal and NTS are two pathogenicity-island encoded type III secretion systems (T3SS): the SPI-1 and SPI-2 T3SS, which are essential for *Salmonella* virulence. In S. Typhi, the SPI-1 T3SS is also required for invasion of nonphagocytic cells [19], but the importance of the SPI-2 T3SS is less clear. Disruption of the SPI-2 T3SS did not influence the survival of *S*. Typhi in THP-1 and human monocyte-derived macrophages [20]; however, S. Typhi strains with transposon insertions in the SPI-2 components ssaQ, ssaP, or ssaN were negatively selected against during competitive growth in human macrophages [21]. The role of SPI-2 during the intracellular lifestyle of typhoidal serovars therefore warrants further investigation.

### 2.3 Somatic O antigen (cell wall Ag or LPS)

The outer L-layer underlying the capsular material has the lipopolysaccharide (LPS) called the '0' antigen. This'L' layer also has certain proteins called outer membrane proteins (OMP) which are antigenic. These OMPs include both porin (OMP F and OMP C) and non-porin substances. Porins are pore-forming channels which help in solute uptake and non-porin proteins are structural proteins (**Figure 2**) [23]. These antigens are highly immunogenic and there is a good antibody response to all these antigens in patients with typhoid fever.". [24, 25]. The somatic antigens represent the side chains of repeating sugar unit projecting outwards from the lipopolysaccharide layer and the surface of the bacterial cell wall; they are hydrophilic and heat stable. It is used for serological diagnosis [26].

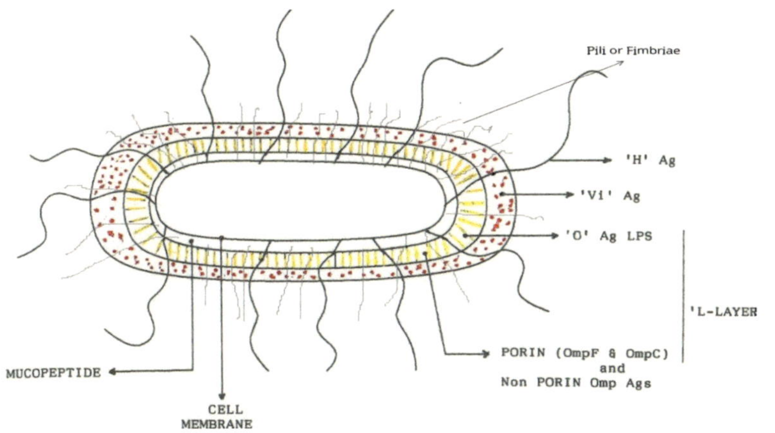


**Figure 2.**
*Antigenic structure of* Salmonella Typhi *[22].*

### 2.4 Flagella(H antigen)

Flagella, while contributing to virulence, are also important activators of innate immune responses via recognition of monomeric flagellin by TLR5 and NAIP receptors [27, 28], H antigen may occur in either or both of two forms, called phase 1 and phase 2. The organism tends to change from one phase to the other. H antigen also provides a useful epidemiologic tool with which to determine the source of infection and its mode of spread [29]; While most NTS display phase variation through the alternate expression of two genes of flagellin (fliC and fljB), most S. Typhi strains are monophasic, expressing FliC of the antigen H: d directly. Interestingly, some Indonesian S. Typhi strains transmit H: j, due to an in-frame deletion in fliC, a variant of H:d. [30], and/or are biphasic, expressing a plasmid-encoded FljB analogue of the H:z66 antigen [31], H:j and H:z66 antigenic variants are thought to have recently emerged during S. Typhi evolution [32], driven by immune selection in this high incidence region [31]. This additional variation seems to play a role in *S*. Typhi interactions with host epithelial cells and macrophages and partly in immune evasion [33].

### 2.5 Fimbriae (adhesion protein) and pili

The significant adhesion factors for S. Typhi are fimbriae and pili. These elements of virulence are employed by S. Typhi during infection and host colonization for its various cellular interactions [34]. The Operon Stg, one of the six Operons Fimbriae found in S. Typhi, But not S. Typhimurium has recently been shown to be involved in cellular invasion and in vitro destruction of epithelial cells [34]. In addition, the STG operon was found to assist S. Typhi targets enterocytes more preferentially than M cells, which promotes S. Typhi By passing the Peyer patches, eludes the innate immune system [35].

### 2.6 Virulence plasmid

Certain *Salmonella* carry a large, low copy number plasmid that contains virulence genes. Virulence plasmids are required to trigger systemic disease; their involvement in the enteric stage of the infection is unclear. *Salmonella* virulence plasmids are heterogeneous (50–90) kb in size, but all share a 7.8 kb area, SPV, necessary for reticuloendothelial system bacterial multiplication [36, 37].

### 2.7 Invasiveness

Unlike most bacteria that rely on endocytosis mediated by receptors in order to invade a target cell, S. Typhi uses a complex mechanism known as bacterial mediated endocytosis, in which bacterial proteins enter the host cell and control signaling cascades that regulate the trafficking of cytoskeletal membrane architecture and gene expression, both of which force endocytosis S. Typhi into the host [37, 38]. The target cell for *S*. Typhi is the macrophage. The ability of S. Typhi to survive in macrophages is due to the development of bacterial proteins that allow the organism to with strand both the oxygen-dependent and the non oxygen-dependent killing mechanisms of these professional phagocytic cells [36, 37].

### 2.8 Biofilm

Biofilm cells manufacture proteinaceous substances that allows synergic growth and protection from possible harsh environments it may encounter [39, 40]. In the seventeenth horn, a Dutch scientist Van Leeuwenhoek was the first individual to discover biofilm cells which he described as "animacules" on his dental plaque. The biofilm development process is initiated with single cells attaching to a surface or to each other, this is then followed by the formation of clustered cells or microcolonies. Over time, the microcolonies are surrounded by a protective layer of protein-rich substances referred to as extracellular polymeric substances (EPS) [37]. The development and genetic signaling pathways involved in a *Salmonella* biofilm formation are complex. There are four major components to the structure of the *Salmonella* biofilm: curli, cellulose, capsular polysaccharides and lipopolysaccharides. Curli fibers, referred to as thin very aggregative fibers (Tafi) are one of the main components of the extracellular polysaccharide (EPS) matrix [40, 37].

Enea et al. [41] were found biofilm production by *S.* Typhi may represent a key factor for the promotion of a persistent infection in the gallbladder, thus sustaining a chronic local inflammatory response and exposing the epithelium to repeated damage cau*sed* by carcinogenic toxins. **Figure 3** demonstrates the potential role of biofilm-producing S. Typhi, in the development of gallbladder cancer. (A) Chronic S. infection. Typhi strains and gallstone presence strongly correlate with the development of gallbladder cancer (GC); The presence of gallstones (B) could provide

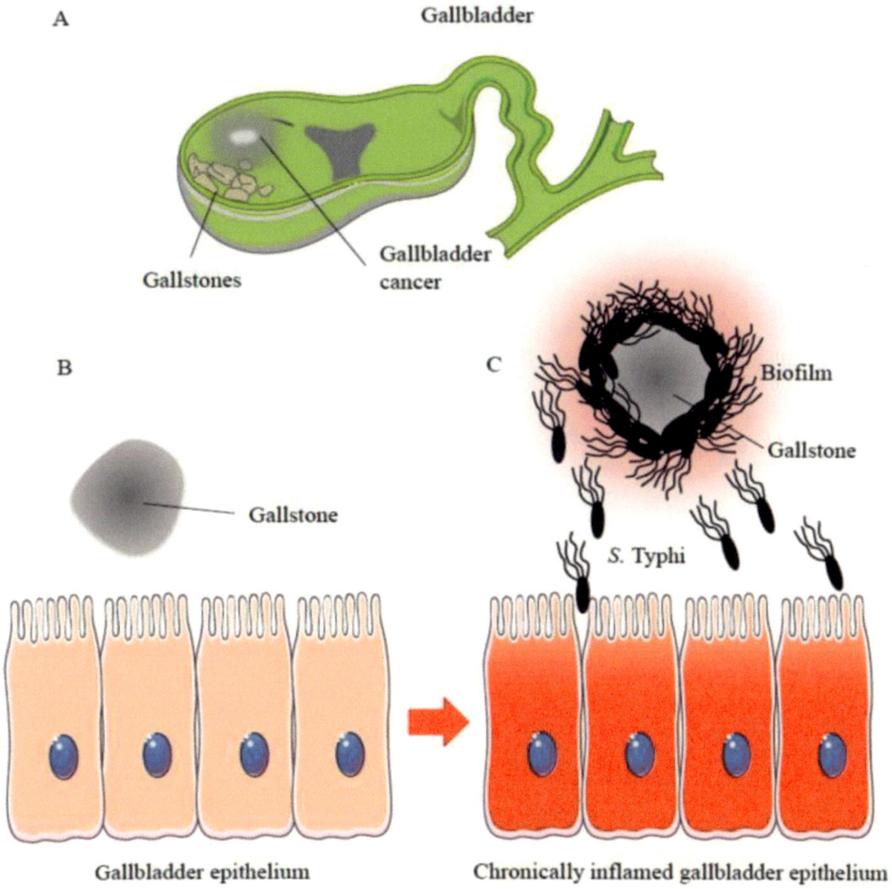


**Figure 3.**
*Showing the potential role of biofilm-producing S. Typhi in the development of gallbladder cancer. (A) Chronic S. infection. Typhi strains and gallstone presence strongly correlate with the development of gallbladder cancer (GC); The presence of gallstones (B) could provide S. Typhi strains with the ideal substrate. (C) Once the biofilm is established, bacterial cells are separated from the gallstones that release carcinogenic molecules [41].*

S. Typhi strains with the ideal substrate. (C) Once the biofilm is established, bacterial cells are separated from the gallstones that release carcinogenic molecules that induce genomic instability and chronic inflammation, which are key prerequisites for the onset of GC. with an increased biofilm forming capacity.

### 2.9 Endotoxin of *Salmonella* Typhi

Endotoxin is a big part of Gram-negative bacteria's outer membrane (OM). Endotoxins have been found to play an important function in the pathogenicity of Gram-negative bacterial infections. It is a powerful mediator of a wide range of pathophysiological effects in humans, mainly in the gasterointestinal tracts. Therefore, these are also known as enterotoxins. These toxic behaviors, as well as many beneficial ones linked to immunostimulation, include lethal toxicity, pyrogenicity and tissue necrotizing activity [42]. Endotoxins are high-molecular weight complexes, of lipopolysaccharides (LPS) which is the major component of bacterial cell wall [42]. It's a heat stable toxic substance released by gram negative bacteria's after disruption of cell envelopes [43, 44]. The role of endotoxins in bacterial pathogenesis and their chemical characterization as lipopolysaccharide (LPS) have been studied earlier [45, 46]. Chemically, LPS consist of a hydrophilic polysaccharide covalently linked to a hydrophobic lipid portion which is termed as lipid A, which anchors the molecules in the outer membrane (OM) [47]. Endotoxins play a major role in human disease states that created interest to investigate the pathogenicity of the producing bacteria [42]. Lipopolysaccharide found to be an important activator for the activation of immune system that leads to non- specific inflammatory immune response [48].

## 3. Conclusions

According to above review we put highlights on the role of the *Salmonella* virulence factors. In addition, we mentioned some strategies that could be explored in order to take control of *Salmonella* infections.

## Conflict of interest

There is no 'conflict of interest' for this work.

## Author details

Noor S.K. Al-Khafaji[1]*, Ali M.K. Al-Bayati[2] and Hussein O.M. Al-Dahmoshi[1]

1 Biology Department, College of Science, University of Babylon, Hilla, Iraq

2 Al-Noor Hospital for Pediatric, Babylon Health Directorate, Ministry of Health, Iraq

*Address all correspondence to: noorvaccine@gmail.com

## References

[1] Parry, C. M. (2006). Epidemiological And Clinical Aspects Of Human Typhoid Fever. In: Matroeni, P. And Maskell, D. (Eds.). Salmonella Infections: Clinical, Immunological And Molecular Aspects, P: 1-18. New York: Cambridge University Press.

[2] Moorhead, R. (2002). William Budd And Typhoid Fever. J R Soc Med. ;95:561-564.

[3] Oldenkamp EP. (2004) Predecessors: Veterinarians From Earlier Times, Daniel Elmer Salmon (1850-1914) Tijdschr Diergeneeskd. 129:554-555.

[4] Marineli, F., Gregory T., Marianna K., and George A. (2013) Mary Mallon (1869-1938) And The History Of Typhoid Fever. Ann Gastroenterol. 26(2): 132-134.

[5] Antillón, M.; & et. al (2017). The Burden Of Typhoid Fever In Lowand Middle-Income Countries: A Meta-Regression Approach. Plos Neglected Tropical Diseases, 11(2).

[6] Shere, K.D.; Goldbreg, M. B.; et.al (1998). Salmonella Infection. In: Gorbach, S. L.; Bartlett, J. G.; And Blacklow, N. R. (Eds.). Infectious Diseases. 2nd Ed. Part1. Philadelphia, Sannders Company. Pp.:699.

[7] Abbas H.H. (2012 ). Detection Of *Salmonella enterica* Serovar Typhi In Patients With Acute And Chronic Typhoid By Some Bacteriological , Immunological And PCR Technique .Msc Thesis Baghdad University.

[8] Su, L. H., Chiu, et. al (2003). In Vivo Acquisition Of Ceftriaxone Resistance In *Salmonella enterica* Serotype Anatum. Antimicrobial Agents And Chemotherapy 7(2), 47-56.

[9] Gal-Mor, O., Boyle, & et. al (2014). Same Species, Different Diseases: How And Why Typhoidal And Non-Typhoidal *Salmonella enterica* Serovars Differ. Frontiers In Microbiology. 2(4), 436-442.

[10] Fàbrega, A., And Vila, J. (2013). *Salmonella enterica* Serovar Typhimurium Skills To Succeed In The Host: Virulence And Regulation. Clinical Microbiology Reviews. 26(2), 308-341.

[11] Browning, G. (2008). Clinical Veterinary Microbiology. Australian Veterinary Journal.

[12] Abdul Aziz T. A. F. AL Jobouri.,(2019). Molecular Characterization of Some Virulence Genes of *Salmonella enterica* in Babylon province Msc Thesis Baghdad University. (2), 47-56.

[13] Popoff M.Y., et. al.(2003). Supplement 2001 (No. 45) To The Kauffmann–White Scheme. Res Microbiol. 154(3):173-174.

[14] Madigan M., Martinko J. (2007). Biologie Des Micro-Organismes. 11 Ème Édition. Pearson, Paris. P : 731-735, 790-792, 943, 947-948.

[15] Robbins 1D, Robbins 18(1984). Reexamination Of The Protective Role Of The Capsular Polysaccharide (Vi Antigen) Of Salmonella Typhi. J Infect Dis;150:436-49.

[16] Hart, P. J., O'Shaughnessy, & et. al. (2016). Differential Killing Of *Salmonella enterica* Serovar Typhi By Antibodies Targeting VI And Lipopolysaccharide O:9 Antigen. Plos One, 11(1).

[17] Wilson, R. P., Winter, S. E., et.al. (2011). The Vi Capsular Polysaccharide Prevents Complement Receptor 3-Mediated Clearance Of *Salmonella enterica* Serotype Typhi. Infection And Immunity, 79(2), 830-837.

[18] Pickard, D., Wain, J., Baker, S., Line, A., Chohan, S., Fookes, M., ... Dougan, G. (2003). Composition, Acquisition, And Distribution Of The Vi Exopolysaccharide-Encoding *Salmonella enterica* Pathogenicity Island SPI-7.

[19] Bishop, A., House, D., Perkins, T., Baker, S., Kingsley, R. A., & Dougan, G. (2008). Interaction Of *Salmonella enterica* Serovar Typhi With Cultured Epithelial Cells: Roles Of Surface Structures In Adhesion And Invasion. Microbiology, 154(Pt 7), 1914-1926. Https://Doi.Org/10.1099/Mic.0.2008/016998-0.

[20] Forest, C. G., et. al. (2010). Intracellular Survival Of *Salmonella enterica* Serovar Typhi In Human Macrophages Is Independent Of Salmonella Pathogenicity Island (SPI)-2. Microbiology, 156(12), 3689-3698.

[21] Sabbagh, S. C., et.al. (2012). Selection Of *Salmonella enterica* Serovar Typhi Genes Involved During Interaction With Human Macrophages By Screening Of A Transposon Mutant Library. Plos One, 7(5), E36643.

[22] V. BALARAJ, G. SRIDHARAN, M. V. JESUDASON(1992), Immunization Against Typhoid Fever, Mahatma Gandhi Institute Of Medical Sciences, Wardha, Sevagram 442102, Maharashtra, India © The National Medical Journal Of India 1992.

[23] Benz R. Structure And Function Of Porins From Gram-Negative Bacteria. Ann Rev Microbiol 1988;42:359-93.

[24] Duguid JP. Salmonella I Typhoid And Paratyphoid Fevers. In: Duguid JP, Marimon BP, Swain RHA (Eds). Medical Microbilogy. Vol I. Microbial Infections London:English Language Book Society, 1978:314-19.

[25] Calderon I, Lobos SR, Rojas HA, Et Al. Antibodies To Porin Antigens Of Salmonella Typhi Induced During Typhoid Infection In Humans. Infect Immun 1986;52:209-12.

[26] Al-Ammar, M. H. (2006). Some Aspects Of Specific Immunity For Salmonella Typhi In Typhoid Patients And Rabbits .Phd Thesis , Alkufa University.

[27] Hayashi, F., Smith, K. D., Ozinsky, A., Hawn, T. R., Yi, E. C., Goodlett, D. R., ... Aderem, A. (2001). The Innate Immune Response To Bacterial Flagellin Is Mediated By Toll- Like Receptor 5. Nature, 410(6832), 1099-1103. Https://Doi.Org/10.1038/35074106 .

[28] Kortmann, J., Brubaker, S. W., & Monack, D. M. (2015). Cutting Edge: Inflammasome Activation In Primary Human Macrophages Is Dependent On Flagellin. The Journal Of Immunology, 195(3), 815-819. Https://Doi.Org/10.4049/Jimmunol.1403100.

[29] Harriet U. And Nandita De.( 2014 ). Mechanisms Of Antibiotic Resistance In Salmonella Typhi. Int.J.Curr.Microbiol.App.Sci (2014) 3(12): 461-476.

[30] Frankel, G., Newton, S. M., Schoolnik, G. K., & Stocker, B. A. (1989). Intragenic Recombination In A Flagellin Gene: Characterization Of The H1-J Gene Of Salmonella Typhi. The EMBO Journal, 8(10), 3149-3152.

[31] Baker, S., Hardy, J., Sanderson, K. E., Quail, M., Goodhead, I., Kingsley, R. A., ... Dougan, G. (2007). A Novel Linear Plasmid Mediates Flagellar Variation In Salmonella Typhi. Plos Pathogens, 3(5), 0605-0610. Https://Doi.Org/10.1371/Journal.Ppat.0030059.

[32] Baker, S., Holt, K., Van Den Vosse, E., Roumagnac, P., Whitehead, S., King, E., Dougan, G. (2008). High-Throughput Genotyping Of *Salmonella enterica* Serovar Typhi Allowing Geographical Assignment Of Haplotypes And Pathotypes Within An Urban District Of Jakarta, Indonesia.

Journal Of Clinical Microbiology, 46(5), 1741-1746. Https://Doi.Org/10.1128/JCM.02249-07.

[33] Schreiber, F., Kay, S., Frankel, G., Clare, S., Goulding, D., Van De Vosse, E., ...Baker, S. (2015). The Hd, Hj, And Hz66 Flagella Variants Of *Salmonella enterica* Serovar Typhi Modify Host Responses And Cellular Interactions. Scientific Reports, 5(1), 7947. Https://Doi.Org/10.1038/Srep07947.

[34] Berrocal, L. Et Al. Stg Fimbrial Operon From S. Typhi STH2370 Contributes To Association And Cell Disruption Of Epithelial And Macrophage-Like Cells. Biol. Res. 48, 1-8 (2015).

[35] Gonzales, A. M., Wilde, S. & Roland, K. L. Insights Into The Roles Of Long Polar Fimbriae And Stg Fimbriae In Salmonella Interactions With Enterocytes. Infect. Immun. 85, 1-12 (2017).

[36] Safia Sultana,(2012). Comparison Of Different Test Methods Including Polymerase Chain Reaction For Early And Reliable Diagnosis Of Typhoid Fever Department Of Microbiology Mymensingh Medical College Mymensingh, Bangladesh

[37] Zahraa Hameed Oda AL-Quraishi (2018), Bacteriological And Molecular Study On Salmonella Typhi Isolated From Gallbladder Chronic Infections In Alnajaf City A Thesis Submitted To The Council Of College Of Science / University Of Kufa.

[38] Ohi, M. E. And Miller, S. I. (2001). Salmonella: A Model For Bacterial Pathogenesis. Ann. Rev. Med. 52:259-274.

[39] Donlan, R. and Costerton, W. (2002). Biofilms: survival mechanisms of clinically relevant microorganisms. Clinical Microbiology Reviews. 15(2): 167-193.

[40] Mary Corcoran, (2013). *Salmonella enterica* - biofilm formation and survival of disinfection treatment on food contact surfaces. A Thesis presented to the National University of Ireland, galway for the degree of Doctor of Philosophy Discipline of Bacteriology, School of Medicine, National University of Ireland Galway.

[41] Enea Gino Di Domenico, ID, Ilaria Cavallo, Martina Pontone, Luigi Toma and Fabrizio Ensoli.(2017). Biofilm Producing Salmonella Typhi: Chronic Colonization and Development of Gallbladder Cancer, International journal, enea.didomenico@ifo.gov.it; Tel.: +39-06-526-652-45

[42] Mahamuni, P.P, Patil, A.R And Ghosh, J.S (2017). Proteolytic And Lipolytic Properties Of Endotoxins (Enterotoxins) Produced By Salmonella Typhi NCIM 5255, Salmonella Typhimurium NCIM 2501 And Shigella Flexneri NCIM 5265. International Food Research Journal 24(6): 2685-2688

[43] Beutler, B. And Rietschel, E.T. (2003). Innate Immune Sensing And Its Roots: The Story Of Endotoxin. Nature Reviews Immunology 3: 169-176

[44] Bishop, R.E.( 2005). Bacterial endotoxin. In Russell, W and Herwald, H. (Eds). Fundamentals of endotoxin structure and function. Basel : Karger medical and Scientific publisher, p.1-27

[45] Raetz C.R.H. And Whitfield C.(2002). Lipopolysaccharide Endotoxins. Annual Reviews Biochemistry 71 : 635-700

[46] Raetz, C.R.H., Reynolds, et.al. (2007). Lipid A Modification Systems In Gram Negative Bacteria. Annual Reviews Biochemestry 76 : 295-329.

[47] Brandenburg, K. and Wiese, A.(2004). Endotoxins: Relationships between structure, function, and

activity. Current Topics in Medicinal Chemistry 4: 1128-1146

[48] Buyse, J.; Swennen, Q.; Niewold, T.A.; Klasing, K.C. ; Janssens, G.P.J., Baumgartner, M. And Gooeeris, B.M. ( 2007 ). Dietary L-Carnitine Supplementation Enhances The Lipopolysaccharide-Induced Acute Phase Protein Response In Broiler Chickens. Veterenary Immunoogy And Immunopathology 118:154-159.

Chapter 2

# Enteric Fever in Primary Care

*Abdulmaleek Idanesimhe Sado and Aduke Oluwambe Sado*

## Abstract

Enteric fever is a bacterial infection caused by *Salmonella typhi* and *paratyphi*. It is endemic in many parts of Africa and South Asia where there is poor access to safe portable water and below par food quality assurance. It is important to ensure prompt recognition, diagnosis and management of symptoms to forestall complications. Due to the rising global burden, significant effort has to be made to improve primary care services like vaccination, antimicrobial stewardship and encouragement of hygiene measures. Hence, it is imperative to be aware of its current burden and options available in primary care for its prevention and treatment.

**Keywords:** *Salmonella*, enteric fever, typhoid, primary care, typhoid-conjugate vaccine

## 1. Introduction

Typhoid fever (now more appropriately called Enteric fever) is a bacterial infectious disease caused by *Salmonella enterica* subspecies *enterica* and serovar *typhi*. It is mainly transmitted through the faeco-oral route via contaminated food, water and asymptomatic carriers [1]. It is endemic in developing countries and low-resource settings where hygiene and sanitation measures are subpar. In developed countries and high-income settings, it is less common but cases still occur in recent travellers to endemic areas [2]. There are a number of factors which contribute to the disease burden including lack of access to clean, portable water, poor food quality control and lack of public health services (e.g well managed public latrine and hand washing facilities); all of which can be attributed to lack of awareness, low political will and sociocultural factors. Symptoms of enteric fever vary significantly and are generally nonspecific. These include pyrexia, headache, myalgia, arthralgia, nausea, rash, abdominal pain, constipation and occasionally diarrhoea [3].

Enteric fever, if left untreated can be life-threatening and result in a myriad of complications including intestinal haemorrhage and perforation, peritonitis, sepsis, meningitis, osteomyelitis, multiorgan failure and death [1, 3]. Hence, it is expedient to ensure early diagnosis and management to mitigate complications.

Central to the actualisation of universal health coverage is an effective primary health care system which is usually the first point of contact for most patients. Hence, the role of the primary care clinician in the prevention, diagnosis and management of enteric fever and its complications cannot be overemphasised. This is what this chapter aims to address.

## 2. Epidemiology

Enteric fever is a global health problem affecting 21.6 million people (incidence of 3.6 per 1000 population) and resulting in just over 216 000 deaths annually [4]. It is endemic in developing and low and middle income countries of Africa, Asia, Latin America, the Caribbean and Oceania mainly due to poor sanitation and environmental hygiene [2, 4]. Bangladesh, Indonesia, China, India, Laos, Nepal, Pakistan and Vietnam account for 80% of cases [4]. Untreated, 10%–30% of patients will die but mortality reduces to 1%–4% with prompt and appropriate treatment. In the pre-antibiotics era, the USA had a case fatality rate of 9%–13% [5]. This illustrates how much the discovery of antibiotics has revolutionised its management, just like most bacterial infections. However, there is an increasing burden of antimicrobial resistant *Salmonella* strains emanating from endemic countries mainly due to poor antimicrobial stewardship and measures must be taken to stem the tide.

Significant intra- and intercountry variation in disease burden exists in many regions of south and south east Asia and parts of Africa. For instance, surveillance performed in two sites in Kenya between 2006 and 2009 found that the incidence of blood-culture proven typhoid fever in rural and urban sites varied from 29 up to 247-cases/100000 person-years [6]. Also, data from the Diseases of Most Impoverished areas have described incidence rates varying from 24.2/100000 in Vietnam to 493.5/100000 in parts of India [7]. However, most disease burden data from low- and middle-income countries are hospital-based which leaves a huge number of cases unaccounted especially in areas of low health-care usage and accessibility. Hence, it is imperative for countries in endemic regions of the world to develop a national and regional surveillance system to identify factors responsible for these variations and adopt guidelines and protocols to improve efficiency in prevention, diagnosis and management. Central to this should be an efficient primary care system where surveillance and data gathering can be co-ordinated and synchronised with hospital-based data providing a broad-based approach and a better reflection of disease burden.

The incidence of enteric fever varies by age. In endemic areas, incidence is higher in younger children but similar across age groups in low burden areas [8]. In general, children are at a higher risk of complications including ileitis and intestinal perforation. When perforation sets in, mortality has been reported to be as high as 62% [9, 10]. Therefore, it is imperative that signs and symptoms of enteric fever are identified and treated early in primary care. Due to the wide disparity in incidence between developed and low and middle-income countries, primary care physicians in the latter will most likely see a lot more cases and have a high pre-test probability. This poses a challenge for a lot of primary care practitioners in developed countries who are less likely to be familiar with its presentation and may result in delay in diagnosis. In England and Wales, any case of *Salmonella* infection is a notifiable disease which must be reported to Public Health England and may require urgent community investigation to forestall an outbreak [11].

## 3. Aeotiopathogenesis

*Salmonella* is a flagellated, non-capsulated facultative anaerobic gram-negative bacilli and non-lactose fermenter of the Enterobacteriaceae family which has flagellar, somatic and outer coat antigens [7, 12]. Its outermost covering is made up of the somatic O antigen while the flagellae are composed of the H antigen. Each

O and H antigen have a unique code number and a varied combination of these form the basis for the determination of serotypes [12]. Of the over 2500 serotypes of *Salmonella* that have been identified, only 100 are thought to be responsible for most human infections [7]. These infections can be broadly divided into nontyphoidal and typhoidal. The typhoidal infection is mainly caused by *S. typhi* and less commonly *paratyphi*. *Salmonella typhi* and *paratyphi* A are thought to be restricted to humans alone. A key virulence factor in most strains of *S.typhi* is the Vi capsular antigen which possesses immunomodulatory properties that are thought to contribute to disease pathogenesis, including limiting complement deposition, reducing immune activation, assisting with phagocytosis evasion, and inhibiting serum bactericidal activity [7, 13, 14]. Without it, *S. typhi* will be more susceptible to attack and destruction by the host immune system. Hence, the Vi antigen has been harnessed as a major component of typhoid vaccines including the new conjugate vaccines [7].

Transmission is through the faeco-oral route from contaminated food, water and unrestricted contact with chronic carriers especially in an unhygienic environment. When *Salmonella typhi* is ingested, it evades degradation by enzymes and gastric acid before entering the host's system primarily through the terminal ileum [15]. At the distal ileum, through specialised structures called fimbrae, they attach to the epithelial cells overlying clusters of lymphoid tissues called Peyer patches. These serve as a relay point for macrophages travelling from the gut to the lymphatic system. Activation of the macrophages at the Peyer's patches release cytokines which attract more macrophages to the site. These macrophages serve as a vehicle by which S. typhi is transported to several parts of the reticuloendothelial system including the liver, spleen and bone marrow where they replicate up to a critical density at this point [16], they break into the bloodstream and invade other parts of the body. One of such places invaded is the gall bladder. The gall bladder is infected haematogenously or through infected bile. Infected bile is then secreted into the gut where it once again comes in contact with the Peyer patches at the distal ileum. This second sensitization of the macrophages at this site results in inflammation and hypertrophy of the lymphoid tissues (typhoid ileitis) [15, 16]. This enlargement encroaches on the blood supply resulting in ischaemic coagulative necrosis and consequently perforation and peritonitis. Some of the *salmonella* is excreted in the stool which is serves as a source of infection spread. This is the source of transmission of *Salmonella* in chronic carriers where the *salmonella* is thought to avoid enzymatic and chemical degradation in the gall bladder for a long time by forming biofilms or entering an intracellular 'comfort zone' in the gall bladder epithelium.

## 4. Clinical presentation

Enteric fever presents with a number of nonspecific symptoms and a wide variation in severity. Symptoms must be correlated with laboratory investigation to reach a diagnosis. Symptoms generally include fever, constipation, diarrhoea, abdominal pain, lethargy, nausea and vomiting, malaise, headache, truncal rash (rose spot), anorexia etc. The incubation period for enteric fever is 1–3 weeks and symptoms progressively get worse over the course of illness if not promptly treated [17].

In the first week, patients may complain of headache, malaise, intermittent fever, cough and constipation. Bradycardia may also be elicited on clinical examination. In the presence of fever, this is termed Faget sign or sphygmothermic dissociation. This is also seen in yellow fever, Brucellosis, Tularaemia and Colorado Tick fever.

In the second week, the patient appears dull with diarrhoea and apathy, sustained pyrexia, distended, tender abdomen and sometimes red macules (rose spots). Splenomegaly may also be present in 75% of cases.

In the third week if still untreated, patient become very ill, delirious and toxic with high pyrexia, intestinal haemorrhage and perforation. Toxic myocarditis may also ensue.

10% of cases relapse within the first 3 weeks of apparent recovery or completion of treatment, hence adequate monitoring and follow up should be arranged.

In the United Kingdom, any *Salmonella* infection is notifiable to Public Health England. Most cases occur in travellers returning from endemic areas. The PHE has developed certain criteria which would serve as an invaluable tool especially for primary care physicians in the early identification of suspicious cases for further escalation, assessment and confirmation.

According to Public Health England (PHE), cases can be classified into confirmed, possible and probable cases based on the following criteria (**Table 1**) [18]:

| Confirmed Case | Probable Case | Possible Case |
|---|---|---|
| • A person *with S. typhi* or *S. paratyphi* infection determined by the Public Health England Gastrointestinal Bacteria Reference Unit<br>OR<br>• A person with documented confirmatory evidence from a recognised overseas reference laboratory | • Local laboratory presumptive identification of *Salmonella typhi* or *paratyphi* on faecal and/or blood culture or culture of another sterile site (e.g. urine), with or without clinical history compatible with enteric fever<br>OR<br>• A returning traveller giving a clinical history compatible with enteric fever and documentation of a positive blood/ faecal culture (or positive PCR for *S. typhi* / *S. paratyphi* on blood) and/or treatment for enteric fever overseas | • A person with a clinical history compatible with enteric fever and where the clinician suspects typhoid or paratyphoid as the most likely diagnosis<br>OR<br>• A person with clinical history of fever and malaise and/or gastrointestinal symptoms with an epidemiological link to a source of enteric fever e.g. if they have 'Warn and inform' information<br>OR<br>• A returning traveller reporting a diagnosis abroad with positive serological testing or *Salmonella* PCR from faeces but no documented evidence of a positive blood or faecal culture positive |

**Table 1.**
*PHE classification of Enteric fever cases*

It must be noted that the typical presentation of course of enteric fever may deviate significantly from that described above. These may include pneumonia, delirium, arthralgias and severe jaundice. Younger children, people living with AIDS and one third of immunocompetent adults may present with diarrhoea instead of the classical constipation. The typical step ladder pyrexia is now only seen I 12% of cases with the fever pattern now mostly of the insidious persistent type [7]. Untreated or poorly treated infections may result in orchitis, intestinal ileitis, haemorrhage and perforation, meningitis, osteomyelitis.

## 5. Diagnosis

The diagnosis of enteric fever is made by correlation of clinical and laboratory investigations. The current gold standard as recommended by the world health organisation (WHO) is blood culture, although this may be culture of bone marrow, stool or urine depending on the time in the course of infection at which the sample was taken [4, 7, 15]. Even blood culture has been found to be an imperfect gold standard, hence there is an advocacy in some quarters for the use of a composite reference standard (CRS) to improve estimation of diagnostic accuracy [19]. The CRS involves combination of several diagnostic tests to increase the sensitivity rather than relying on individual tests. However, at present there is no consensus as to which tests should be included in the CRS [19]. This may be the future gold standard but further research is needed.

In low- and middle-income countries where the disease is endemic, access to contemporary diagnostic tests may be a challenge and a lot of patient in these countries pay out of pocket for health service delivery which they may not be able to afford. Hence, there is a case for empirical treatment based on clinical symptoms. However, this should be seen as a last resort and priority should be given to improving access to a simple, effective rapid diagnostic test (RDT) which is both reliable and valid. At present, although there are RDTs available commercially in endemic areas such as Typhidot, TUBEX and Test-it, their diagnostic accuracy is uncertain [20].

A lot of laboratories in low resource settings are still very much dependent on the Widal test. The Widal test is a serologic agglutination test developed by F Widal in 1896 [21]. The test is based on the presence of antibodies against the flagellar H and somatic O antigens of *Salmonella* typhi. Over the years, it has become a lot more controversial and largely abandoned in developed countries [21]. The main limitations with the test include a high cross- reactivity with other infectious agents (like nontyphoidal *salmonella*, plasmodium and tuberculosis), past enteric fever and BCG vaccination history. Other limitations include poor performance technique and result interpretation. Therefore, its use should be restricted to situations where there is no other supportive confirmatory test [21].

There are lots of other tests in development which hold promise for the future of enteric fever diagnosis. The antibody-in lymphocyte-supernatant (ALS) test has demonstrated good sensitivity and specificity in endemic settings [7, 22]. Others include PCR-based assays and high through-put technologies on clinical specimens using mass spectrometry [23].

## 6. Clinical management

In primary care, a thorough history and clinical examination could be suggestive. Unstable patients or those at high risk of deterioration should be referred to secondary care for same day hospital assessment and treatment. The mainstay of enteric fever treatment is antibiotics. The route of administration is often oral and parenteral in primary and secondary care respectively. There are several options of antibiotics and first line choice is usually determined by national and local guidelines according to sensitivities and antibiotics resistance pattern. In most places, fluoroquinolones such as ciprofloxacin and ofloxacin are first line. Second line antibiotics include third generation cephalosporins such as ceftriaxone, ampicillin, co-amoxiclav and trimethoprim-sulphamethoxazole [7]. Over the years, several options of antibiotics have been preferred but have changed based on resistance patterns. Decades ago, top on the list of antibiotics were chloramphenicol, ampicillin and co-trimoxazole [7]. The resistance to these traditional antibiotics resulted in multidrug resistance (MDR) typically conferred via IncHI1 plasmids, harbouring resistance genes such as *catA*, *sul1*, *sul2*, *dfrA*, $bla_{\text{TEM-1}}$, *strA*, *strB*, *tetA*, *tetB*, *tetC*, and *tetD* on composite transposons [7]. MDR strains were responsible for several outbreaks of enteric fever in the 1980/1990s and led to the widespread use of fluoroquinolones as first-line therapy [7, 24]. In the event of MDR and fluoroquinolone resistance, third generation cephalosporins provided respite and an effective alternative. Unfortunately, there are now emerging resistant strains to fluroquinolone and cephalosporin especially in Africa, south-east Asia and the Indian subcontinent resulting in extreme drug-resistance [7]. Fluoroquinolone resistance occurs mainly via chromosomal mutations in the *gyrA*, *gyrB*, *parC,* and *parE* genes. The local pharmacologist or microbiologist should be involved in discussion of the treatment of such cases where recommendations can be made for the use of other options. Such options would likely include Azithromycin, tigecycline or the monobactam, Aztreonam. On a positive note, re-emerging sensitivity to the traditional antibiotics of chloramphenicol, ampicillin and co-trimoxazole is being reported after the prolonged decline in their use [25]. This makes a case for strict adherence to antibiotics stewardship and similar trend may be the case for lot of other infectious diseases which is worth exploring. It is inevitable that various forms of MDR may emerge in future and antibiotic guidelines have to evolve to reflect the trend.

Also, chronic carriers may be treated with a combination of medical and surgical interventions. About 80% clearance rate can be achieved with a 28- day course of ciprofloxacin 750 mg twice daily or norfloxacin 400 mg twice daily. Azithromycin may be beneficial in those with floroquinolone resistance. In chronic carriers with cholethiasis, cholecystectomy under antibiotic cover is indicated and those with schistosomal infection should be covered with praziquantel [7, 24].

## 7. Prevention

As earlier discussed, central to the transmission and pathogenesis of enteric fever is poor sanitation and hygiene standards and lack of access to safe drinking water. Therefore, public health interventions targeted at addressing these will go a long way in reducing the global burden of the disease.

Another factor responsible for high global burden is the emergence of antimicrobial resistance strains which result in treatment failure and increased carriage rate [7, 14, 16]. In light of this, it is totally rational that the development of a highly efficacious vaccine will significantly reduce global burden. This is highly important especially for endemic areas with high disease burden and those at high risk of complications especially young children. A number of typhoid vaccines have been developed over the years and others are still in various stages of development with varying degrees of efficacy. Examples include Ty21a and Vi-polysaccharide vaccines which have shown efficacy at 2 years of 58% (95% CI 40–71%) and 59% (95% CI 45–69%), respectively [7]. Typhoid conjugate vaccines (TCVs) have been developed using the Vi-polysaccharide vaccine covalently linked to a protein to enhance immunogenicity, antibody quality, magnitude and duration [7]. Recent studies have shown better and long-lasting immunogenicity from TCV than Vi alone. For instance a prototype Vi-rEPA vaccine made up of Vi covalently linked to rEPA, a recombinant exoprotein A from *Pseudomonas aeruginosa* demonstrated efficacy of up to 91% (95% CI 77–97%) at 2 years, when given as a two dose schedule in 2–5 year-old children and protection lasted at least 4 years [7, 26, 27].

In England, the two main vaccines available are the Vi vaccine given as a single injection and the Ty21a vaccine available in the form of three capsules to be taken on alternate days. It is also available in combination with hepatitis A vaccine with protection lasting 1 year and 3 years for hepatitis A and Typhoid respectively [28]. Vaccination is highly recommended for people who are travelling to high risk areas including the Indian subcontinent, Africa, South America, South and South-east

Asia [28]. The Ty21a being live-attenuated, should not be given to immunocompromised patients or children below six years of age. However, the Vi vaccine can be given from the age of 2 years [28].

The World Health Organisation (WHO) strategic Group of Experts on immunisation, in October, 2017, recommended the inclusion of TCVs in vaccination programme schedules in endemic countries from 6 month of age and catch-up vaccinations in children and adolescents up to 15 years old where it is feasible and appropriate [7, 29]. However, this is yet to be implemented in most of these countries due to several factors including lack of political will, poor funding and in some cases poor uptake due to local cultural beliefs.

In 2008, the Vaccine Alliance (Gavi) made TCVs a priority as part of the typhoid investment initiative but did not make any financial commitments due to unavailability of a suitable vaccine [29], however, with the development of promising vaccine candidates with clinically appreciable efficacy albeit in the short- to-medium term, in November 2017, Gavi committed an $85 million funding window to support the roll out of these vaccines in eligible countries between 2019 and 2020 [7, 29] and in In January 2018, the WHO prequalified the TYPBAR-TCV. Since then, three countries have applied for support from Gavi, which includes a request for TCV use in response to an outbreak. The first Gavi-supported introduction of the TCV began in 2019 with Pakistan being the first country to request the vaccine in response to widespread transmission of an of extreme drug resistant strain (XDR) of *Salmonella typhi*. Following on this in the same year, Zimbabwe also applied for the TCV to combat an outbreak of drug resistant strains in Harare and was the first non-research use of the TCV in sub-Saharan Africa [29]. It is hoped that many more countries will apply for vaccine support and increase coverage to enhance reduction in global disease burden.

Another important factor driving an increase in disease burden is antibiotics resistance. This is mainly in endemic countries where the implementation of antibiotic stewardship is still a huge challenge. In these areas, a lot of antibiotics can be bought over-the-counter and patients often get them from the local chemist or patent medicine store without having to see a clinician. This has fostered the propagation of drug resistant strains of *Salmonella typhi* and *paratyphi* resulting in treatment failure, increased morbidity and mortality and increased tendency for chronic carriage. To mitigate this, there has to be a deliberate policy in these countries to better control access to antimicrobials, improved access to rapid diagnostic tests and public sanitation measures like clean, safe water, running pipe-borne water, clean toilet and waste disposal facilities.

## 8. Conclusion and recommendation

In conclusion, enteric fever caused by *Salmonella typhi* and *paratyphi* is still a huge global challenge and remains of major public health concern especially in low resource settings. The fundamental reasons for this unrelenting disease burden are multifactorial. Major factors include poor hygiene and sanitation measures, lack of access to portable drinking water, antibiotics resistance and poor access to vaccines. There has to be a strong political will for disease surveillance and primary care interventions such as ensuring the typhoid vaccines are included in routine immunisation schedules in endemic countries especially for children. The traditional typhoid vaccines held some challenges including convenience of administration and poor immunogenicity. However, the TCVs with better administration convenience especially in children as young as six months and longer-term protection, hold huge promise for the future of typhoid vaccine prophylaxis. Also, more countries need to seize the opportunity for vaccine support provided by Gavi to improve vaccine uptake. There also has to be increased investment in research and development into novel vaccines and diagnostic tools which are accessible, available, reliable and affordable. With the aforementioned and improved commitment to environmental, food and hygiene status, we will hopefully combat this scourge and forge a better, healthier future with enteric fever disease burden reduced to the barest minimum.

## Author details

Abdulmaleek Idanesimhe Sado[1*] and Aduke Oluwambe Sado[2]

1 Penwortham St. Mary's Group, Penwortham, Preston, Lancashire, United Kingdom

2 Department of Stroke Medicine, Royal Preston Hospital, Preston, Lancashire, United Kingdom

*Address all correspondence to: sadoabdulmalik@gmail.com

## References

[1] Kirchhelle C, Pollard AJ, Vanderslott S. Typhoid-From Past to Future. Clin Infect Dis [Internet]. 2019;69(Suppl 5):S375-6. Available from: http://www.ncbi.nlm.nih.gov/pubmed/31612936

[2] Chau TT, Campbell JI, Galindo CM, Van Minh Hoang N, Diep TS, Nga TTT, et al. Antimicrobial drug resistance of *Salmonella enterica* serovar typhi in asia and molecular mechanism of reduced susceptibility to the fluoroquinolones. Antimicrob Agents Chemother [Internet]. 2007 Dec;51(12):4315-23. Available from: http://www.ncbi.nlm.nih.gov/pubmed/17908946

[3] Kumar P, Kumar R. Enteric Fever. Indian J Pediatr [Internet]. 2017 Mar;84(3):227-30. Available from: http://www.ncbi.nlm.nih.gov/pubmed/27796818

[4] Crump JA, Luby SP, Mintz ED. The global burden of typhoid fever. Bull World Health Organ [Internet]. 2004 May;82(5):346-53. Available from: http://www.ncbi.nlm.nih.gov/pubmed/15298225

[5] Crump JA, Ram PK, Gupta SK, Miller MA, Mintz ED. Part I. Analysis of data gaps pertaining to *Salmonella enterica* serotype Typhi infections in low and medium human development index countries, 1984-2005. Epidemiol Infect [Internet]. 2008 Apr;136(4):436-48. Available from: http://www.ncbi.nlm.nih.gov/pubmed/17686194

[6] Breiman RF, Cosmas L, Njuguna H, Audi A, Olack B, Ochieng JB, et al. Population-Based Incidence of Typhoid Fever in an Urban Informal Settlement and a Rural Area in Kenya: Implications for Typhoid Vaccine Use in Africa. Gosling RD, editor. PLoS One [Internet]. 2012 Jan 19;7(1):e29119. Available from: https://dx.plos.org/10.1371/journal.pone.0029119

[7] Gibani MM, Britto C, Pollard AJ. Typhoid and paratyphoid fever. Curr Opin Infect Dis [Internet]. 2018 Oct;31(5):440-8. Available from: http://journals.lww.com/00001432-201810000-00010

[8] Radhakrishnan A, Als D, Mintz ED, Crump JA, Stanaway J, Breiman RF, et al. Introductory Article on Global Burden and Epidemiology of Typhoid Fever. Am J Trop Med Hyg [Internet]. 2018 Sep 6;99(3_Suppl):4-9. Available from: http://www.ajtmh.org/content/journals/10.4269/ajtmh.18-0032

[9] Atamanalp SS, Aydinli B, Ozturk G, Oren D, Basoglu M, Yildirgan MI. Typhoid Intestinal Perforations: Twenty-six Year Experience. World J Surg [Internet]. 2007 Aug 20;31(9):1883-8. Available from: http://link.springer.com/10.1007/s00268-007-9141-0

[10] Sharma A, Sharma R, Sharma S, Sharma A, Soni D. Typhoid intestinal perforation: 24 Perforations in one patient. Ann Med Health Sci Res [Internet]. 2013;3(5):41. Available from: http://www.amhsr.org/text.asp?2013/3/5/41/121220

[11] Public Health England. NOIDs Weekly Report. Statutory Notifications of INfectious Diseases in England and Wales. Week 2020/53 Week ending 03/01/2021 [Internet]. 2021. Available from: https://assets.publishing.service.gov.uk/government/uploads/system/uploads/attachment_data/file/949867/NOIDS-weekly-report-week53-2020.pdf

[12] Crump JA, Sjölund-Karlsson M, Gordon MA, Parry CM. Epidemiology, Clinical Presentation, Laboratory Diagnosis, Antimicrobial Resistance, and Antimicrobial Management of Invasive Salmonella Infections. Clin Microbiol Rev [Internet]. 2015 Oct

15;28(4):901-37. Available from: https://cmr.asm.org/content/28/4/901

[13] Hart PJ, O'Shaughnessy CM, Siggins MK, Bobat S, Kingsley RA, Goulding DA, et al. Differential Killing of *Salmonella enterica* Serovar Typhi by Antibodies Targeting Vi and Lipopolysaccharide O:9 Antigen. Skurnik M, editor. PLoS One [Internet]. 2016 Jan 7;11(1):e0145945. Available from: https://dx.plos.org/10.1371/journal.pone.0145945

[14] Raffatellu M, Chessa D, Wilson RP, Dusold R, Rubino S, Bäumler AJ. The Vi Capsular Antigen of *Salmonella enterica* Serotype Typhi Reduces Toll-Like Receptor-Dependent Interleukin-8 Expression in the Intestinal Mucosa. Infect Immun [Internet]. 2005 Jun;73(6):3367-74. Available from: https://iai.asm.org/content/73/6/3367

[15] Raffatellu M, Chessa D, Wilson RP, Tükel C, Akçelik M, Bäumler AJ. Capsule-mediated immune evasion: a new hypothesis explaining aspects of typhoid fever pathogenesis. Infect Immun [Internet]. 2006 Jan;74(1):19-27. Available from: http://www.ncbi.nlm.nih.gov/pubmed/16368953

[16] Parry CM, Hien TT, Dougan G, White NJ, Farrar JJ. Typhoid fever. N Engl J Med [Internet]. 2002 Nov 28;347(22):1770-82. Available from: http://www.ncbi.nlm.nih.gov/pubmed/12456854

[17] Crump JA. Progress in Typhoid Fever Epidemiology. Clin Infect Dis [Internet]. 2019 Feb 15;68(Supplement_1):S4-9. Available from: https://academic.oup.com/cid/article/68/Supplement_1/S4/5320155

[18] PHE. Recommendations for the Public Health Management of Gastrointestinal Infections. 2019; Available from: https://assets.publishing.service.gov.uk/government/uploads/system/uploads/attachment_data/file/614875/Public_Health_Operational_Guidelines_for_Typhoid_and_Paratyphoid.pdf

[19] Storey HL, Huang Y, Crudder C, Golden A, de los Santos T, Hawkins K. A Meta-Analysis of Typhoid Diagnostic Accuracy Studies: A Recommendation to Adopt a Standardized Composite Reference. Schallig HDFH, editor. PLoS One [Internet]. 2015 Nov 13;10(11):e0142364. Available from: https://dx.plos.org/10.1371/journal.pone.0142364

[20] Wijedoru L, Donegan S, Parry C. Rapid Diagnostic Tests for Typhoid and Paratyphoid (Enteric) Fever. In: Wijedoru L, editor. Cochrane Database of Systematic Reviews [Internet]. Chichester, UK: John Wiley & Sons, Ltd; 2010. Available from: http://doi.wiley.com/10.1002/14651858.CD008892

[21] Olopoenia LA. Classic methods revisited: Widal agglutination test - 100 years later: still plagued by controversy. Postgrad Med J [Internet]. 2000 Feb 1;76(892):80-4. Available from: https://pmj.bmj.com/lookup/doi/10.1136/pmj.76.892.80

[22] Islam K, Sayeed MA, Hossen E, Khanam F, Charles RC, Andrews J, et al. Comparison of the Performance of the TPTest, Tubex, Typhidot and Widal Immunodiagnostic Assays and Blood Cultures in Detecting Patients with Typhoid Fever in Bangladesh, Including Using a Bayesian Latent Class Modeling Approach. Baker S, editor. PLoS Negl Trop Dis [Internet]. 2016 Apr 8;10(4):e0004558. Available from: https://dx.plos.org/10.1371/journal.pntd.0004558

[23] Kuhns M, Zautner AE, Rabsch W, Zimmermann O, Weig M, Bader O, et al. Rapid Discrimination of *Salmonella enterica* Serovar Typhi from Other Serovars by MALDI-TOF Mass Spectrometry. Chakravortty D,

editor. PLoS One [Internet]. 2012 Jun 29;7(6):e40004. Available from: https://dx.plos.org/10.1371/journal.pone.0040004

[24] Parry CM, Hien TT, Dougan G, White NJ, Farrar JJ. Typhoid Fever. N Engl J Med [Internet]. 2002 Nov 28;347(22):1770-82. Available from: http://www.nejm.org/doi/abs/10.1056/NEJMra020201

[25] Zellweger RM, Basnyat B, Shrestha P, Prajapati KG, Dongol S, Sharma PK, et al. A 23-year retrospective investigation of Salmonella Typhi and Salmonella Paratyphi isolated in a tertiary Kathmandu hospital. PLoS Negl Trop Dis [Internet]. 2017 Nov;11(11):e0006051. Available from: http://www.ncbi.nlm.nih.gov/pubmed/29176850

[26] Lin FYC, Ho VA, Khiem HB, Trach DD, Bay P Van, Thanh TC, et al. The Efficacy of a Salmonella typhi Vi Conjugate Vaccine in Two-to-Five-Year-Old Children. N Engl J Med [Internet]. 2001 Apr 26;344(17):1263-9. Available from: http://www.nejm.org/doi/abs/10.1056/NEJM200104263441701

[27] Szu SC. Development of Vi conjugate – a new generation of typhoid vaccine. Expert Rev Vaccines [Internet]. 2013 Nov 9;12(11):1273-86. Available from: http://www.tandfonline.com/doi/full/10.1586/14760584.2013.845529

[28] NHS England. Typhoid Vaccination [Internet]. Available from: https://www.nhs.uk/conditions/typhoid-fever/vaccination/

[29] Gavi. Typhoid Vaccine [Internet]. Available from: https://www.gavi.org/types-support/vaccine-support/typhoid

Chapter 3

# In-Silico Approach in the Development of *Salmonella* Epitope Vaccine

*Hidajah Rachmawati, Raditya Weka Nugraheni and Firasti A.N. Sumadi*

## Abstract

In the case of infection control, one of our primary concerns is typhoid fever. According to WHO, typhoid prevalence in Indonesia is highly endemic. There is also the problem with the low efficacy of the available vaccine to prevent the disease. Therefore, there is an urgent need to develop a highly effective typhoid vaccine. One of the phases in vaccine development is an exploratory phase, a research-intensive phase of the vaccine development process designed to identify natural or synthetic antigens that might help prevent or treat a disease through computer in silico prediction targets. The vaccines developed through epitope peptide are designed to be safer, more efficacious, and less expensive than traditional vaccines. A thorough understanding of the disease agent, particularly critical epitopes to induce the appropriate immunological reaction, is required to achieve these aims. Mapping epitope sequences or antigenic peptides from pathogenic proteins recognized by B cells and T cells is crucial for vaccine development. Once the epitopes were identified, the polypeptide production could be produced through protein recombinant technology. The polypeptide vaccine, in the end, could be delivered using a liposomal delivery system.

**Keywords:** epitope, vaccine, typhoid, infection, control

## 1. Introduction

### 1.1 The urgency of infection control

Research about vaccines is urgently required because vaccinations are still the most effective way to prevent illness, disability, and death from vaccine-preventable diseases, such as Diphtheria, Pertussis, and Tetanus. WHO reported that global immunization successfully averts 2–3 million deaths of children every year. However, this achievement can still be improved, which means an additional 1,5 million deaths can be prevented. 19,4 million children worldwide are still missing out on essential vaccines. One of the critical strategies to improve global vaccine coverage is to provide the vaccine at all times and all places, in the best quality.

In the case of infection control, one of the major concerns is typhoid fever. According to WHO, typhoid prevalence in Indonesia is highly endemic. Typhoid disease still has to get serious attention because of its increasingly complex problems, making it difficult to manage, treat, and prevent [1]. This problem becomes even more difficult with the increasing resistance to commonly used antibiotic drugs. At present, there have even been reported cases of resistance to many drugs (multidrug resistance) spread throughout the world [2]. There is also a problem with the low efficacy of the available vaccine to prevent the disease. Therefore, there is an urgent need to develop a highly effective typhoid vaccine.

## 2. Six stage vaccine development

According to the CDC [3], there are six vaccine development stages: exploratory, preclinical, clinical development, regulatory review, approval, manufacturing, and quality control.

**Exploratory:** This research-intensive phase of the vaccine development process is designed to identify "natural or synthetic antigens that might help prevent or treat a disease." Antigens might include weakened strains of a particular virus/bacteria.

**Preclinical:** During this phase, researchers use tissue-culture or cell-culture systems and animal testing to determine whether the candidate vaccine will produce immunity. Many candidate vaccines do not move on to the next stage of development because they fail to produce that immunity or prove harmful to test subjects.

**Clinical development:** At this point, a sponsor, usually a private company, submits an application for an Investigational New Drug (IND) to the U.S. Food and Drug Administration (FDA) or BPOM (*Badan Pengawas Obat dan Makanan,* National Food, and Drug Agency) in Indonesia. This step summarizes findings to date and describes how the drug will be tested and created. An institution that will host the clinical trial holds a review board for approval of the application. The FDA has 30 days to approve the application. Once the proposal has been approved, the vaccine must pass three trial stages of human testing.

**Regulatory review and approval**: If a vaccine passes through all three clinical development phases, the vaccine developer submits the registration documents to the regulatory board.

**Manufacturing:** Major drug manufacturers provide the infrastructure, personnel, and equipment necessary to create mass quantities of vaccines. They also reap the profits of successful or widely distributed drugs.

**Quality control:** The approval and distribution are far from the end of the line. Stakeholders must adhere to procedures that allow them to track whether a vaccine is performing as anticipated.

## 3. Improving safety and efficacy of typhoid vaccine using epitope vaccine

On the other hand, there are vaccine safety and efficacy issues that cannot be ignored. There are three different types of vaccination developed for *S. typhi*: live-attenuated pathogens, inactivated pathogens, and sub-unit vaccines. From the safety point of view, sub-unit vaccines provide better safety profiles because they only use specific proteins and could not be reverted into a virulent form. However, conventional protein isolation usually results in a minimal yield; thus, we need to develop an epitope vaccine. Epitope vaccine is a part of the subunit vaccine, which only uses the antigen's epitope area. The interaction of epitopes and antibodies are particular, and the peptides are well characterized. Therefore, we can produce the peptides for the epitope vaccine using the recombinant technique.

Until now, vaccines for typhoid fever that have been available and show the safety and effectiveness of several clinical trials and are recommended by the CDC

(Center for Disease Control and Prevention, USA) are oral Ty21a vaccine and ViCPS vaccine (Vi capsular polysaccharide) given parenterally [4]. Ty21a is a vaccine that uses a weakened organism (oral attenuated vaccine). This orally administered vaccine is technically more comfortable to use because it does not cause pain but can be virulent if given to an immunocompromised individual. ViCPS vaccine is a parenterally administered independent T-cell antigen that gives uncomfortable pain to the patients [5].

The development of bioinformatics tools and advances in recombinant DNA technology (rDNA) and the knowledge on the host immune response and the genetic background of the pathogen will lead to new vaccines against diseases that currently have few or no control measures in just 1 or 2 years. Through computer in silico predictions to define targets. The vaccines developed through rDNA technologies are safer, more efficacious, and less expensive than traditional vaccines. A thorough understanding of the disease agent, particularly critical epitopes to induce the appropriate immunological reaction, is required to achieve these aims [6].

The epitope is part of the antigen that would be recognized by the antibody [7]. Different epitopes of protein antigens can be identified based on sequences from amino acids or different conformational forms. Some epitopes are hidden in antigen molecules and exposed as a result of physicochemical changes. Epitope vaccine is part of the subunit/peptide vaccine. Peptide vaccines can be used to induce broad-spectrum immunity against some serological variants (serovar) or certain pathogenic strains by formulating several non-contiguous immunodominant epitopes and conserved epitopes between different serovars/pathogenic strains.

On the other hand, due to the relatively small peptides, they are often immunogenic weak on their own and therefore require carrier molecules to add chemical stability and adjuvants to induce a robust immune response. Allergenicity and molecular reactogenicity of the carrier itself increases the complexity of the peptide vaccine design. Making peptide vaccines are generally considered safe and cost-effective when compared to conventional vaccines [8].

## 4. Stages development of epitope vaccine

In principle, a material's antigenic nature shows how much the antigen's ability to bind to antibodies and cause different reactions in human immunity formation. Antibodies or immunoglobulins are specialized proteins that are products of differentiated B lymphocytes or plasma cells. The bond between antibodies and antigens induces systemic immunity and activates the complement to process further activate the humoral immune system [9].

Specific antibodies can be made on an individual's body by immunizing selected peptides that present epitopes of these proteins. Epitopes play an essential role in vaccine development. Mapping epitope sequences or antigenic peptides from pathogenic proteins recognized by B cells and T cells is crucial for vaccine development. Epitope mapping provides useful information for designing peptide-based vaccines and as libraries to monitor specific cellular immunity in protected individuals, patients, and vaccines [8].

B cell epitope mapping is divided into linear and nonlinear B cell epitope mapping. Although minor, linear B cell mapping has further attention in vaccine research because linear epitopes are epitopes ready to replace antigens in immunization. There are various epitope mapping method using different approaches and algorithms; for instance Kolaskar and Tongaonkar [10], Bepipred, Preditop, ABCPred, LBtope, and many others [11].

The IEDB (The Immune Epitope Database) maps experiments identify and characterize epitopes and epitope-specific receptors with related details, including host organisms, immune exposures, and induced immune responses. The genes that encode polypeptides are then cloned into pRSETA, a bacterial vector for high-level expression of proteins, plasmid vectors. Transformation of the recombinant plasmid in *E. coli* host was followed by induction with IPTG (Isopropyl ß-D-1-thiogalactopyranoside), which resulted in a polypeptide expression as observed in SDS-PAGE (Sodium dodecyl sulfate polyacrylamide gel electrophoresis) and Western blot analysis. Purified polypeptides were subjected to step dialysis. The final concentration was quantified and adjusted to 2.5 mg/ml by Lowry's calorimetric assay against BSA (Bovine Serum Albumin) standard [12].

## 5. In silico experiments as tools to design epitopes for vaccine

Epitope-based vaccine design using this computational method silico is an effective strategy that can lead to vaccine development to induce the necessary immunogenicity without the emergence of a cytokine storm or immune tolerance. Based on several in vitro studies and in vivo, if scientifically and critically designed, epitope-based vaccines offer several advantages over other types of vaccines, including their fast design and accurate, time/cost effective formulations, and desired immunogenicity with minimal side effects [13].

There is no denying that vaccination is beneficial in promoting a healthy global population. This act has saved countless lives, reduced healthcare costs, and improved the quality of human life. Accidental discoveries in immunology are augmented by knowledge about bioinformatics tools for epitope prediction, resulting in the emergence of a pattern new vaccine design. The art and science of efficient and comprehensive information extraction and analysis of data stored in relevant databases is of increasing importance in research related to immunology [14].

Fortunately, although research in experimental immunology are expensive and highly intensive, usually large amounts of data are generated. Such data can only be analyzed with high precision and high-speed using bioinformatics tools. For example, genome sequencing as well as in vitro T-cell confirmation takes place within a few months. With conventional vaccine designs, computational immunology methods drastically reduce time and labor requirements in epitope screening. With computational immunological techniques, it is possible to find vaccine candidate epitopes only by scanning deep protein sequences of the desired pathogen. Many of these proteins have not been isolated or at least cloned into specific and unique pathogens, and they present a ready candidate in vaccine construction [14].

This in silico strategy also helps in selecting better molecules before testing conditions in vitro or in vivo. In this early stage, the use of in silico methods can direct and thus significantly shorten the next experimental work. Besides, proper use of the silico method can replace, reduce and improve the usage of animal experiments that are often misleading and time-consuming [15].

The position of the in-silico method in the vaccine development process lies at the preclinical development stage. In vaccine development, the first steps are to do is identify a vaccine candidate. The preclinical stage aims to determine the safety profile of the vaccine. During this stage, the researchers will carefully select the appropriate antigen and technology, and in vitro and in vivo will do. The information gathered from this study will be essential to continue with the subsequent clinical trials in humans [16].

There are some advantages of using the In silico method [17]. This technique offers an advantage in giving new drug candidates faster and at a lower cost. It is also Increases the chances of success in the many stages of the discovery process and facilitates access to the large amount of data generated. In silico experiments, it was also turning massive complex biological data into useful knowledge.

Revolution in information technology and molecular biology, together with growth in genome data storage, has provided the basis for vaccine design using computational and bioinformatics tools. These tools are used for silico mapping of the most precise and immunogenic components for the manufacture of a hypothetical protein. The vaccine that is designed can then be simulated and evaluated prior to experimental validation, enabling the process research and development (R&D) to be carried out efficiently, and leads to the development of vaccines with few adverse effects [13]. Here is some software that can be used to help in silico design of epitope vaccine.

### 5.1 Mega x™

Over the last decade, genome sequencing has become an efficient and efficient way to investigate a wide variety of biological systems, ranging from diversity studies large-scale biology to tracking the evolution and origin of pathogenic microbes. The steps needed to gather interpretable and interpretable results actionable from raw sequence data always require a comparative analysis of molecular sequences to find differences in functional and adaptive genomes. The Molecular Evolutionary Genetics Analysis (MEGA) software provides the tools to do that analysis. MEGA includes many programs for assembling sequence alignments, inferring evolutionary trees, estimating genetic distances and diversities, inferring ancestral sequences, computing time trees, and testing selection [18]. MEGA software or other software allows alignment of *salmonella* typhi genome sequences from various countries to determine which sequences are considered sustainable. The selection of sustainable sequences is vital for the manufacture of vaccines so that later the vaccine can be used in many countries.

### 5.2 IEDB

The Immune Epitope Database (IEDB, iedb.org) [19] captures experimental data limited in figures, texts, and tables of the scientific literature, making them available online free and easily searchable by the public. The scope of the IEDB includes data on immune epitopes associated with all studied species and includes the antibody binding context, T cells, and MHCs associated with infection, allergy, autoimmune, and related diseases transplant [20]. The IEDB is a website that provides tools computations that focus on the prediction and analysis of B and T cell epitopes. The IEDB maps experiments identify and characterize epitopes and epitope-specific receptors with related details, including host organisms, exposures immunity, and induced immune response. The site associated with the IEDB is the ISDBA Analysis Resource, which is a predictor of various B cell and T cell epitopes. Using a trained and validated algorithm [21]. An antigenicity analysis was carried out using the Kolaskar and Tongaonkar principles followed by epitope mapping using Bepipred. Both are done based on a database owned by IEDB (The Immune Epitope Database).

IEDB web server using some parameters, including Kolaskar-Tongaonkar as the antigenicity scale parameter, which is the standard for epitope prediction [22]. The Kolaskar and Tongaonkar antigenicity scales based on the amino acid residues' physicochemical properties and the known tendency frequency as an experimental epitope have an accuracy of 75% [10]. The BepiPred-2.0 server predicts B cell epitopes from protein sequences, using a Random Forest algorithm trained on epitopes and amino acids non-epitope determined from the crystal structure [21, 23]. The Immune Epitope Database (IEDB) retrieves experimental data through pictures, texts, and tables of scientific literature, even the scope of the IEDB extends to the immune epitope, where data are linked to all studied species including antibodies, T cells, and the binding context MHC is associated with infections, allergies, autoimmune, and transplant-related diseases [20].

### 5.3 VAXIJEN 2.0™

VaxiJen is the first server for the prediction of protective antigens, antigens tumor, and vaccine subunits. It is also a grade-free bioinformatics tool first for silico immunogen identification. VaxiJen uses the Z-scale Wold to explain the Physicochemical properties of the primary amino acids building protein. The Amino acids are tested, converting the derived string to a uniform vector with auto cross-covariance (ACC). The next step is to select the relevant variable with a genetic algorithm (G.A.) or gradual regression and classify protein as a protective or non-antigen antigen by discriminant analysis least-squares based (PLS). Initially, Algorithm files are trained to identify the protective immunogens of bacteria. A model for immunogens viruses and tumors was then included, and VaxiJen was developed to provide access free to the model. The latest version of VaxiJen (VaxiJen 2.0) also includes a model for identifying parasitic and fungal immunogens [15]. Vaxijen used to validate the results of the epitope sequences found from the IEDB whether they are immunogenic or not.

## 6. Epitope mapping model using IEDB

The IEDB was first published in 2004 using data that limited numbers, text, and tables of scientific literature. Finally, in 2015, the IEDB experiments' number increased by 140% to exceed 1.6 million, and receptor sequence data in the IEDB scheme. Previously, this device could only capture antibodies and the T cell receptor (TCR) but can now capture antibodies and complementarity determining regions (CDR), which is essential for the antigen's specificity of its diversity. The scope of this device extends throughout the epitope for data relating to all species includes immunity body/antibody, the context of T cell binding, and MHC-related infections, allergies, autoimmune, and transplantation of certain diseases and features to access also summarize data in terms of quantity and complexity [20].

An antigenicity analysis is used, and epitope mapping is also carried out (**Figure 1**) to analyze areas that have antigenicity potential against B-cell, for example, using Bepipred and Emini methods. Both are done based on a database owned by IEDB (The Immune Epitope Database). Selection of epitopes according to the score above the threshold. Epitope mapping was performed using Bepipred software from the immune epitope database (HTTP://toolsiedb.ofg/bcell/) to find linear B cell epitopes from a sustainable region with an average threshold value of 0.030. This method is classically used to measure propensity.

Moreover, using hidden Markov programming [21], we show the epitope mapping of the OMP28 protein, a typhus vaccine candidate researched by Saxena et al. (2012). OMP28 is outer membrane protein 28 from *Salmonella enterica subsp. enterica* serovar Typhi with Accession number NCBI: ACX42427. The OMP28 protein sequence is MNKFSLATAGIIVAALVTSVSVN
AATDTTKTNVTPKGMSCQEFVDLNPQTMAPVAFWVLNEDEDFKGGD
YVDFQETETTAVPLAVELCKKNPQSELSKIKDEIKKELSK. Preferably, before starting epitope mapping, we need to make sure that the sequence that will be

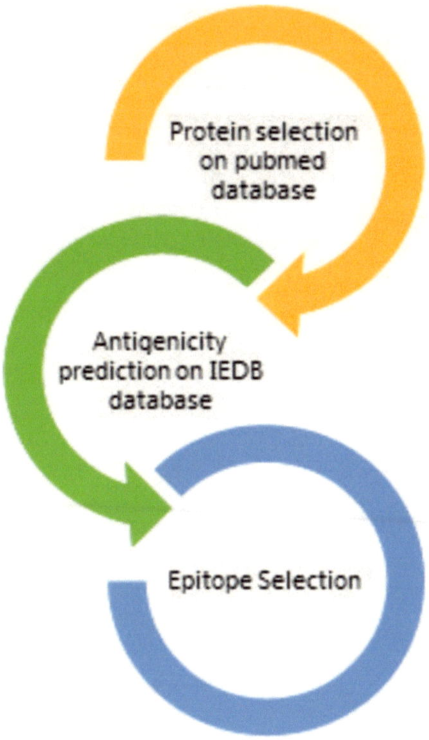


**Figure 1.**
*Steps of epitope mapping.*

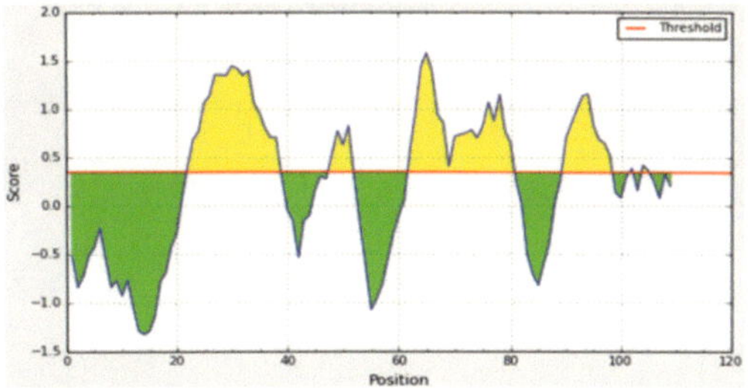


**Figure 2.**
*B cell epitope mapping using the IEDB Bepipred method (IEDB.org) [19].*

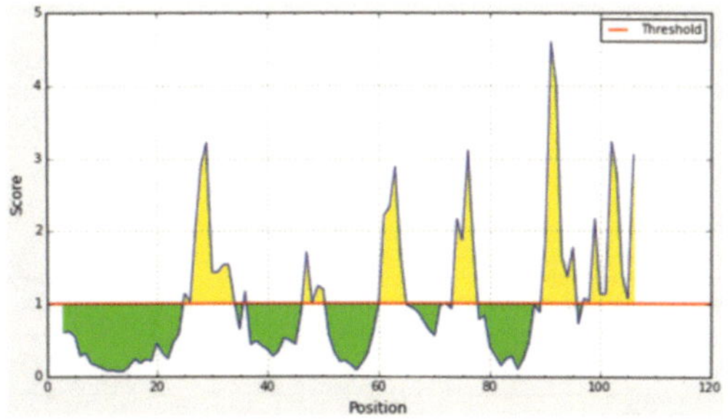


**Figure 3.**
*B cell epitope mapping using the IEDB Emini method (IEDB.org) [19].*

analyzed is a conserved region. This step can be done by aligning many of the same protein sequences from different strains of *salmonella*.

The epitope mapping results are shown in **Figures 2** and **3**; the yellow area is considered to have high antigenicity potential. In bepipred linear epitope prediction, a value equal to or greater than the threshold value of 0.030 is said to have a strong potential to bind to B cells. This is also done for other methods, such as the emini surface method with a threshold limit of 1.

From the **Tables 1** and **2**, it can be seen that the highest epitope potential is in positions 62th–80th because it has the highest score 1.581 with a long sequence of 19 amino acid sequences, and 22–38 positions with a score of 1.449 along with

| **Bepipred Method Analysis (Treshold = 1,030)** | | | |
|---|---|---|---|
| **Score Epitope Prediction** | | | |
| **Position** | **Residue** | **Score** | **Assignment** |
| 65 | F | 1.581 | E |
| 30 | K | 1.449 | E |
| 64 | D | 1.436 | E |
| 31 | T | 1.422 | E |
| 33 | V | 1.399 | E |

**Sequence Epitope Pediction**

| **No.** | **Start** | **End** | **Peptide** | **Length** |
|---|---|---|---|---|
| 1 | 22 | 38 | VNAATDTTKT NVTPKGM | 17 |
| 2 | 48 | 51 | PQTM | 4 |
| 3 | 62 | 80 | DEDFKGGDYV DFQETETTA | 19 |
| 4 | 90 | 98 | KNPQSELSK | 9 |
| 5 | 102 | 102 | E | 1 |
| 6 | 104 | 105 | KK | 2 |

**Table 1.**
*Resume of Bepipred analysis on OMP28 sequence.*

**Emini Analysis (Treshold = 1) Score Epitope Prediction**

| **Position** | **Residue** | **Start** | **End** | **Peptide** | **Score** |
|---|---|---|---|---|---|
| 91 | N | 89 | 94 | KKNPQS | 4.605 |
| 92 | P | 90 | 95 | KNPQSE | 3.988 |
| 102 | E | 100 | 105 | KDEIKK | 3.235 |
| 29 | T | 27 | 32 | DTTKTN | 3.221 |
| 76 | T | 74 | 79 | QETETT | 3.115 |

**Sequence Epitope Prediction**

| **No.** | **Start** | **End** | **Peptide** | **Length** |
|---|---|---|---|---|
| 1 | 25 | 34 | ATDTTKTNVT | 10 |
| 2 | 90 | 95 | KNPQSE | 6 |

**Table 2.**
*Resume of Emini analysis on OMP28 sequence.*

the 17 amino acid sequences. As a comparison, the antigenicity analysis was also performed using the emini surface technique, and the most significant potential was obtained at positions 25–34 with a sequence length of 10 amino acids, namely ATDDTKTNVT.

Based on the analysis of bepipred and emini surface, the sequences that have the potential to provide the greatest immunogenicity can be identified at positions 22–38 and positions 62–80. Furthermore, this peptide can be produced by the recombinant protein method. The multi-epitope candida vaccine is considered more promising than the single epitope vaccine.

## 7. Epitope vaccine delivery

One of the drawbacks of using peptide fragments as vaccine antigens is the weak immunogenicity generated compared to inactive and live-attenuated vaccines [24]. A formulation with the addition of adjuvant ingredients is needed to increase the immune response in the subjects. One example of adjuvants that have been used commercially for a long time in vaccine formulations is aluminum salts or "alum" [25]. Several publications report alum mechanisms such as creating a gel depot that prolonged exposure of the immune system to antigen, forming particulate structures that promote antigen uptake by APCs (Antigen Presenting Cells) via phagocytosis, and inducing inflammation and secretion of chemokines. However, its inability to induce Th1 cell-mediated immune responses also becomes a limitation for this adjuvant [26], Since the previous study indicated that the typhoid sub-unit vaccine-induced cellular immunity through TCD4+ [27].

The benefit of liposome as carrier and adjuvant for antigen has been known for quite a long time [28]. Liposomes can boost immune response due to its mechanism to create depot effect by causing antigen retention and slowly releasing them to the immune cells [29]. Liposome delivery of antigen is influenced by lipid bilayer components such as lipid choice and the role of cholesterol [30]. We can either choose neutral lipid or charged lipid as the membrane constructing unit. However, previous research showed that positively charged lipid like DDA (Dimethyldioctadecyl ammonium) has an advantage over neutral lipid [31–33]. Cationic DDA can effectively interact with the APC cell membrane and increase the number of antigens delivered into the cell [34]. To put it simply, the more antigens delivered, the more significant immune responses are. However, DDA could not produce a stable vesicle due to its positive charge due to electro-repulsive force. Therefore, previous publications mentioned that DDA should be formulated with phosphatidylcholine as a major lipid component and cholesterol to maintain membrane integrity [35, 36].

## 8. Conclusion

There is an urgent need to develop a highly effective typhoid vaccine, especially in a highly endemic region. One of the reasons above is that this paper's rationale is to find a faster, cheaper, and more efficient vaccine candidate design. The vaccine design used is the Peptide vaccine, and the design is done in silico. Peptide vaccine has several advantages when compared to conventional vaccines. The advantages of peptide vaccines can be seen in specifications, disease, purity, production capacity, and production cost efficiency.

## Acknowledgements

The authors are grateful to the University of Muhammadiyah Malang authorities, especially the Faculty of Health Sciences and Medical Faculty, for the technical facilitation to conduct the research related to the epitope vaccine.

## Conflict of interest

The authors declare no conflict of interest.

## Author details

Hidajah Rachmawati*, Raditya Weka Nugraheni and Firasti A.N. Sumadi
University of Muhammadiyah Malang, Malang, Indonesia

*Address all correspondence to: hidajah@umm.ac.id

## References

[1] Buckle GC, Walker CLF, Black RE. Typhoid fever and paratyphoid fever: Systematic review to estimate global morbidity and mortality for 2010. J Glob Health [Internet]. 2012 Jun [cited 2020 Dec 31];2(1). Available from: https://www.ncbi.nlm.nih.gov/pmc/articles/PMC3484760/

[2] Ugboko H, Nandita D. Mechanisms of Antibiotic resistance in Salmonella typhi. Int J Curr Microbiol Appl Sci. 2014 Aug 8;3:461-76.

[3] CDC. Vaccine Testing and Approval Process | CDC [Internet]. 2020 [cited 2020 Dec 31]. Available from: https://www.cdc.gov/vaccines/basics/test-approve.html

[4] Slayton RB, Date KA, Mintz ED. Vaccination for typhoid fever in Sub-Saharan Africa. Hum Vaccines Immunother. 2013 Apr 1;9(4):903-6.

[5] MacLennan CA, Martin LB, Micoli F. Vaccines against invasive Salmonella disease: current status and future directions. Hum Vaccines Immunother. 2014;10(6):1478-93.

[6] Soria-Guerra RE, Nieto-Gomez R, Govea-Alonso DO, Rosales-Mendoza S. An overview of bioinformatics tools for epitope prediction: implications on vaccine development. J Biomed Inform. 2015 Feb;53:405-14.

[7] Abbas AK, Lichtman AH, Pillai S. Cellular and Molecular Immunology E-Book. Elsevier Health Sciences; 2011. 557 p.

[8] Li Pira G, Ivaldi F, Moretti P, Manca F. High Throughput T Epitope Mapping and Vaccine Development. J Biomed Biotechnol [Internet]. 2010 [cited 2020 Dec 31];2010. Available from: https://www.ncbi.nlm.nih.gov/pmc/articles/PMC2896667/

[9] Wang Y, Wu W, Negre NN, White KP, Li C, Shah PK. Determinants of antigenicity and specificity in immune response for protein sequences. BMC Bioinformatics. 2011 Jun 21;12:251.

[10] Kolaskar AS, Tongaonkar PC. A semi-empirical method for prediction of antigenic determinants on protein antigens. FEBS Lett. 1990 Dec 10;276(1-2):172-4.

[11] Sanchez-Trincado JL, Gomez-Perosanz M, Reche PA. Fundamentals and Methods for T- and B-Cell Epitope Prediction [Internet]. Vol. 2017, Journal of Immunology Research. Hindawi; 2017 [cited 2020 Dec 31]. p. e2680160. Available from: https://www.hindawi.com/journals/jir/2017/2680160/

[12] Babu L, Uppalapati SR, Sripathy MH, Reddy PN. Evaluation of Recombinant Multi-Epitope Outer Membrane Protein-Based *Klebsiella pneumoniae* Subunit Vaccine in Mouse Model. Front Microbiol. 2017;8:1805.

[13] Parvizpour S, Pourseif MM, Razmara J, Rafi MA, Omidi Y. Epitope-based vaccine design: a comprehensive overview of bioinformatics approaches. Drug Discov Today. 2020 Jun;25(6):1034-42.

[14] Oli AN, Obialor WO, Ifeanyichukwu MO, Odimegwu DC, Okoyeh JN, Emechebe GO, et al. <p>Immunoinformatics and Vaccine Development: An Overview</p> [Internet]. Vol. 9, ImmunoTargets and Therapy. Dove Press; 2020 [cited 2020 Dec 31]. p. 13-30. Available from: https://www.dovepress.com/immunoinformatics-and-vaccine-development-an-overview-peer-reviewed-article-ITT

[15] Flower DR, Doytchinova I, Zaharieva N, Dimitrov I.

Immunogenicity Prediction by VaxiJen: A Ten Year Overview. J Proteomics Bioinform [Internet]. 2017 [cited 2020 Dec 31];10(11). Available from: https://www.omicsonline.org/open-access/immunogenicity-prediction-by-vaxijen-a-ten-year-overview-jpb-1000454-95687.html

[16] International Federation of Pharmaceutical Manufacturers & Associations. The Complex Journey of a Vaccine – The Steps Behind Developing a New Vaccine – IFPMA [Internet]. IFPMA Resources; 2014 [cited 2020 Dec 31]. Available from: https://www.ifpma.org/resource-centre/the-complex-journey-of-a-vaccine-final/

[17] Gangrade D, Sawant G, Mehta A. Re-thinking drug discovery: In silico method. 2016;8.

[18] Kumar S, Stecher G, Li M, Knyaz C, Tamura K. MEGA X: Molecular Evolutionary Genetics Analysis across Computing Platforms. Mol Biol Evol. 2018 Jun 1;35(6):1547-9.

[19] IEDB. IEDB.org: Free epitope database and prediction resource [Internet]. [cited 2020 Dec 31]. Available from: http://www.iedb.org

[20] Vita R, Mahajan S, Overton JA, Dhanda SK, Martini S, Cantrell JR, et al. The Immune Epitope Database (IEDB): 2018 update. Nucleic Acids Res. 2019 Jan 8;47(D1):D339-43.

[21] Dhanda SK, Mahajan S, Paul S, Yan Z, Kim H, Jespersen MC, et al. IEDB-AR: immune epitope database-analysis resource in 2019. Nucleic Acids Res. 2019 Jul 2;47(W1):W502-6.

[22] Baidya S, Das R, Kabir MdG, Arifuzzaman Md. Epitope design of L1 protein for vaccine production against Human Papilloma Virus types 16 and 18. Bioinformation. 2017 Mar 31;13(3):86-93.

[23] Jespersen MC, Peters B, Nielsen M, Marcatili P. BepiPred-2.0: improving sequence-based B-cell epitope prediction using conformational epitopes. Nucleic Acids Res. 2017 Jul 3;45(W1):W24-9.

[24] Gause KT, Wheatley AK, Cui J, Yan Y, Kent SJ, Caruso F. Immunological Principles Guiding the Rational Design of Particles for Vaccine Delivery. ACS Nano. 2017 Jan 24;11(1):54-68.

[25] Shardlow E, Mold M, Exley C. Unraveling the enigma: elucidating the relationship between the physicochemical properties of aluminium-based adjuvants and their immunological mechanisms of action. Allergy Asthma Clin Immunol. 2018 Nov 7;14(1):80.

[26] Bobbala S, Hook S. Is There an Optimal Formulation and Delivery Strategy for Subunit Vaccines? Pharm Res. 2016 Sep 1;33(9):2078-97.

[27] Rachmawati H, Winarsih S, Prawiro SR, Barlianto W, Santoso S, Djunaedi D, et al. AdhO36 Liposomes from Salmonella Typhi in Combination With β-Glucan Immuno-adjuvant From *Candida albicans* Cell Wall as Oral Vaccine Against Typhoid Fever in Mice Model. Open Access Maced J Med Sci. 2020 May 25;8(A):441-8.

[28] Allison AG, Gregoriadis G. Liposomes as immunological adjuvants. Nature. 1974 Nov 15;252(5480):252.

[29] Wilkinson A. Nanotechnology for the delivery of vaccines [Internet] [phd]. Aston University; 2014 [cited 2016 Mar 12]. Available from: http://eprints.aston.ac.uk/21411/

[30] Giddam AK, Zaman M, Skwarczynski M, Toth I. Liposome-based delivery system for vaccine candidates: constructing an effective formulation. Nanomed. 2012 Dec;7(12):1877-93.

[31] Kaur R, Henriksen-Lacey M, Wilkhu J, Devitt A, Christensen D, Perrie Y. Effect of Incorporating Cholesterol into DDA:TDB Liposomal Adjuvants on Bilayer Properties, Biodistribution, and Immune Responses. Mol Pharm. 2014 Jan 6;11(1):197-207.

[32] Hamborg M, Jorgensen L, Bojsen AR, Christensen D, Foged C. Protein Antigen Adsorption to the DDA/TDB Liposomal Adjuvant: Effect on Protein Structure, Stability, and Liposome Physicochemical Characteristics. Pharm Res. 2013 Jan;30(1):140-55.

[33] Liu X, Da Z, Wang Y, Niu H, Li R, Yu H, et al. A novel liposome adjuvant DPC mediates *Mycobacterium tuberculosis* subunit vaccine well to induce cell-mediated immunity and high protective efficacy in mice. Vaccine. 2016 Mar 8;34(11):1370-8.

[34] Smith Korsholm K, Agger EM, Foged C, Christensen D, Dietrich J, Andersen CS, et al. The adjuvant mechanism of cationic dimethyldioctadecylammonium liposomes. Immunology. 2007 Jun;121(2):216-26.

[35] Kett V, Yusuf H, McCarthy H, Chen KH. Liposomal delivery system [Internet]. 2015 [cited 2016 Aug 25]. Available from: http://www.google.ch/patents/US20150079156

[36] Yusuf H, Ali AA, Orr N, Tunney MM, McCarthy HO, Kett VL. Novel freeze-dried DDA and TPGS liposomes are suitable for nasal delivery of vaccine. Int J Pharm. 2017 Nov;533(1):179-86.

Chapter 4

# Computational Identification of the Plausible Molecular Vaccine Candidates of Multidrug-Resistant *Salmonella enterica*

*Rohit Mishra, Yong Chiang Tan,*
*Amr Adel Ahmed Abd El-Aal and Chandrajit Lahiri*

## Abstract

*Salmonella enterica* serovars are responsible for the life-threatening, fatal, invasive diseases that are common in children and young adults. According to the most recent estimates, globally, there are approximately 11–20 million cases of morbidity and between 128,000 and 161,000 mortality per year. The high incidence rates of diseases like typhoid, caused by the serovars Typhi and Paratyphi, and gastroenteritis, caused by the non-typhoidal Salmonellae, have become worse, with the ever-increasing pathogenic strains being resistant to fluoroquinolones or almost even the third generation cephalosporins, such as ciprofloxacin and ceftriaxone. With vaccination still being one of the chosen methods of eradicating this disease, identification of candidate proteins, to be utilized for effective molecular vaccines, has probably remained a challenging issue. In our study here, we portray the usage of computational tools to analyze and predict potential vaccine candidate(s) for the multi-drug resistant serovars of *S. enterica*.

**Keywords:** typhoid, *Salmonella* Typhi, multidrug resistance, computational identification, vaccine candidates

## 1. Introduction

With a current worldwide prevalence of around twenty-seven million cases [1, 2] and hundreds of thousands of deaths every year [2, 3], salmonellosis remains the second most common food/water-borne illness. It constitutes a disease caused due to the systemic infection of human and animal hosts by the facultatively anaerobic, Gram-negative rod-shaped bacterial species of *Salmonella enterica* from the family *Enterobacteriaceae*. Clinically, several serologic variants (serovars) of *S. enterica* exist, which differ with respect to their different antigenic variation in lipopolysaccharide and flagella [4, 5]. They include Typhi and Paratyphi A, besides the non-typhoidal serotypes like Typhimurium and Enteriditis [4]. Among these, the enteric fever termed typhoid, caused by *S.* Typhi and Paratyphi, is typically a more severe illness than those caused by other non-typhoidal serovars [5].



IntechOpen

Being contagious in nature, salmonellosis, like typhoid, can spread through feces, water and the hands of those caring for the sick while, for non-typhoidal serovars, through the consumption of raw or undercooked contaminated food of animal origin such as meat, poultry, eggs and milk by humans [1, 6, 7]. Salmonellosis begins with ingestion of a dose for the bacterium enough to broach the first-line host defenses and colonize the gastrointestinal tract. The onset symptoms for typhoid are usually accompanied with fever, headache, myalgia, anorexia and sometimes diarrhea or constipation [6, 7], moving onto remittent fever, with a stepwise increment in the daily peak temperature, reaching 40°C by the end of the first week [6]. Slow recovery after 3–4 weeks is the normal case, though, for untreated patients with complications, major fatalities occur due to intestinal hemorrhage or perforation [6, 7].

Drugs available for the treatments are mostly ineffective due to the resistance developed with the emergence of multidrug-resistance (MDR) *Salmonella* strains [8]. These new strains are ineffective to the older generations of drugs including ampicillin, chloramphenicol, ciprofloxacin, trimethoprim as well as co-trimoxazole and their derivatives, thereby necessitating the newer classes of cephalosporins and quinolone derivatives to be greatly explored to combat such MDR threats [1, 8]. Moreover, dating as early as the 1890s, whole-cell vaccines with parenteral administration of killed suspensions of *S.* Typhi [9] has several problems having: a) high-reactivity with 20–25% fever and 40–50% local reactions, b) moderate efficacy with protection rates of 51–88% insufficient to halt disease transmission in endemic area and c) logistical and safety problems having the need for needles and two doses. Approaches with recent vaccines, like, single-dose Typhim Vi® containing purified Vi capsular polysaccharide, or, the live attenuated vaccine *S.* Typhi Ty21a (Vivotif®), confer around 50% protection in adults, and very poor immunogenicity among young children, without any license for under two years old, besides being considered to be expensive for low-middle income areas [10, 11]. Thus, the urgency, for new and specific vaccines and/or drugs to combat the disease, is evident and indeed, proteins of the pathogen-specific biochemical and biosynthetic pathways, involved in the virulence of *S.* Typhi, has already begun to be targeted with a view to developing novel vaccines/drugs.

While the two afore-mentioned vaccines are for *S.* Typhi, those for other serovars including Paratyphi, Typhimurium and Enteritidis were largely unavailable until some few years back [11]. Of late, efforts to confer protective immunity for serovars of Typhimurium has been reported with the *lppA* and *lppB* Braun lipoprotein genes with and without the *msbB* gene, encoding an acetyltransferase enzyme required for modification of the lipid A of lipopolysaccharide [12]. Other candidate genes proposed for effective vaccines for different serovars include *rpoS*, *phoPQ*, *ssaV*, *htrA* [13], besides the proteins of SseBI, OmpACDFL and SopB being used as antigens in other vaccination studies [14]. Such recombinant attenuated *Salmonella* vaccines (RASV) are considered to be same or more effective than the whole wild-type strains [15]. RASV can persistently colonize internal lymphoid tissues to produce recombinant antigens having their maximum abilities to elicit mucosal and systemic antibody along with those of the cell mediated immune responses [15]. Thus, development of such recombinant vaccines is considered to be the cost-effective and most promising strategy against the pressing antibiotic resistance threats. In this regard, several strategies have been adopted in other drug resistant bacteria including reverse vaccinology through comparative genome analysis and *in vitro* proteomics [16, 17]. These become especially effective keeping in mind the new and emerging threats of multidrug resistance strains of *Salmonella*. Such strains might possibly arise form immune selection leading to antigen

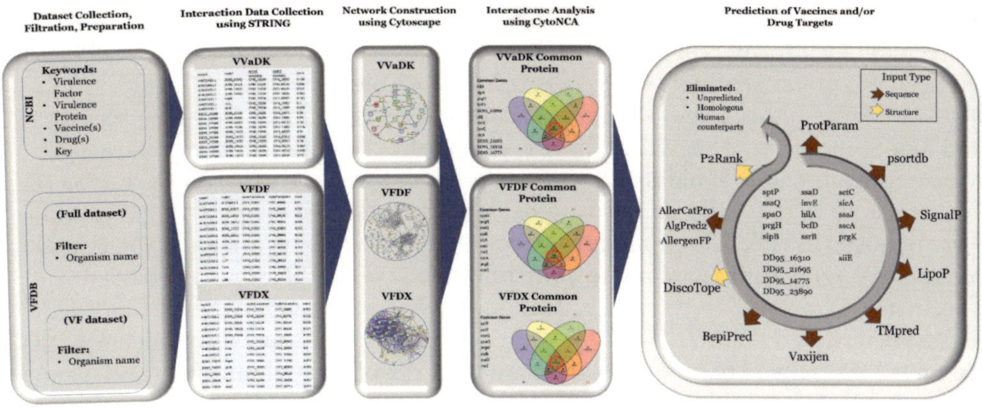

**Figure 1.**
*Graphical summary of the methods adopted in vaccine candidates and druggability prediction. This comprises a network-based approach to identify the key players in* Salmonella *virulent proteome coupled with downstream predictions of vaccine candidates and druggable pockets among the top rankers.*

sequence variability followed by a down-regulation of the target antigens, thereby conferring poor "cross-protective efficacy" as reported for MDR *Acinetobacter baumannii* [18]. Therefore, identification of new and effective vaccine candidates is, probably, the current need of the hour.

With an availability of different virulent proteins, reported from different experimental verification and predictive databases, selection of the most plausible vaccine candidates can be confusing. To cater to the need of simplifying this complex problem of selection, graph theoretical analysis of the interacting networks of such virulent proteins, involved in the disease scenario, might be poised to be quite useful. Such virulent protein interaction networks (PIN) can be utilized to find out the most central or sought-after proteins for such cases [19]. Ideally, the centrality of any biological networks is efficiently analyzed through global parameters like betweenness, closeness, degree and eigen-vector centralities, referred to as the BC, CC, DC and EC, respectively [19–21]. Among them, BC has been regarded to be efficient enough to impart central character of a network above CC and DC for long until EC gained some prominence and can be quite effective as reported through recent studies [22–25].

In this study, we proposed the vaccine candidates for *Salmonella* serovars (**Figure 1**) as explained in the next section. Essentially, we utilized the four different centrality measures for analyzing three different virulent PINs denoted as VVaDK, VFDF and VFDX. Among the top 20 rankers of each of the different centralities, the unanimously present unique candidates were finally collected for further downstream analyses. These shortlisted candidate virulent proteins were rigorously analyzed through different bioinformatic tools to determine their antigenic and allergenic potential besides revealing the epitopes for efficient vaccines or molecular crevices for good drug targets.

## 2. Approach

### 2.1 Dataset collection

We have initiated our study with the proteins collected for *Salmonella enterica* serovar Typhimurium str. LT2 (NCBI txid: 99287) on the 19th of December 2020.

They were retrieved from two different sources namely, the National Center for Biotechnology Information (NCBI) and the Virulence Factor Database (VFDB) [26]. From NCBI, protein datasets were collected through literature search using various keywords such as Virulence, Virulence Factor, Virulence Protein, Drug(s), Vaccine(s) and Key. Some of these keywords, having essentially the same meaning, were used to get more hits and to avoid missing of any possible candidates thereby reducing the false-negative hits. Finally, all the candidates of the lists were merged, and duplicates were removed to yield 120 proteins to be considered for further analysis. They were termed as VVaDK for easy reference, where V stands for Virulence, Va represents Vaccine(s), D means Drug(s) and K denotes Key. Moreover, two types of candidates' lists were retrieved from VFDB. They comprised the Full dataset which covers all the proteins (261) related to unknown and predicted VFs of *S.* Typhimurium and were referred as VFDF. Additionally, 117 experimentally verified candidates were retrieved for *S.* Typhimurium and termed as VFDX.

All the afore-mentioned proteins for the different categories of VVaDK, VFDF and VFDX were fed as queries to the biological meta-database of protein interaction, STRING version 11.0 [27] to retrieve all the possible interactions of a particular protein [date and time of access: Dec 22, 2020, from 17 hours IST onwards]. Detailed protein links file under the accession number 90371 in STRING v11 was used to collect all the interactions of the whole genome proteins of *S.* Typhimurium. In each case, a database dictated default medium confidence value of 0.4, for the combined scores from different parameters of interaction, was used. Accordingly, the total number of protein interactions obtained were 138, 3501 and 2464 for VVaDK, VFDF and VFDX listed candidates, respectively.

### 2.2 Interactome construction

The protein interaction data for all individual sets for VVaDK, VFDF and VFDX, having medium confidence values, were imported into Cytoscape version 3.8.2 [28] to integrate and build the respective interactomes of protein interactions. Care was taken to remove duplicate and bidirectional interactions from each dataset. In essence, such interactome of proteins or the protein interaction network (PIN) has been constructed as an undirected graph, $G = (V, E)$, consisting of E edges and a finite set of V vertices (or nodes) where, edge, $e = (u, v)$, is connected to two vertices u and v. Each vertex/node in our PIN represents a protein. The number of connections/interactions/associations/links, a protein has with other proteins, reflects its degree, d [29].

### 2.3 Network analysis

All the constructed 3 PINs have been viewed by Cytoscape v 3.8.2 in the form of interactomes of aforementioned interconnected proteins. They were subsequently analyzed through the integrated java plugin CytoNCA version 2.1.6 [30] to compute values for BC, CC, DC and EC as the four different global network centrality parameters. The different parametric combined scores from STRING were considered as edge weights for computing the CytoNCA scores of the 4 centrality parameters. Upon sorting these 4 measures from largest to smallest, top 20 proteins for each of the categories of centrality were picked to create Venn diagrams using Venny 2.0 [31] for finding the common proteins from each of the measures. This resulted in 12, 10, 7 proteins from VVaDK, VFDF and VFDX, respectively. Among these 29 candidates, 9 duplicates were removed to yield a total of 20 proteins. Through a

BLASTp alignment, these Typhimurium proteins were unanimously found in the serovars of Typhi and Paratyphi, and thus, considered for further analyses.

## 2.4 Vaccine and/or drug candidature prediction

### *2.4.1 Basic analysis*

The 20 shortlisted protein candidates from VVaDK, VFDF and VFDX PIN analysis were subjected to further analyses for predicting the plausible vaccine and/or drug candidates. All such proteins were explored for their molecular weight calculation, cellular localization, signal peptide prediction followed by antigenicity prediction. ProtParam was used to find the molecular weight and number of amino acids [32] and cellular localization was analyzed by PSORTb v3.0.2 [33]. Location of signal peptides was predicted using the server called SignalP 4.1 [34]. Lipoprotein signal peptides were predicted using the LipoP 1.0 [35]. Finally, Vaxijen was used to predict the possible antigenicity of the proteins [36].

### *2.4.2 Mapping of available 3D structures in PDB*

For the top ranked proteins, the respective crystallized protein 3D structures available in Protein Data Bank (PDB) were retrieved (**Table 1**). The seleno-methionine in PDB structures were changed back into methionine using Dock Prep in Chimera [37].

| Protein | Structural Information | | | |
|---|---|---|---|---|
| | PDB ID | Chain ID | Structure Coverage | Resolution |
| SptP | 1G4W | R | 161–543 | 2.20 |
| | 1JYO | E | 35–139 | 1.90 |
| SpaO | 4YX1 | A | 232–297 | 1.35 |
| | 4YX7 | A | 145–213 | 2.00 |
| | 4YX7 | B | 232–297 | 2.00 |
| PrgH | 4G1I | A | 170–392 | 1.85 |
| | 4G2S | A | 11–119 | 1.86 |
| | 6UOT | A | 1–392 | 3.30 |
| SipB | 3TUL | A | 81–237 | 2.79 |
| SsrB | 2JPC | A | 133–193 (First 19 N-terminal amino acids missing) | NMR |
| SctC | 4G08 | A | 22–178 | 1.80 |
| | 6PEE | A | 1–562 | 3.42 |
| PrgK | 6UOT | Y | 1–252 | 3.30 |
| | 4OYC | A | 96–200 | 2.60 |
| SiiE | 2YN5 | A | 5078–5365 | 1.85 |

*SsaQ, SsaD, InvE, HilA, BcfD, SicA, SsaJ, SscA, DD95_23890, DD95_21695, DD95_16310, and DD95_14775 have no structures available in PDB.*

**Table 1.**
*PDB structure availability among top rankers.*

*2.4.3 B-cell epitope prediction*

Unlike viral pathogens, most bacterial pathogens are not intracellular parasites, especially *Salmonella*. Thus, the humoral immune response, which involves B cells and antibodies, will be of great focus in this study. Herein, BepiPred v2.0 and DiscoTope v2.0 were utilized in predicting linear and discontinuous B-cell epitopes, respectively [38, 39]. For BepiPred, the default threshold score of 0.5 was applied for epitope recognition. For DiscoTope, the propensity score radius was 22 Angstrom, upper half sphere radius was 14 Angstrom, window size was 1, and alpha was 0.115. An in-house script (DiscoTope2ChimeraAttr) has been utilized to convert DiscoTope result into Chimera attributes for visualization in 3D, with a default threshold DiscoTope score of −3.7 [40]. These analyses were done to pinpoint the specific immunogenic regions within the full-length proteins. Thus, the immunogenically insignificant regions can be trimmed out, resulting in shorter peptides which can confer higher specificity and ease the peptide synthesis process.

*2.4.4 Allergenicity prediction*

The ability of proposed immunogen to potentially evoke allergic reactions can usually fail clinical trials due to the severe adverse effects arising upon vaccination. Herein, we utilized AllerCatPro, AlgPred2, and AllergenFP v1.0 to predict possible allergic reactions raised by the query proteins, which were the top rankers in this case. For AlgPred2, the hybrid algorithm was selected and the default threshold value of 0.3 was selected. AllerCatPro predicts allergenicity by comparing the protein structural and sequential information to known allergens [41]. Besides, the hybrid algorithm of AlgPred2.0 utilizes the random forest, BLAST, and MERCI algorithms to predict the allergenicity of the query proteins [42]. Moreover, the allergenicity prediction of AllergenFP v1.0 utilizes an alignment-independent fingerprint-based approach [43].

*2.4.5 Druggable pocket prediction*

P2Rank was being utilized to predict the presence of druggable pockets in the available 3D structures of proteins [44]. P2Rank utilizes a template-independent machine learning algorithm in predicting potential ligand-binding sites on the query proteins. Herein, the topmost ranked predicted pockets were selected for further analyses. Thus, besides being utilized in vaccination, the potential druggability of the top rankers can be discovered.

*2.4.6 Detecting human counterparts*

Peptide vaccines that contain regions of high sequence similarity to human proteome counterparts can lead to ineffective vaccination due to recognition as "self" by the immune system, which can result in low antigenicity or adverse effects that arise from potential self-reactivity. Thus, the top rankers were screened for human counterparts via sequence alignment approach using BLASTp against non-redundant proteins (nr) database with *Homo sapiens* as the specified organism [45].

## 3. Interactome analyses of three virulent PINs

Three different interactomes of virulent proteins of *Salmonella* were built using the method described above. The first of them comprised those available through literature search using different keywords comprising Virulence, Virulence Factor, Virulence Protein, Drug(s), Vaccine(s) and Key. This was named as VVaDK. The other two PINs were made of the full and experimentally verified datasets of virulent proteins from *Salmonella*, listed in VFDB and were named as VFDF and VFDX, respectively. The four centrality measures were applied for analyzing each of these PINs and twenty top rankers from each of the measures were initially segregated. Among them, the proteins present unanimously for all the measures were noted as 12, 10 and 7 for VVaDK, VFDF and VFDX, respectively, and a removal of duplicates from them finally yielded 20 candidates for further downstream analysis.

Our unique way of streamlining the candidates is based upon the following facts. Under pathological conditions, the virulent proteins are expected to be working in unison to render the final disease phenotype. Thus, their connectivity could be perceived in terms of the said PINs. Among these proteins, some can be master regulators and connecting to others more frequently thereby having higher order of connectivity. This renders them degree centrality (DC). Alternatively, there could be different types of such regulators for carrying out different subfunctions of the main disease phenotype and they form the bridge between the other proteins. These could impart the betweenness centrality (BC) of such proteins. Moreover, among such conglomerate of different proteins, certain numbers could connect to others faster to sequentially carry out their function, leading to a concept of closeness for them and having higher closeness centrality (CC). Furthermore, certain proteins could be more important to render the final disease phenotype and they are only connected to other important proteins to carry out their functions. These could bring out their character of eigen vector centrality (EC). Finally, from the top-ranking proteins of all these centrality measures, those, appearing unanimously, are expected to play a major role in virulence and could be segregated to scan for further analysis. These are 20 unique virulent proteins, mostly belonging to the *Salmonella* Pathogenicity Islands (SPI) from three different PIN analyses and reflected in **Figure 1** and **Table 2**. These are discussed in the next section.

## 4. Features of the twenty virulent proteins

All the virulent proteins from different serovars of *Salmonella* are discussed here, with their characteristic features along with a note on their existing vaccine potential.

**SptP** is one of the most important SPI-1 Type III Secretion System (T3SS) effector proteins which facilitates the bacterial translocation and survival into the host non-phagocytic cells by inhibition of the extracellular-regulated kinase (ERK) mitogen-activated protein kinase (MAP) pathways [46]. It requires SicP as a chaperone protein for its secretion and stabilization [46]. Moreover, SptP is directly responsible for the reversal of the actin cytoskeletal changes in the host cells by acting as a GTPase-activating protein (GAP) for Rac-1 and Cdc42. In fact, the efficacy of *sptP* deletion mutation of *S.* Enteriditis has been shown to be effective for live attenuated vaccine (LAV) in chickens [47].

**SsaQ** is a member of FliN/YscQ/Spa33/HrcQ family of both T3SS and flagellum proteins [48]. The gene *ssaQ* is encoded in the *ssaMVNOPQ* operon within the SPI-2 and transcribes to two products namely, SsaQL of 322 residues and SsaQS of 106 residues. SsaQS acts as a chaperone-like protein for SsaQL and optimize its function. SsaQ interact with SsaK and SsaN to form the C-ring complex, which have a crucial role in secretion by acting as a cytoplasmic sorting platform at the base of T3SS as well as rotation and direction switching of the flagella [49].

| Protein | ProtParam | | PsortB | SignalP & LipoP | TMpred | | | Vaxijen | |
|---|---|---|---|---|---|---|---|---|---|
| | # amino acids | Molecular Weight | Localization | Prediction: CSPosition | Prediction score | Status (Predicted) | # of TM Helices | Position | Score (Orientation) |
| SptP | 543 | 60047.68 | E | — | 0.5192 | A | 1 | 477–496 | 570 (o-i) |
| SsaQ | 322 | 36009.35 | C | — | 0.3857 | NA | 1 | 186–209 | 611 (i-o) |
| SpaO | 303 | 33793.74 | C | — | 0.5073 | A | 1 | 62–86 | 600 (o-i) |
| PrgH | 392 | 44459.53 | C | — | 0.5122 | A | 1 | 142–163 | 2551 (o-i) |
| SipB | 593 | 62450.71 | E | — | 0.4855 | A | 2 | 320–343<br>409–428 | 2293 (i-o)<br>2875 (o-i) |
| SsaD | 403 | 44849.66 | CM | — | 0.4319 | A | 1 | 119–135 | 2978 (o-i) |
| InvE | 372 | 42421.49 | CM | — | 0.3335 | N | 2 | 255–273<br>317–337 | 634 (o-i)<br>612 (i-o) |
| HilA | 553 | 63040.96 | C | — | 0.3985 | N | 1 | 340–361 | 523 (o-i) |
| BcfD | 335 | 35928.51 | U | SP(SPI): 21–22 | 0.6728 | A | 2 | 1–21<br>204–222 | 1342 (o-i)<br>582 (i-o) |
| SsrB | 212 | 24354.49 | C | — | 0.4053 | A | 0 | — | — |
| SctC | 562 | 61765.81 | OM | SP(SPI): 24–25 | 0.371 | N | 1 | 5–25 | 1725 (i-0) |
| SicA | 165 | 19220.72 | C | — | 0.5954 | A | 0 | — | — |
| SsaJ | 249 | 28521.47 | U | LIPO(SPII): 18–19 | 0.529 | A | 2 | 6–24<br>225–245 | 1371 (i-o)<br>2963 (o-i) |
| SscA | 157 | 18134.88 | C | — | 0.3731 | N | 0 | — | — |
| PrgK | 252 | 28210.3 | OM | LIPO(SPII): 17–18 | 0.5132 | A | 1 | 208–225 | 2812 (i-o) |

| Protein | ProtParam | | PsortB | SignalP & LipoP | TMpred | | | Vaxijen | |
|---|---|---|---|---|---|---|---|---|---|
| | # amino acids | Molecular Weight | Localization | Prediction: CSPosition | Prediction score | Status (Predicted) | # of TM Helices | Position | Score (Orientation) |
| DD95_23890 (BigA) | 1721 | 176526.46 | U | — | 0.8358 | A | 5 | 193–213<br>466–487<br>800–820<br>952–971<br>1141–1165 | 641 (i-o)<br>610 (o-i)<br>513 (i-o)<br>558 (o-i)<br>504 (i-o) |
| SiiE | 5559 | 594451.38 | U | — | 0.76 | A | 3 | 39–57<br>4349–4367<br>5013–5036 | 692 (o-I)<br>512 (i-o)<br>1827 (o-i) |
| DD95_21695 (SspH2) | 788 | 87222.97 | E | — | 0.382 | N | 0 | — | — |
| DD95_16310 (TorS) | 911 | 100273.65 | CM | — | 0.4384 | A | 3 | 10–29<br>332–350<br>622–641 | 2340 (i-o)<br>2738 (o-i)<br>1048 (i-o) |
| DD95_14775 (MarT_1) | 147 | 16757.61 | U | — | 0.4706 | A | 1 | 127–144 | 1622 (o-i) |

*TM: Transmembrane. For Localization, E: Extracellular, C: Cytoplasmic, CM: Cytoplasmic Membrane, OM: Outer Membrane, U: Unknown. For TMpred status, A: Antigen, N: Non-antigen.*

**Table 2.**
*Basic screening of plausible vaccine candidates.*

**SpaO** is a major invasion factor of *S. enterica spp.* and the core component of the sorting platform in *S.* Typhimurium. SpaO is comprised of 303 residues of two translated products with SpaOS (the shorter product) encompassing the last 101 amino acids of SpaOL (full length protein) [50]. It is a highly conserved element in T3SS that shares similarity with limited residues with flagellar C-ring substructure [51]. In fact, SpaO, along with H1a, has been suggested to be promising new vaccine candidates to prevent typhoid fever caused by *S.* Paratyphi A infection [52].

**PrgH** is a 55 kDa protein encoded within *prgHIJK* operon in the SPI-1. All the genes of *prg* operon are essential for the formation of T3SS needle complex (NC) and known to share sequence similarity with the flagellar protein, FliF [53]. PrgH inserts in the inner membrane by its hydrophobic domain where it forms the MS-ring of the flagellar basal body as well as provides the structural foundation required for *prgK* oligomerization for further assembly of the NC [53].

**SicA** is a wide acting chaperone protein (18 KDa) which aids in the secretion process of all T3SS proteins through the invasion of host cells. Accordingly, it is encoded upstream to the *Sip/SspABCD* operon in SPI-1. SipB and SipC proteins are responsible for the translocon formation in the host cell membrane to facilitate the injection of Type III effector proteins into the host cell to manipulate it [54]. Moreover, SicA is essential for the expression of the most virulence genes that encode T3SS effector proteins and is identified as a co-regulator with InvF for *SigDE* and *SptP* [55].

**HilA** is a member of the OmpR/ToxR regulator protein family and the central activator of SPI-1 genes, belonging to T3SS. The *hilA* gene is encoded within SPI-1 and is the key factor in SPI-1-T3SS regulation, starting from the expression of downstream genes *sicA* and *invF* to ultimate regulation of the effector genes *sipA* and *sipB* [56]. The upregulation of *hilA* results in the high expression of all genes encoded within the SPI-1 which are necessary for the invasion of epithelial cells. Moreover, the expression of *hilA* is controlled by many different activators and suppressors in response to specific environmental changes during invasion of the host cells, such as, temperature, bile, fatty acids, osmolarity, pH, oxygen concentrations and growth state [57]. Additionally, certain studies considered HilA as a promising drug target to inhibit the activity of T3SS without affecting the growth of *Salmonella* [58].

**SiiE** is the largest protein in *Salmonella* proteome, with the size of 595 kDa. It consists of 53 repetitive bacterial immunoglobulin domains, each containing several conserved residues [59]. The protein helps to contact the host cell membrane and positions the SPI T3SS, to initiate the translocation of effector proteins. A study states that *Salmonella* SiiE-mediated entry of enterocytes via the apical route requires transmembrane mucin MUC1 [60]. Moreover, it is shown that, *siiE* is required for the prevention of efficient humoral immune response against the pathogen and it induces the high tires of specific *Salmonella*-specific IgG [61].

**PrgK** is a component from the inner membrane of *Salmonella* SPI-1 T3SS basal body, in its N-terminus. It possess the canonical lipoproteins which acts as anchor for the hydrophilic proteins onto the surface of the bacterial cell membranes [62]. In addition, C-terminus of PrgK is found in the cytoplasm which confirms that the protein traverses the inner membrane. A study observed reduced fever in swine which were vaccinated with *prgK* gene attenuated *S.* Typhimurium in comparison with mock-vaccinated swine [63].

**SscA** is a chaperone protein of about 18 KDa size. It is an independent α-helical protein, that consists of eight α-helices and repeated large tetratricopeptide domain from 36 to 137 amino acids. SscA is a virulence factor which encodes the chaperonin of SseC and the translocon is involved during the adaptation and survival to desiccation [64]. A huge effect of the gene expression level of *sscA*, has been noted on treatment of the samples with ciprofloxacin [65].

**SsaJ** is a core encoding component of the T3SS. It is required for SpvB, in-order to induce the actin depolymerization, especially inside the human macrophages. *Salmonella* depends on SsaJ effector protein as it prevents the interaction of NADPH oxidase subunit Cytb558 with the *Salmonella* containing vesicle (SCV) thereby helping to avoid the oxidative burst [66]. An *in vivo* study, conducted with the peptide of SsaJ, however, showed its inability to provide antigen specific immunity when compared with the other chosen peptides [67].

**SctC** is a layer of outer membrane anchor forming two distinct outer rings namely, OR1 and OR2. It is homologous to a protein of Type II Secretion System (T2SS) which requires pilotin lipoprotein for its optimal assembly and localization [68]. SctC serves as a midline between the inner and outer membrane, with evidence showing that the translocation of foreign antigens can induce potent immune response against pathogens [69].

**SsrB** is responsible for the survival and replication of *Salmonella* in the host cell and plays an important role in the transcription of multiple genes of SPI-2. SsrB has been claimed as one of the most important factors for *Salmonella*'s virulence by the fact that, a mutated *ssrB*, resulted in reduced ability of colonization on comparing with the wild type [70]. Moreover, one alteration in the gene *ssrB*, preferentially silencing the acquired DNA, can have a high contribution towards low transcription in the virulence factors of *Salmonella* [71].

**BcfD** is a fimbrial protein and part of the operon Bcf [72]. BcfD is a surface molecule, which helps in the adherence through specific receptors on the host cell. This step of adhesion is considered to be an important course during infection as it allows bacteria to initiate the colonization [73]. A research shows that the knock-out of this gene influenced in the low adhesion capacity of *Salmonella* to the host cell [74].

**InvE**, encoded within SPI-1, is a protein located in the cell membrane and said to be essential for the translocation of *Salmonella* proteins into the host cells by regulating the functions of the Sip protein translocases [75]. An investigation of finding the region of InvE, as the T3SS regulator protein, indicates that it may have two functional domains which are responsible for regulating the secretion of translocases as N-terminal secretion signal and C-terminal regulatory domain [76]. An *in-vivo* study conducted with the BALB/c mice, showed less pathogenicity when it is injected with the mutated *invE* gene *Salmonella* on comparing with the wild strain [77].

**SipB** is one of the effector proteins of SPI-1 T3SS which facilitates the entry of *Salmonella* into the host cell. It is also called as an invasion protein as it initiates the bacterial entry process. It forms a complex along with the SipC to assemble into plasma membrane-integral structure which mediates the effectors delivery [78]. It also affects the membrane fluidity and bacterial osmotolerance and hence a small alteration of this gene will pave a huge way to prevent *Salmonella* entry into the host cell [79]. In fact, a study evaluating the effect of *sipB* deleted mutants, showed significant decrease in the virulence of *sipB* mutants when compared with the wild-type strains [80].

**SsaD** is an important cellular component which is responsible for the virulence of *Salmonella.* It is found to be in the transmembrane of the bacteria. The gene *ssaD* encodes for the proteins related to the basal body, cytoplasmic rings and export apparatus and it is also involved in the ATPase complex, regulation and translocation of T3SS [81]. A study shows that there is an important defect in the intercellular survival with the mutant *ssaD* strains on comparing with the wild-type *Salmonella* [82].

**DD95_23890** refers to the computationally predicted protein, mapping to the autotransporter adhesin BigA protein. The BigA protein in *Salmonella* has recently been identified via automated genome annotation in 2015. Thus, studies on this protein has been scarce. Inferring from its homolog in *Brucella*, the cell surface BigA protein promotes adhesion of bacteria on host epithelial cells [83, 84]. The adhesive properties of the BigA protein can be established by binding onto the cell adhesion molecules on the host epithelial cellular surface [85].

**DD95_21695** maps to the RING-type E3 ubiquitin transferase (SspH2) protein. The SspH2 protein aids in *Salmonella* pathogenicity by conferring anti-inflammatory properties, hence delaying the host immune response in reaction to bacterial invasion [86]. Moreover, the ability of SspH2 to ubiquitinate host NOD1 protein, through an essential interaction with host SGT1 protein, can result in NOD1-mediated IL-8 secretion in host [87].

**DD95_16310** maps to the *Salmonella* TorS histidine kinase sensor. The TorS protein comprises the two-component systems along with the TorT response regulator [88]. Upon stimulation by Trimethylamine-N-oxide, TorS, along with TorT, carry out osmoregulation and protect the cellular proteins against low-pH induced denaturation in urea [88].

**DD95_14775** refers to the putative transcriptional regulator *marT_1* in *Salmonella*. The MarT protein mainly regulates the expression of MisL autotransporter protein, which is a fibronectin-binding protein that is involved in the cell adhesive properties of *Salmonella* [89]. Moreover, MarT has also been reported to regulate the expression of genes related to bacterial biofilm formation [90].

## 5. Initial screening of the candidate proteins

All the twenty proteins were screened to ascertain their potential for plausible candidatures as vaccines (**Table 2**). Proteins were localized in extracellular matrix (3), cytoplasm (7), cytoplasmic membrane (3) and outer membranes (2), besides some of them being predicted with unknown cellular location (5). Of these, surface/outer membrane proteins and vesicles have been deployed for prospective vaccinations against bacterial pathogens [91–94]. Again, extracellular proteins have been potentiated as drugs for prospects against disease management, *albeit,* in a different scenario [95, 96]. Our results predict the proteins namely, SptP, SipB, SsaD, PrgK and TorS to be potentially antigenic except InvE, SctC and SspH2. Notably, the five proteins of unknown location, namely, BcfD, SsaJ, BigA, SiiE, and MarT_1 are all potentially antigenic. Of the two signal peptides BcfD and SctC, the latter was predicted to be non-antigenic while SsaJ and PrgK belongs to another category of signal peptides (lipoproteins) with good antigenic potential. Of these, SsaJ has been predicted with two transmembrane (TM) spanning helices and poses itself a good candidate for vaccines. Other candidates with more TM helices are BigA (5), SiiE (3) and TorS (3). Furthermore, a BLASTp alignment of these 20 proteins revealed SptP and SspH2 to have 40–50% similarity for 101 and 106 hits, respectively, against human counterparts, thereby completely ruling out their candidature as potential vaccines.

## 6. Selection of potential vaccine candidates

The 20 top ranked proteins were further screened for B cell epitopes. Therein, InvE, SsrB, SicA, and SscA were omitted from being considered as vaccine candidates due to the absence of predicted epitopes that fall within the normal range of peptide length (**Table 3**). Moreover, in allergenicity prediction, HilA,

| Protein | Start | End | Peptide | Length | Average Score |
|---|---|---|---|---|---|
| SptP | 5 | 25 | EERKLNNLTLSSFSKVGVSND | 21 | 0.5922 |
| | 59 | 78 | FKNTEVVQKHTENIRVQDQK | 20 | 0.5383 |
| | 378 | 396 | EDQMQAKQLPPYFRGSYTF | 19 | 0.5593 |
| SsaQ | 5 | 23 | ANEERPWVEILPTQGATIG | 19 | 0.5839 |
| | 58 | 75 | WQRWCEGLIGTANRSAID | 18 | 0.5467 |
| | 93 | 113 | ASDATLCQNEPPTSCSNLPHQ | 21 | 0.5675 |
| SpaO | 21 | 37 | ECQRHGREATLEYPTRQ | 17 | 0.5256 |
| PrgH | 84 | 104 | LHELKEGNSESRSVQLNTPIQ | 21 | 0.5941 |
| | 115 | 138 | ESEPWVPEQPEKLETSAKKNEPRF | 24 | 0.6261 |
| | 164 | 182 | NSPQRQAAELDSLLGQEKE | 19 | 0.5371 |
| | 261 | 277 | SRQRNTMSKKELEVLSQ | 17 | 0.5634 |
| SipB | 5 | 22 | ASSISRSGYTQNPRLAEA | 18 | 0.5777 |
| | 232 | 253 | GTANAASQNQVSQGEQDNLSNV | 22 | 0.5356 |
| | 545 | 569 | MDQIQQWLKQSVEIFGENQKVTAEL | 25 | 0.5389 |
| SsaD | 18 | 38 | GHVLQGREVWLNEGNLSLGEK | 21 | 0.5384 |
| | 155 | 170 | LDKSNIHYVRAQWKED | 16 | 0.5236 |
| | 257 | 274 | IPGLLHWQISHSHQSQGD | 18 | 0.5234 |
| | 331 | 345 | QDIAPSHDESKYLPA | 15 | 0.5794 |
| HilA | 206 | 227 | VKGYHLLHQESIKLIEHQPASL | 22 | 0.5358 |
| | 242 | 256 | GLRWDTKQISELNSI | 15 | 0.5672 |
| BcfD | 121 | 144 | PMNNVLMGYDENVKAGQPFYVRDS | 24 | 0.5932 |
| | 214 | 232 | LYSGNFNHAGQKPEGVRAK | 19 | 0.6014 |
| | 282 | 303 | NALIPNDVQSVAPFITDSAGRA | 22 | 0.5514 |
| SctC | 430 | 449 | DGNDKTPQSDTTTSVDALPE | 20 | 0.6323 |
| SsaJ | 20 | 34 | DVDLYRSLPEDEANQ | 15 | 0.5328 |
| | 91 | 111 | NQLVVSPQEEQQKINFLKEQR | 21 | 0.5596 |
| PrgK | 20 | 34 | DKDLLKGLDQEQANE | 15 | 0.5634 |
| | 188 | 207 | SERSDAQLQAPGTPVKRNSF | 20 | 0.5906 |
| | 229 | 249 | YYKNHYARNKKGITADDKAKS | 21 | 0.6077 |
| BigA | 1429 | 1448 | RVLSNRFTMLADAAPQIKDG | 20 | 0.5452 |
| | 1456 | 1475 | KGDPRAELGNDTQYDMLALR | 20 | 0.5380 |
| | 1689 | 1709 | SSNDTALHLDAYQWKEDGISD | 21 | 0.5748 |
| SspH2 | 72 | 95 | FELLRTLAYAGWEESIHSGQHGEN | 24 | 0.5541 |
| | 431 | 446 | RNQLTRLPESLIHLSS | 16 | 0.5176 |
| | 574 | 594 | TEATSSCEDRVTFFLHQMKNV | 21 | 0.5220 |
| | 620 | 639 | FRLGKLEQIAREKVRTLALV | 20 | 0.5312 |
| TorS | 133 | 148 | TLRAQQQQLSRQIAEA | 16 | 0.5421 |
| | 172 | 192 | AGIYDLIESGKGDQAERALDR | 21 | 0.5473 |
| | 663 | 677 | SKPASKSAFREPINL | 15 | 0.5645 |

| Protein | Start | End | Peptide | Length | Average Score |
|---|---|---|---|---|---|
| MarT_1 | 106 | 126 | ITIIATDSETKGRKKQIVRQT | 21 | 0.5981 |

*Only predicted peptides of length between 15 to 25 amino acids were selected [97]. SiiE protein were omitted from prediction because of its overly huge sequence.*

**Table 3.**
*BepiPred v2.0 prediction of linear B-cell epitopes.*

| Protein | AllerCatPro | AlgPred2 | | AllergenFP v1.0 |
|---|---|---|---|---|
| | | Hybrid Score | Prediction | |
| SptP | No Hits | 0.04 | Non-Allergen | Probable Non-Allergen |
| SsaQ | No Hits | 0.08 | Non-Allergen | Probable Non-Allergen |
| SpaO | No Hits | 0.03 | Non-Allergen | Probable Non-Allergen |
| PrgH | No Hits | 0.03 | Non-Allergen | Probable Non-Allergen |
| SipB | No Hits | 0.24 | Non-Allergen | Probable Non-Allergen |
| SsaD | No Hits | 0.09 | Non-Allergen | Probable Non-Allergen |
| InvE | No Hits | 0.02 | Non-Allergen | Probable Non-Allergen |
| HilA | No Hits | 0.54 | **Allergen** | Probable Non-Allergen |
| BcfD | No Hits | 0.85 | **Allergen** | Probable Non-Allergen |
| SsrB | No Hits | −0.43 | Non-Allergen | Probable Non-Allergen |
| SctC | No Hits | 0.09 | Non-Allergen | Probable Non-Allergen |
| SicA | No Hits | 0.31 | **Allergen** | **Probable Allergen** |
| SsaJ | No Hits | −0.45 | Non-Allergen | Probable Non-Allergen |
| SscA | No Hits | 0.18 | Non-Allergen | Probable Non-Allergen |
| PrgK | No Hits | −0.48 | Non-Allergen | Probable Non-Allergen |
| BigA | No Hits | 0.75 | **Allergen** | Probable Non-Allergen |
| SiiE | No Hits | 0.86 | **Allergen** | N/A |
| SspH2 | No Hits | −0.48 | Non-Allergen | Probable Non-Allergen |
| TorS | No Hits | 0.02 | Non-Allergen | Probable Non-Allergen |
| MarT_1 | No Hits | 0.55 | **Allergen** | Probable Non-Allergen |

*For AllergenFP v1.0, N/A refers to Not Available because of overly large protein size.*

**Table 4.**
*Allergenicity assessment through different predictive tools. Potential allergens are in bold case.*

BcfD, SicA, BigA, SiiE, and MarT_1 were predicted to be potential allergens (**Table 4**), and thus, were excluded from consideration as well. Hence, we report SptP, SsaQ, SpaO, PrgH, SipB, SsaD, SctC, SsaJ, PrgK, SspH2, and TorS to be potentially utilized as B cell epitopes. Moreover, in discontinuous B cell epitope prediction, the localizations of the highly antigenic regions were illustrated in 3D (**Figure 2**). For successful vaccination, these regions should be prioritized and retained as much as possible due to their important roles in antigenicity.

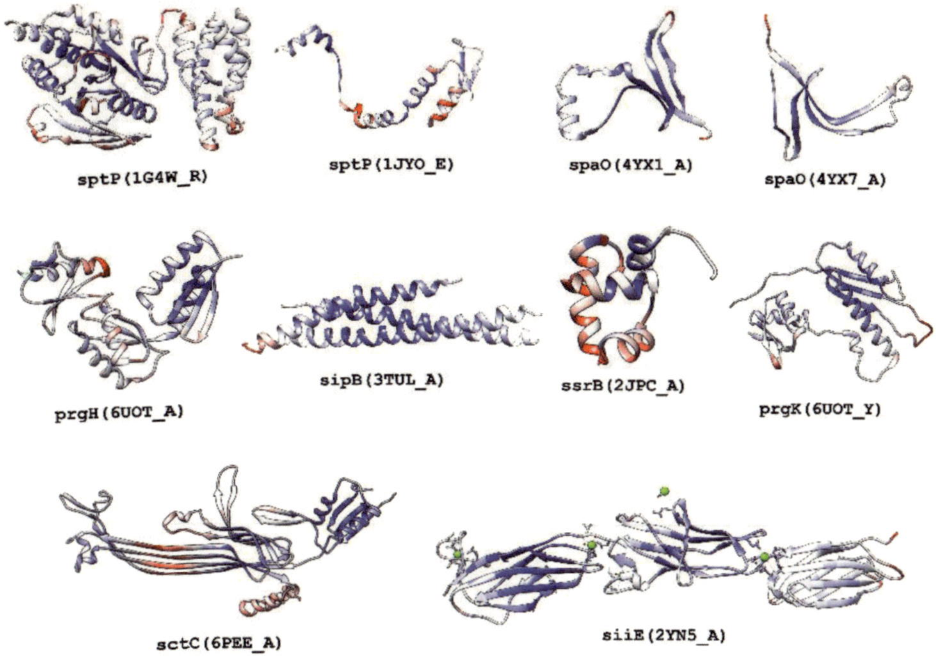


**Figure 2.**
*DiscoTope v2.0 prediction of discontinuous B-cell epitopes. The residues are colored according to their respective DiscoTope scores (red: high, white: threshold of -3.7 and blue: low).*

## 7. Potential druggable proteins

Besides potential vaccine candidates, we have conducted predictions on the druggability and druggable sites of the 20 top ranked proteins which have their 3D crystallized structures available in PDB. Eventually, the localization of the top ranked druggable pockets of SptP, SipB, SctC, SpaO, SsrB, PrgK, PrgH, and SiiE were illustrated in 3D (**Figure 3**). This can help future research in structure-aided drug discovery, by designing drugs specific for the druggable pockets to suppress the virulence of *Salmonella*.

## 8. Conclusions

The study depicted here essentially delineates a schematic approach of shortlisting the most probable virulent proteins as potential vaccine and/or drug candidates from the proteome of *Salmonella* spp. It starts with the building of the theoretical PIN comprising the known and predicted virulent proteins followed by the graph theoretical parametric analyses for identifying a probable set of them. These were further screened through different essential tools enabling the prediction of cellular localisation, signal peptides, transmembrane helices, antigenicity, epitopes, allergenicity and molecular crevices besides comparing with any human homologs. A thorough analysis revealed SsaJ and PrgK to come to the forefront among those already known to be virulent. PrgK even has nice druggable pocket to be targeted through potential drugs. Our approach can pave the way for screening such effective molecular vaccines and/or drug targets for such pathogens. Newer candidates, however, could be unraveled through other effective methods.

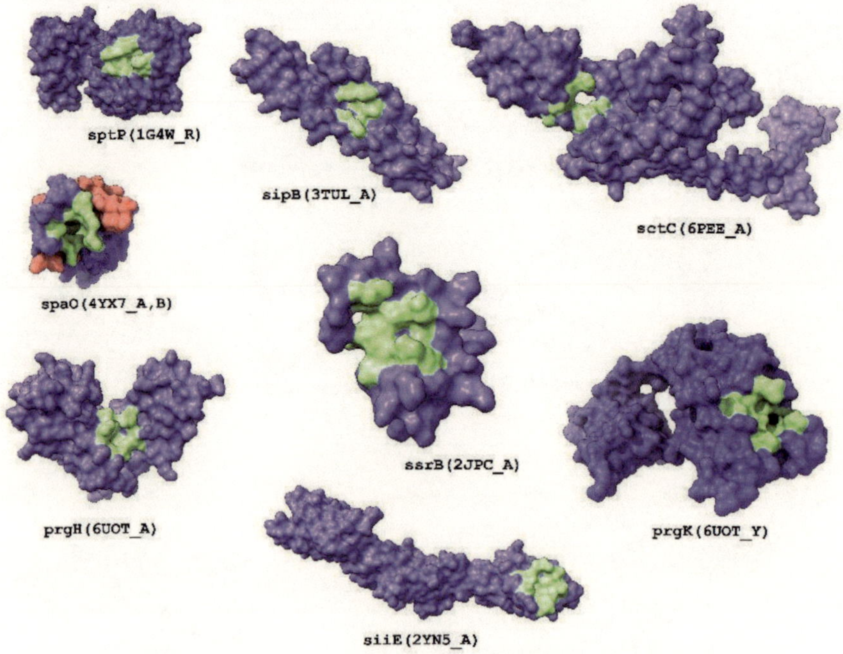


| Protein | PDB ID | Chain ID | Druggable Residues in Rank 1 Pocket |
|---|---|---|---|
| SptP | 1G4W | R | 310, 314, 315, 441, 442, 482, 483, 485, 486, 487, 519, 522, 526 |
| | 1JYO | E | NONE |
| SpaO | 4YX7 | A | 19, 20, 21, 25, 34, 35, 36, 38 |
| | | B | 7, 9 |
| PrgH | 4G2S | A | NONE |
| | 6UOT | A | 226, 229, 230, 233, 245, 246, 247, 347, 352, 353 |
| SipB | 3TUL | A | 89, 92, 93, 96, 206, 209, 210 |
| SsrB | 2JPC | A | 25, 26, 27, 28, 29, 30, 60, 61, 64, 68, 73 |
| SctC | 6PEE | A | 187, 188, 189, 207, 297, 299, 301, 376, 378, 379 |
| PrgK | 6UOT | Y | 102, 106, 109, 133, 143, 144, 146, 175, 176 |
| | 4OYC | A | NONE |
| SiiE | 2YN5 | A | 5304, 5306, 5307, 5311, 5313, 5327, 5328, 5329, 5330, 5331, 5335 |

**Figure 3.**
*P2Rank predicted druggable pockets colored in light green. For SpaO, chain A is in blue, while B is in red. Residues contributing to druggability are tabulated.*

## Acknowledgements

The authors wish to acknowledge the support of Sunway University, Malaysia for the provision of computational facilities.

## Conflict of interest

The authors declare no conflict of interest.

## Author details

Rohit Mishra[1†], Yong Chiang Tan[2†], Amr Adel Ahmed Abd El-Aal[2]
and Chandrajit Lahiri[2*]

1 Advanced Centre for Treatment, Research and Education in Cancer,
Tata Memorial Centre, Mumbai, India

2 Department of Biological Sciences, Sunway University, Petaling Jaya, Selangor,
Malaysia

*Address all correspondence to: chandrajitl@sunway.edu.my

† These authors have contributed equally to the work.

## References

[1] Miller S. Salmonella species, including S. typhi [Internet]. [place unknown]: InfectiousDiseaseAdvisor; [date unknown; cited 2021 Jan 01]. Available from: https://www.infectiousdiseaseadvisor.com/home/decision-support-in-medicine/infectious-diseases/salmonella-species-including-s-typhi/

[2] World Health Organization. Typhoid [Internet]. [place unknown]: World Health Organization; 2018 [updated 2018; cite 2021 Jan 01]. Available from: https://www.who.int/news-room/fact-sheets/detail/typhoid

[3] World Health Organization. Salmonella (non-typhoidal) [Internet]. [place unknown]: World Health Organization; 2018 [updated 2018; cite 2021 Jan 01]. Available from: https://www.who.int/news-room/fact-sheets/detail/salmonella-(non-typhoidal)

[4] Giannella RA. Salmonella. 4th edition. Galveston (TX): Medical Microbiology; 1996. Chapter 21.

[5] Ashurst JV, Truong J, Woodbury B. Salmonella Typhi. InStatPearls [Internet] 2019 Jun 11. StatPearls Publishing.

[6] Mandell GL. Mandell, Douglas, and Bennett's Principles and Practice of Infectious Diseases. 7th Edition. United States: Churchill Livingstone; 2009.

[7] Bennett JE, Dolin R, Blaser MJ. Mandell, douglas, and bennett's principles and practice of infectious diseases: 2-volume set. Elsevier Health Sciences; 2014 Aug 28.

[8] Liao YS, Chen BH, Hong YP, Teng RH, Wang YW, Liang SY, Liu YY, Tu YH, Chen YS, Chang JH, Tsao CS. Emergence of multidrug-resistant *Salmonella enterica* serovar Goldcoast strains in Taiwan and international spread of the ST358 clone. Antimicrobial agents and chemotherapy. 2019 Oct 1;63(10):e01122-19.

[9] Tacket CO, Levine MM. Human Typhoid Vaccines-Old and New. In: Ala'Aldeen AA, Hormaeche CE, eds. Molecular and Clinical Aspects of Bacterial Vaccine Development. Chichester: John Wiley & Sons, 1995. p. 155-178.

[10] Papadimitropoulos V, Vergidis PI, Bliziotis I, Falagas ME. Vaccination against typhoid fever in travellers: a cost-effectiveness approach. Clinical Microbiology and Infection. 2004 Aug;10(8):681-3.

[11] MacLennan CA, Martin LB, Micoli F. Vaccines against invasive Salmonella disease: current status and future directions. Human vaccines & immunotherapeutics. 2014 Jun 4;10(6):1478-93.

[12] Erova TE, Kirtley ML, Fitts EC, Ponnusamy D, Baze WB, Andersson JA, Cong Y, Tiner BL, Sha J, Chopra AK. Protective immunity elicited by oral immunization of mice with *Salmonella enterica* serovar Typhimurium Braun lipoprotein (Lpp) and acetyltransferase (MsbB) mutants. Frontiers in cellular and infection microbiology. 2016 Nov 10;6:148.

[13] Galen JE, Buskirk AD, Tennant SM, Pasetti MF. Live attenuated human Salmonella vaccine candidates: tracking the pathogen in natural infection and stimulation of host immunity. EcoSal Plus. 2016 Nov;7(1).

[14] Gayet R, Bioley G, Rochereau N, Paul S, Corthésy B. Vaccination against Salmonella infection: the mucosal way. Microbiology and Molecular Biology Reviews. 2017 Sep 1;81(3).

[15] Clark-Curtiss JE, Curtiss R. Salmonella vaccines: conduits for

protective antigens. The Journal of Immunology. 2018 Jan 1;200(1):39-48.

[16] Moriel DG, Beatson SA, Wurpel DJ, Lipman J, Nimmo GR, Paterson DL, Schembri MA. Identification of novel vaccine candidates against multidrug-resistant Acinetobacter baumannii. PloS one. 2013 Oct 8;8(10):e77631.

[17] Singh R, Capalash N, Sharma P. Immunoprotective potential of BamA, the outer membrane protein assembly factor, against MDR Acinetobacter baumannii. Scientific reports. 2017 Sep 29;7(1):1-1.

[18] Chiang MH, Sung WC, Lien SP, Chen YZ, Lo AF, Huang JH, Kuo SC, Chong P. Identification of novel vaccine candidates against Acinetobacter baumannii using reverse vaccinology. Human vaccines & immunotherapeutics. 2015 Apr 3;11(4):1065-73.

[19] Pan A, Lahiri C, Rajendiran A, Shanmugham B. Computational analysis of protein interaction networks for infectious diseases. Briefings in bioinformatics. 2016 May 1;17(3):517-26.

[20] Jeong H, Mason SP, Barabási AL, Oltvai ZN. Lethality and centrality in protein networks. Nature. 2001 May;411(6833):41-2.

[21] Lahiri C, Pawar S, Sabarinathan R, Ashraf MI, Chand Y, Chakravortty D. Interactome analyses of Salmonella pathogenicity islands reveal SicA indispensable for virulence. Journal of theoretical biology. 2014 Dec 21;363:188-97.

[22] Ashraf MI, Ong SK, Mujawar S, Pawar S, More P, Paul S, Lahiri C. A side-effect free method for identifying cancer drug targets. Scientific reports. 2018 Apr 27;8(1):1-9.

[23] Pawar S, Ashraf M, Mehata K, Lahiri C. Computational identification of indispensable virulence proteins of Salmonella Typhi CT18. Curr. Top. Salmonella Salmonellosis. 2017 Apr 5.

[24] Pawar S, Ashraf M, Mujawar S, Mishra R, Lahiri C. In silico identification of the indispensable quorum sensing proteins of multidrug resistant *Proteus mirabilis*. Frontiers in cellular and infection microbiology. 2018 Aug 7;8:269.

[25] Lahiri C, Mujawar S, Mishra R, Ashraf M, Pawar S, Gatherer D. Delineating the plausible molecular vaccine candidates and drug targets of multidrug resistant Acinetobacter baumannii. Frontiers in Cellular and Infection Microbiology. 2019;9:203.

[26] Liu B, Zheng D, Jin Q, Chen L, Yang J. VFDB 2019: a comparative pathogenomic platform with an interactive web interface. Nucleic acids research. 2019 Jan 8;47(D1):D687-92.

[27] Szklarczyk D, Gable AL, Lyon D, Junge A, Wyder S, Huerta-Cepas J, Simonovic M, Doncheva NT, Morris JH, Bork P, Jensen LJ. STRING v11: protein–protein association networks with increased coverage, supporting functional discovery in genome-wide experimental datasets. Nucleic acids research. 2019 Jan 8;47(D1):D607-13.

[28] Shannon P, Markiel A, Ozier O, Baliga NS, Wang JT, Ramage D, Amin N, Schwikowski B, Ideker T. Cytoscape: a software environment for integrated models of biomolecular interaction networks. Genome research. 2003 Nov 1;13(11):2498-504.

[29] Diestel R. Graph Theory. Berlin and Heidelberg: Springer-Verlag, 2000.

[30] Tang Y, Li M, Wang J, Pan Y, Wu FX. CytoNCA: a cytoscape plugin for centrality analysis and evaluation of protein interaction networks. Biosystems. 2015 Jan 1;127:67-72.

[31] Oliveros JC. Venny. An interactive tool for comparing lists with Venn's diagrams [Internet]. Venny 2.1.0. 2007 [cited 2020]. Available from: https://bioinfogp.cnb.csic.es/tools/venny/index.html

[32] Gasteiger E, Hoogland C, Gattiker A, Wilkins MR, Appel RD, Bairoch A. Protein identification and analysis tools on the ExPASy server. InThe proteomics protocols handbook 2005 (pp. 571-607). Humana press.

[33] Yu NY, Wagner JR, Laird MR, Melli G, Rey S, Lo R, Dao P, Sahinalp SC, Ester M, Foster LJ, Brinkman FS. PSORTb 3.0: improved protein subcellular localization prediction with refined localization subcategories and predictive capabilities for all prokaryotes. Bioinformatics. 2010 Jul 1;26(13):1608-15.

[34] Armenteros JJ, Tsirigos KD, Sønderby CK, Petersen TN, Winther O, Brunak S, von Heijne G, Nielsen H. SignalP 5.0 improves signal peptide predictions using deep neural networks. Nature biotechnology. 2019 Apr;37(4):420-3.

[35] Juncker AS, Willenbrock H, Von Heijne G, Brunak S, Nielsen H, Krogh A. Prediction of lipoprotein signal peptides in Gram-negative bacteria. Protein Science. 2003 Aug;12(8):1652-62.

[36] Doytchinova IA, Flower DR. VaxiJen: a server for prediction of protective antigens, tumour antigens and subunit vaccines. BMC bioinformatics. 2007 Dec 1;8(1):4.

[37] Pettersen EF, Goddard TD, Huang CC, Couch GS, Greenblatt DM, Meng EC, Ferrin TE. UCSF Chimera—a visualization system for exploratory research and analysis. Journal of computational chemistry. 2004 Oct;25(13):1605-12.

[38] Jespersen MC, Peters B, Nielsen M, Marcatili P. BepiPred-2.0: improving sequence-based B-cell epitope prediction using conformational epitopes. Nucleic acids research. 2017 Jul 3;45(W1):W24-9.

[39] Kringelum JV, Lundegaard C, Lund O, Nielsen M. Reliable B cell epitope predictions: impacts of method development and improved benchmarking. PLoS Comput Biol. 2012 Dec 27;8(12):e1002829.

[40] Tan YC. DiscoTope2ChimeraAttr_v1.0.0 [Internet]. Malaysia: Zenodo; 2021 [updated 2021, cited 2021 Jan 01]. Available from: https://github.com/TanYongChiang/DiscoTope2ChimeraAttr/tree/v1.0.0. DOI: 10.5281/zenodo.4408680

[41] Maurer-Stroh S, Krutz NL, Kern PS, Gunalan V, Nguyen MN, Limviphuvadh V, Eisenhaber F, Gerberick GF. AllerCatPro—prediction of protein allergenicity potential from the protein sequence. Bioinformatics. 2019 Sep 1;35(17):3020-7.

[42] Sharma N, Patiyal S, Dhall A, Pande A, Arora C, Raghava GP. AlgPred 2.0: an improved method for predicting allergenic proteins and mapping of IgE epitopes. Briefings in Bioinformatics. 2020 Nov 17.

[43] Dimitrov I, Naneva L, Doytchinova I, Bangov I. AllergenFP: allergenicity prediction by descriptor fingerprints. Bioinformatics. 2014 Mar 15;30(6):846-51.

[44] Kriváк R, Hoksza D. P2Rank: machine learning based tool for rapid and accurate prediction of ligand binding sites from protein structure. Journal of cheminformatics. 2018 Dec 1;10(1):39.

[45] Altschul SF, Gish W, Miller W, Myers EW, Lipman DJ. Basic local alignment search tool. Journal

of molecular biology. 1990 Oct 5;215(3):403-10.

[46] Ehrbar K, Friebel A, Miller SI, Hardt WD. Role of the Salmonella pathogenicity island 1 (SPI-1) protein InvB in type III secretion of SopE and SopE2, two Salmonella effector proteins encoded outside of SPI-1. Journal of bacteriology. 2003 Dec 1;185(23):6950-67.

[47] Guo Y, Xu Y, Kang X, Gu D, Jiao Y, Meng C, Tang P, Wang X, Huang C, Geng S, Jiao X. Immunogenic potential and protective efficacy of a sptP deletion mutant of Salmonella Enteritidis as a live vaccine for chickens against a lethal challenge. International Journal of Medical Microbiology. 2019 Dec 1;309(8):151337.

[48] Yoshida Y, Miki T, Ono S, Haneda T, Ito M, Okada N. Functional characterization of the type III secretion ATPase SsaN encoded by Salmonella pathogenicity island 2. PLoS One. 2014 Apr 10;9(4):e94347.

[49] Sakai T, Miyata T, Terahara N, Mori K, Inoue Y, Morimoto YV, Kato T, Namba K, Minamino T. Novel insights into conformational rearrangements of the bacterial flagellar switch complex. Mbio. 2019 Apr 30;10(2).

[50] Lara-Tejero M, Qin Z, Hu B, Butan C, Liu J, Galán JE. Role of SpaO in the assembly of the sorting platform of a Salmonella type III secretion system. PLoS pathogens. 2019 Jan 22;15(1):e1007565.

[51] Notti RQ, Bhattacharya S, Lilic M, Stebbins CE. A common assembly module in injectisome and flagellar type III secretion sorting platforms. Nature communications. 2015 May 21;6(1):1-1.

[52] Ruan P, Xia XP, Sun D, Ojcius DM, Mao YF, Yue WY, Yan J. Recombinant SpaO and H1a as immunogens for protection of mice from lethal infection with Salmonella paratyphi A: Implications for rational design of typhoid fever vaccines. Vaccine. 2008 Dec 2;26(51):6639-44.

[53] Kimbrough TG, Miller SI. Contribution of Salmonella typhimurium type III secretion components to needle complex formation. Proceedings of the National Academy of Sciences. 2000 Sep 26;97(20):11008-13.

[54] Myeni SK, Wang L, Zhou D. SipB-SipC complex is essential for translocon formation. PLoS One. 2013 Mar 27;8(3):e60499.

[55] Darwin KH, Miller VL. Type III secretion chaperone-dependent regulation: activation of virulence genes by SicA and InvF in Salmonella typhimurium. The EMBO journal. 2001 Apr 17;20(8):1850-62.

[56] Lv Q, Li S, Wei H, Wen Z, Wang Y, Tang T, Wang J, Xia L, Deng X. Identification of the natural product paeonol derived from peony bark as an inhibitor of the *Salmonella enterica* serovar Typhimurium type III secretion system. Applied Microbiology and Biotechnology. 2020 Feb 1;104(4):1673-82.

[57] Queiroz MH, Madrid C, Paytubi S, Balsalobre C, Juarez A. Integration host factor alleviates H-NS silencing of the *Salmonella enterica* serovar Typhimurium master regulator of SPI1, hilA. Microbiology. 2011 Sep 1;157(9):2504-14.

[58] Zhang Y, Liu Y, Wang T, Deng X, Chu X. Natural compound sanguinarine chloride targets the type III secretion system of *Salmonella enterica* Serovar Typhimurium. Biochemistry and biophysics reports. 2018 Jul 1;14:149-54.

[59] Barlag B, Hensel M. The giant adhesin SiiE of *Salmonella enterica*. Molecules. 2015 Jan;20(1):1134-50.

[60] Li X, Bleumink-Pluym NM, Luijkx YM, Wubbolts RW, van Putten JP, Strijbis K. MUC1 is a receptor for the Salmonella SiiE adhesin that enables apical invasion into enterocytes. PLoS pathogens. 2019 Feb 4;15(2):e1007566.

[61] Männe C, Takaya A, Yamasaki Y, Mursell M, Hojyo S, Wu TY, Sarkander J, McGrath MA, Cornelis R, Hahne S, Cheng Q. Salmonella SiiE prevents an efficient humoral immune memory by interfering with IgG+ plasma cell persistence in the bone marrow. Proceedings of the National Academy of Sciences. 2019 Apr 9;116(15):7425-30.

[62] Gan Q, Lehman BP, Bobik TA, Fan C. Expanding the genetic code of Salmonella with non-canonical amino acids. Scientific reports. 2016 Dec 23;6:39920.

[63] Bearson BL, Bearson SM, Brunelle BW, Bayles DO, Lee IS, Kich JD. Salmonella DIVA vaccine reduces disease, colonization and shedding due to virulent S. Typhimurium infection in swine. Journal of medical microbiology. 2017 May;66(5):651.

[64] Maserati A, Fink RC, Lourenco A, Julius ML, Diez-Gonzalez F. General response of *Salmonella enterica* serovar Typhimurium to desiccation: a new role for the virulence factors sopD and sseD in survival. PLoS One. 2017 Nov 8;12(11):e0187692.

[65] Askoura M, Hegazy WA. Ciprofloxacin interferes with Salmonella Typhimurium intracellular survival and host virulence through repression of Salmonella pathogenicity island-2 (SPI-2) genes expression. Pathogens and Disease. 2020 Feb;78(1):ftaa011.

[66] Wang M, Qazi IH, Wang L, Zhou G, Han H. Salmonella Virulence and Immune Escape. Microorganisms. 2020 Mar;8(3):407.

[67] Singh S, Pratap Cb, Nath G. Vaccine Development Against Salmonella Typhi: A Search is Still On. Journal of Clinical & Diagnostic Research. 2019 Aug 1;13(8).

[68] Notti RQ, Stebbins CE. The structure and function of type III secretion systems. Virulence Mechanisms of Bacterial Pathogens. 2016 Jun 22:241-64.

[69] Seyrek K, Ivanisenko NV, Richter M, Hillert LK, König C, Lavrik IN. Controlling Cell Death through Post-translational Modifications of DED Proteins. Trends in Cell Biology. 2020 Mar 29.

[70] Cox CE, Wright AC, McClelland M, Teplitski M. Influence of *Salmonella enterica* Serovar Typhimurium ssrB on Colonization of Eastern Oysters (*Crassostrea virginica*) as Revealed by a Promoter Probe Screen. Applied and environmental microbiology. 2016 Jan 1;82(1):328-39.

[71] Sanapala S, Mosca L, Wang S, Curtiss R. Comparative evaluation of Salmonella Typhimurium vaccines derived from UK-1 and 14028S: Importance of inherent virulence. PloS one. 2018 Sep 7;13(9):e0203526.

[72] De Masi L, Yue M, Hu C, Rakov AV, Rankin SC, Schifferli DM. Cooperation of adhesin alleles in Salmonella-host tropism. Msphere. 2017 Apr 26;2(2).

[73] Azinheiro S, Carvalho J, Prado M, Garrido-Maestu A. Evaluation of different genetic targets for *Salmonella enterica* serovar Enteriditis and Typhimurium, using Loop-Mediated Isothermal AMPlification for detection in food samples. Frontiers in Sustainable Food Systems. 2018 Feb 21;2:5.

[74] Hu Y, Cong Y, Li S, Rao X, Wang G, Hu F. Identification of in vivo induced protein antigens of *Salmonella enterica*

serovar Typhi during human infection. Science in China Series C: Life Sciences. 2009 Oct 1;52(10):942-8.

[75] Du X, Liu Y, Sun X, Zhao J, Wang J, Li J, Liu X, Wang J, Li Y. The mRNA expression of ompF, invA and invE was associated with the ciprofloxacin-resistance in Salmonella. Archives of Microbiology. 2020 Jun 13:1-6.

[76] Kim JS, Im Jang J, Eom JS, Oh CH, Kim HG, Kim BH, Bang IS, Bang SH, Park YK. Molecular characterization of the InvE regulator in the secretion of type III secretion translocases in *Salmonella enterica* serovar Typhimurium. Microbiology. 2013 Mar 1;159(Pt_3):446-61.

[77] Velasco-Carrillo J, Araque-Granadillo M, Ayala-Serrano J, Dávila-Vera D, Peña-Contreras Z, Mendoza-Briceño R. Patogenia de mutantes de Salmonella Typhimurium en dos modelos experimentales in vivo. VacciMonitor. 2019;28(1):1-8.

[78] Gong H, Su J, Bai Y, Miao L, Kim K, Yang Y, Liu F, Lu S. Characterization of the expression of Salmonella Type III secretion system factor PrgI, SipA, SipB, SopE2, SpaO, and SptP in cultures and in mice. BMC microbiology. 2009 Dec 1;9(1):73.

[79] Birhanu BT, Park NH, Lee SJ, Hossain MA, Park SC. Inhibition of Salmonella Typhimurium adhesion, invasion, and intracellular survival via treatment with methyl gallate alone and in combination with marbofloxacin. Veterinary research. 2018 Dec 1;49(1):101.

[80] Chen S, Liao C, Zhang C, Cheng X. Roles of the crp and sipB genes of *Salmonella enterica* serovar Typhimurium in protective efficacy and immune responses to vaccination in mice. Canadian Journal of Veterinary Research. 2018 Apr 1;82(2):102-5.

[81] Seribelli AA, Cruz MF, Vilela FP, Frazão MR, Paziani MH, Almeida F, Medeiros MI, Rodrigues DD, Kress MR, Allard MW, Falcão JP. Phenotypic and genotypic characterization of Salmonella Typhimurium isolates from humans and foods in Brazil. Plos one. 2020 Aug 18;15(8):e0237886.

[82] Riquelme S, Varas M, Valenzuela C, Velozo P, Chahin N, Aguilera P, Sabag A, Labra B, Álvarez SA, Chávez FP, Santiviago CA. Relevant genes linked to virulence are required for Salmonella Typhimurium to survive intracellularly in the social amoeba Dictyostelium discoideum. Frontiers in microbiology. 2016 Aug 23;7:1305.

[83] Guard J, Cao G, Luo Y, Baugher JD, Davison S, Yao K, Hoffmann M, Zhang G, Likens N, Bell RL, Zheng J. Genome sequence analysis of 91 Salmonella Enteritidis isolates from mice caught on poultry farms in the mid 1990s. Genomics. 2020 Jan 1;112(1):528-44.

[84] Czibener C, Merwaiss F, Guaimas F, Del Giudice MG, Serantes DA, Spera JM, Ugalde JE. BigA is a novel adhesin of Brucella that mediates adhesion to epithelial cells. Cellular microbiology. 2016 Apr;18(4):500-13.

[85] Bialer MG, Sycz G, Muñoz González F, Ferrero MC, Baldi PC, Zorreguieta A. Adhesins of Brucella: Their Roles in the Interaction with the Host. Pathogens. 2020 Nov;9(11):942.

[86] Shappo MO, Li Q, Lin Z, Hu M, Ren J, Xu Z, Pan Z, Jiao X. SspH2 as anti-inflammatory candidate effector and its contribution in Salmonella Enteritidis virulence. Microbial Pathogenesis. 2020 May 1;142:104041.

[87] Bhavsar AP, Brown NF, Stoepel J, Wiermer M, Martin DD, Hsu KJ, Imami K, Ross CJ, Hayden MR, Foster LJ, Li X. The Salmonella type III

effector SspH2 specifically exploits the NLR co-chaperone activity of SGT1 to subvert immunity. PLoS Pathog. 2013 Jul 25;9(7):e1003518.

[88] Moore JO, Hendrickson WA. An asymmetry-to-symmetry switch in signal transmission by the histidine kinase receptor for TMAO. Structure. 2012 Apr 4;20(4):729-41.

[89] Tükel Ç, Akçelik M, De Jong MF, Şimşek Ö, Tsolis RM, Bäumler AJ. MarT activates expression of the MisL autotransporter protein of *Salmonella enterica* serotype Typhimurium. Journal of bacteriology. 2007 May 15;189(10):3922-6.

[90] Eran Z, Akçelik M, Özcengiz G, Akçelik N. Regulation of biofilm formation by marT in Salmonella Typhimurium. Molecular Biology Reports. 2020 Jun 11:1-0.

[91] Grandi G. Bacterial surface proteins and vaccines. F1000 biology reports. 2010;2.

[92] Baliga P, Shekar M, Venugopal MN. Potential outer membrane protein candidates for vaccine development against the pathogen Vibrio anguillarum: a reverse vaccinology based identification. Current microbiology. 2018 Mar 1;75(3):368-77.

[93] Maiti B, Dubey S, Munang'andu HM, Karunasagar I, Karunasagar I. Application of outer membrane protein-based vaccines against major bacterial fish pathogens in india. Frontiers in Immunology. 2020;11:1362.

[94] Hu R, Li J, Zhao Y, Lin H, Liang L, Wang M, Liu H, Min Y, Gao Y, Yang M. Exploiting bacterial outer membrane vesicles as a cross-protective vaccine candidate against avian pathogenic *Escherichia coli* (APEC). Microbial Cell Factories. 2020 Dec;19(1):1-7.

[95] Cudic M, Fields GB. Extracellular proteases as targets for drug development. Current Protein and Peptide Science. 2009 Aug 1;10(4):297-307.

[96] Ahmad V. Prospective of Extracellular Matrix and Drug Correlations in disease management. Asian Journal of Pharmaceutical Sciences. 2020 Aug 30.

[97] Potocnakova L, Bhide M, Pulzova LB. An introduction to B-cell epitope mapping and in silico epitope prediction. Journal of immunology research. 2016 Oct;2016.

Section 2

# An Overview of Non-Typhoidal *Salmonella*

Chapter 5

# Non-Typhoidal Salmonellosis: A Major Concern for Poultry Industry

*Mamta Pandey and Emmagouni Sharath Kumar Goud*

## Abstract

*Salmonella* is the most important gastrointestinal pathogen distributed ubiquitously. The major serovars involved in Non-typhoidal salmonellosis are *S.* Typhimurium and *S.* Enteritidis. In the viewpoint of ban in the export and import of the *Salmonella* contaminated poultry food and poultry products, the need for rapid detection and mitigation of *Salmonella* has increased mani-folds. The major problem associated with its control is the growing incidence of antimicrobial resistance, which has been reported worldwide in the recent years. From causing self limiting gastroenteritis they have found to be responsible for several fatal diseases like endocarditis, meningitis, lung infestations, appendicitis, pneumonia, and cerebral abscess in human beings. Targeting several proteins such as adhesive proteins, lipoproteins, outer membrane proteins (Omps) etc. as vaccine candidates may pave a way in its control. So, continuous monitoring using one health approach and development of effective treatment and control strategies are critical.

**Keywords:** non typhoidal *Salmonella*, gastroenteritis, multiple drug resistance, poultry, vaccine

## 1. Introduction

Non-typhoidal Salmonellosis is caused by bacteria belonging to Enterobacteriaceae family. In poultry, *Salmonella* is known to be present in the gastrointestinal tract without showing any symptoms [1]. This leads to an undetected condition at farm level and after consumption of such poultry products like meat and eggs, humans gets infected at fork end. Domestic animals act as a reservoir for the food-borne spread of host-generalist serovars, which accounts for worldwide incidence of non-typhoidal *Salmonella* (NTS) infections. The range of symptoms varies from self limiting gastroenteritis to various dreadful diseases like endocarditis, meningitis etc. Generally condition becomes severe in children, geriatric and immunocompromised individuals [2]. NTS accounts for 93 million enteric infections and 155,000 deaths globally on annual basis [3]. The two factors contributing to majority of NTS infections are its broad host range and multiple drug resistance (MDR), which has been reported universally in recent years [4]. In developing countries, the situation is grimmer due to poor hygienic conditions. Near about 100 cells of virulent *Salmonella* are sufficient to cause infection in humans, which will further depend upon the health condition of an individual [5], hence it is critical to adopt multiple intervention strategies. Vaccination is considered as an effective tool to control the disease [6], but the available vaccines has their own restrictions such as short term immunity etc. which limits their applicability. So, there is a must requirement to develop a suitable vaccine against NTS. Several proteins such as lipoproteins, outer membrane proteins (Omps) and polysaccharides have been targeted to evaluate their potential as suitable vaccine candidates. This chapter aims to present a brief overview on some such valuable information on NTS.

## 2. Non typhoidal *Salmonella* strains and its transmission

Till now, more than 2500 serotypes of *Salmonella* have been identified [7]. Non typhoidal salmonellosis is caused by all serotypes of *Salmonella* except for Typhi, Paratyphi A, Paratyphi B and Paratyphi C. Poultry can get infected either with host-specific *Salmonella* serovars, like *S.* Pullorum and *S.* Gallinarum, which cause a typhoid-like systemic disease or wide ranged NTS. Wide range NTS represents *Salmonella enterica* subspecies *enterica* serovar Enteritidis and Typhimurium together with serovars such as *S.* Newport, *S.* Heidelberg and *S.* Javiana etc. Broad host ranged *S.* serotypes get colonize [8] in host and carry infection asymptomatically. *Salmonella enterica* serotype Enteritidis and Typhimurium are the two most important NTS serotypes transmitted from animals to humans in most parts of the world [9].

NTS transmission usually occurs through consumption of contaminated food i.e. chicken, eggs, pork, beef, dairy products, and water contaminated with animal feces. However, contact with animals such as reptiles and animal environment are equally important sources [10]. Majority of reptiles are known to carry *Salmonella* as part of their natural intestinal flora. In poultry, mode of transmission can be vertical or horizontal. Vertical transmission occurs when parent poultry is suffering from systemic infection or transovarian infection which results in infection of infants. *S.* Enteritidis serovar have a particular preference to this mode of transmission. Polluted feed and drinking water, dirty cages, fomites etc. includes horizontal mode of transmission. Colonization of *Salmonella* in poultry without showing any sign and symptoms is common, hence, its transmission in layers (vertical) and broilers (horizontal) can occur at primary production level [11]. Transmission through eggs and meat from such healthy poultry with colonized NTS is common [12]. The degree of *Salmonella* colonization depends on parameters specific to *Salmonella* and effects of environmental stimuli on gene expression. Factors such as age, environmental and physiological stress, diet, and survival of *Salmonella* through gastric barrier, use of antimicrobials in the farm, chicken health, and genetic background of the chicks could possibly influence the colonization [13]. In poultry farms transmission can also occur through workers, vehicles, clothing, footwear, garbage, insects, rodents, wild birds, pets, equipment, and many other factors. In humans, factors contributing to susceptibility of NTS infections include pernicious anemia, any previous gastric surgery, excessive use of medications responsible for gastric barrier reduction etc. [14]. Other associated susceptibilities include homozygosity for sickle cell anemia [15], HIV [16], malaria [17], malnourished infants, and young adults [18].

## 3. Global disease epidemiology

In many countries, over the past years, the incidence of NTS has increased markedly. In western countries, the predominant serotypes are *S.* Typhimurium and *S.* Enteritidis. In United States, each year NTS causes approximately 1.35 million illnesses, 26,500 hospitalizations, and 420 deaths [19]. As per the fact sheets of World Health Organization, NTS is 1 of 4 key global causes of diarrhea. The burden is so substantial that every year 33 millions of lives are lost. In Europe, NTS is the second most investigated zoonosis responsible for causing gastrointestinal infections in humans. As per the European Food Safety Authority (EFSA) and European Centre for Disease Prevention and Control (ECDPC) reports, the number of confirmed cases of salmonellosis in Europe still remains high, with a total of 91,857 cases reported in 2018 [20]. The estimation of the total number of NTS infections is a difficult task in developing and under-developed countries, a possible reason to this may be non-reporting of the diseased cases to hospitals. The epidemiological pattern has been variable over the past decade in African countries. Sub-Saharan African region is principally affected region in Africa. According to the hospital based studies of Africa, NTS is the second most frequently occurring pathogen in children and is leading cause of bacteremia in adults [21]. A population-based surveillance data reported the incidence to be between 0 and 54 cases per 100,000 person-years of observation in 13 surveillance sites [22]. The disease incidence ranged from 1.4/100,000 population/year in South Africa (all ages) to 2,520/100,000 population/year in Ghana (<5 years of age) [18]. The community acquired NTS bacteremia prevalence varied from 8% in Nigeria to 45% in Central African Republic [18]. From Kenya, documented incidences were found to be 4134/100,000 person-years [23]. According to the Statistical Committee of the Republic of Armenia, a total of 4,392 cases of salmonellosis were reported during the period 2010–2019, comprising at least 50% of patients below 6 years of age [24]. There is a scarcity in data related to NTS infections from several regions of Asia, limited reports are available from India [25], and Taiwan [26]. A hospital-based multicenter study from Indonesia, Thailand, and Vietnam, investigated NTS positivity rates of 27.5% and 11.7% in children and adults respectively from bacteremia cases [27]. From 2009 to 2013, the prevalence rate of NTS was found to be limited 20/12,940 in bacteremia patients with 25% case fatality report in Bangladesh [28]. In Malaysia, reported prevalence was 16.2%, among which most of the affected cases were from children below 1 year of age [29]. A variety of NTS serovars are known to be present in South-East Asia [30], even some less common serovars are also known to be prevalent such as occurrence of *S. enterica* Weltevreden from the farms of Vietnam [31].

## 4. Clinical manifestations

NTS infections can cause several clinical symptoms depending on the type of serovar and host factors in humans. NTS symptoms are generally non-specific and hence their identification is a challenging task particularly in areas where laboratory diagnosis facilities are not accessible. Most commonly, *Salmonella* causes self limiting gastroenteritis in human beings. After an incubation period of 6–72 h (mean 24 h), there is sudden onset of nausea, vomiting, abdominal pain and tenderness, followed by mild to severe watery diarrhea and sometimes diarrhea may contain blood and mucus. The stool examination reveals a moderate number of polymorphonuclear leukocytes and blood. Fever is seen in about 70% of patients. Usually, symptoms subside within 2–7 days in healthy children. In certain high-risk groups, like in neonates, young infants, and immunodeficient individuals symptoms may persist for several weeks. As a complication of gastroenteritis, transient bacteremia may occur in some patients (reported incidences in approximately 5% of the patients) [32]. Certain serotypes i.e. *S.* Choleraesuis and *S.* Dublin show a higher predisposition for bacteremia in humans [33]. After gaining entry to the bloodstream, *Salmonella* get metastasize to different organs and cause focal suppurative infection. In sickle cell anemic patients a common finding as a result of NTS is osteomyelitis [34]. Less frequent occurrence of meningitis has been observed specially in infants [35]. Despite of antibiotic therapy, patients may develop rapid neurological deterioration. Other feared lethal complications include development of endarteritis [36], endocarditis [37], meningitis [38], lung infestations [39], appendicitis [40], pneumonia [41], bone and joint defects [15] and cerebral abscess [42].

## 5. Multiple drug resistance

In current scenario, there has been an extensive increase in documentation of antimicrobial resistance in NTS. Multiple drug resistance (MDR) is the antimicrobial resistance shown by the microorganism to at least three different groups of antimicrobials. Some *Salmonella* strains are characterized by carrying several antimicrobial resistance. The possibility of having MDR in bacteria is due to the presence of several different resistance genes or a single resistance gene that shows resistance to more than one antibiotic. Some important factors that could cause MDR in microorganisms include selective pressures, proliferation of multiple resistant clones, and inability to detect emerging phenotypes. The overuse or misuse of antimicrobials for the treatment of human disease, in agriculture, and in-home disinfectants comes under selective pressure [43]. The development of the antimicrobial resistance in bacteria is as a result of the genetic modifications of a microorganism for its own survival either spontaneously or acquired. In spontaneous mutation, a genetic modification occurs naturally which helps to survive from the lethal effects of antimicrobials. The reason behind the occurrence of spontaneous mutations is unknown, but the exposure to the antimicrobials may provide selective pressure for antimicrobial resistance [44]. Acquired resistance eventuate from gene transfer from other bacteria [45]. High resistance rate have been reported from *S.* Typhimurium DT104, resistant to five antimicrobial agents i.e. ampicillin, chloramphenicol, streptomycin, sulphonamide, and tetracycline [46]. Resistance to some extended spectrum antibiotics like cephalosporins and fluoroquinolones have been increasingly reported [47]. The emergence of *S.* Choleraesuis resistance to multiple antibiotics including ciprofloxacin has posed serious public health concerns [48].

Bacteria develop MDR by three different mechanisms. In first mechanism of resistance, the bacteria are known to produce certain specific proteins such as hydrolytic enzymes, which destroy the antimicrobials present in their surroundings. An example to this is penicillin resistance, where *Salmonella* produces β-lactamases enzymes which cleave the β-lactam ring of active penicillin and convert it into its inactive form [49]. The second mechanism of resistance is the presence of an active efflux pump system in the cell which actively pumps out the antimicrobials before they become effective [45]. *Salmonella* have energy-dependent efflux pumps for tetracycline and chloramphenicol which inhibit protein synthesis in bacteria by binding to tRNA to the A-site of the 30S subunit of the ribosome [50]. The third mechanism of resistance is called as the receptor modification in which the bacteria tends to chemically modify or mutate the target of the antimicrobial agent. For example vancomycin is the antibiotic which binds with D-Ala-D-Ala on the cell wall and inhibits the peptidoglycan synthesis of the cell wall of bacteria. But vancomycin-resistant enterococcus mutates its terminal peptide to D-Ala-D-Lac that has a lower affinity for vancomycin [45]. The mechanism of action adopted by *Salmonella* for different antimicrobial classes has been enlisted in **Table 1**. Plasmid mediated

| Antimicrobial class | Antimicrobial mechanism of action by *Salmonella* | Reference |
|---|---|---|
| Aminoglycosides | Decreased drug uptake, drug modification, and modification of the ribosomal target of the drug | [51] |
| Beta-lactams | Secretion of β-lactamase enzymes | [50] |
| Phenicols | Enzymatic inactivation of the antibiotic by chloramphenicol O-acetyl-transferase and removal of the antibiotic by an efflux pump | [52] |
| Quinolones | Inducing mutations in the quinolone resistance determining region, increased efflux pumps expression and decreased outer membrane permeability. | [53] |
| Tetracycline | Produce energy dependent efflux pumps to remove tetracycline out from the bacterial cell | [54] |
| Sulfonamides and trimethoprim | Expression of *sul* genes i.e. *sul1* or *sul2* for sulfamethoxazole resistance and dihydrofolate reductase (*dfr*) genes for trimethoprim | [55] |

**Table 1.**
*Antimicrobial mechanisms of* Salmonella *for different antimicrobial class.*

resistance determinants (genes) to antimicrobials have been known to be responsible for the worldwide dissemination of several *Salmonella* serotypes i.e. Enteritidis, Heidelberg, Typhimurium, Infantis, Virchow, Kentucky. The most common genes found in poultry and its meat products are β-lactamases, *CTX-M* (*CTX-M-1, -2, -9* and *- 15*), *TEM-52*, *AmpC-type CMY-2*. The transmission of these genes is associated with diverse plasmid families such as Incl 1 ($bla_{CTX-M-1}$, $bla_{TEM-52}$, $bla_{CMY-2}$), Incl A/C ($bla_{CMY-2}$), Incl H12 ($bla_{CTX-M-2}$, $bla_{CTX-M-9}$). Plasmid mediated quinolone resistance is governed by *QnrB2*, *QnrB19*, *QnrS1* genes. The genes mediating R-type ACSSuT in NTS are commonly clustered together in *Salmonella* genomic island 1 (SGI-1), a chromosomal genetic element.

## 6. Antimicrobial resistance in poultry food chain

The practice of using antimicrobials in food animals is rigorous, it may be either for growth promotion, prophylactic, therapeutic or metaphylactic reasons and this results in MDR. In poultry sector, the use of antimicrobials as growth promoters, such as bambermycin, bacitracin, chlortetracycline, penicillin, tylosin, fluoroquinolones and cephalosporins is concerning [56]. Consumption of low doses of antibiotics in poultry feed for rapid poultry growth is a general practice. Use of antibiotics not only kill majority of the gut microbiota, but, some resilient bacteria survive and become resistant. Over time, these resistant bacteria transfer antibiotic resistant genes to other susceptible microbial population. The situation is crucial in developing countries where laws to control the sale and use of antibiotics are not strict. For therapy, antimicrobials like erythromycin, fluoroquinolones, gentamycin, neomycin, penicillin, spectinomycin, tetracyclines and tylosin are commonly used in poultry [56]. The minimum time period from administering the last dose of medication to the production of meat or other animal-derived products for consumption purpose is referred to as withdrawal period. The withdrawal period for antimicrobials should be followed strictly, in order to prevent the detrimental effects of drug residues in food. There are numerous programmes to reduce the flow of foodborne pathogens from animals to humans, for instance programs for meat and poultry inspection, Hazard Analysis Critical Control Point (HACCP) system and standard operating procedures for sanitation.

The drug-resistant bacteria can be present anywhere, in various environmental samples, farms, and retail meat products. *S.* Enteritidis (88%) isolated from hatching eggs, litter, feed, drinkers, bird rinse, and ceca, were reported to be resistant to drugs, ampicillin, nalidixic acid, and tetracycline [57]. There is a frequent isolation of *Salmonella* serovars such as *S.* Enteritidis, *S.* Infantis, *S.* Typhimurium, and *S.* Heidelberg from broiler carcasses. The antibiotic-resistant *Salmonella* isolates has been found from poultry chiller water and carcasses. The isolated *Salmonella* were resistant to antibiotics including tetracycline, ampicillin, amoxicillin-clavulanic acid, ceftiofur, streptomycin, and sulfisoxazole. Broiler farms with *Salmonella* isolates resistant to multiple antibiotics, i.e. streptomycin (30.9%), gentamicin (12.6%), sulfadimethoxine (20.9%), tetracycline (13.9%), and trimethoprim-sulfamethoxazole combination (8.6%) were recovered. Among these isolates, 67% of *S.* Heidelberg and 54% of *S.* Kentucky isolates showed resistance to five or more antibiotics [11]. These serovars have a high resistance towards ceftriaxone and ceftiofur antibiotics. It has been observed that conversion of conventional farms to organic farms can reduce the prevalence of antibiotic resistant *Salmonella* from 44% to 6% [58]. A comparison between the *Salmonella* isolates obtained from poultry samples of Maryland retail shop from conventional and organic farms revealed that conventional carcass samples were resistant for five to seven antimicrobials, whereas 79% of the isolates from organic carcass samples were susceptible to all 17 tested antimicrobials. Assessment of the status of poultry retail shops is necessary, as in many countries, people prefer to procure the freshly slaughtered chicken. In such retail shops, the likelihood of cross-contamination of poultry carcasses is high and MDR-*Salmonella* has been isolated from retail meat shops as well. The Food and Drug Administration (FDA) have reported the National Antimicrobial Resistance Monitoring System (NARMS), regarding retail meat interim report for *Salmonella*, which includes the antibiotic resistance profile of *Salmonella* sp. in retail poultry meat [59]. The retail meats show high resistance to the common antibiotics such as tetracycline, streptomycin, sulfamethoxazole, and ampicillin.

## 7. Diagnosis

*Salmonella* diagnosis requires isolation of bacterium from the clinical samples and its culture in suitable culture media. The most common selective media used for *Salmonella* are SS agar, bismuth sulfite agar, Hektoen Enteric (HE) medium, Brilliant Green agar and Xylose-Lysine-Deoxycholate (XLD) agar. To further confirm diagnosis, biochemical, and serological tests are employed. The biochemical tests include sugar fermentation test, decarboxylation and dehydrogenation reactions, and hydrogen sulphide production. Serological examinations are usually carried out in outbreaks. Suffering from any other ailments makes diagnosis more cumbersome such as in cases of HIV-infected adults [60]. Hence, development of a rapid and sensitive diagnostic test is the need of the hour. A multiplex PCR has been found to be useful to identify NTS i.e. *S.* Typhimurium and variants, *S.* Enteritidis, *S.* Dublin and *S.* Stanleyville with 100% sensitivity and specificity [61]. Presence of low number of bacilli in clinical specimen is a limitation to this. So, to detect low infective loads of NTS, a microwave-accelerated metal-enhanced fluorescence (MAMEF) technique has been developed [62], which is well efficient enough to detect as little as 1 CFU/ml in less than 30 seconds. But, this still needs wider field applicability. A well defined ELISA with a definitive cut-off has not yet being commercialized for detection of NTS. But several researchers have suggested the use of lipopolysaccharide antigens from *S.* Enteritidis (serogroup D) and *S.* Typhimurium (serogroup B) for NTS detection [63].

## 8. Prevention and control measures

The fundamental basis for the control of NTS is food safety at every step from farm to fork. Even antibiotic treatment is not recommended in uncomplicated gastroenteritis cases as this condition is self-limiting. The list for preventive and control measures include good sanitation practices, safer food, and water handling methods, vaccination, public awareness, malaria control, and antiretroviral therapy programmes. To limit the number of infections arising as a result of animal contact it is advisable to wash hands properly after each animal contact, as in many cases the organism is in colonized state in animals without showing any sign and symptoms. Proper food cooking contributes to limit infections. Although irradiation technology has been approved by several health agencies like WHO, CDC, and European commission's Scientific Committee on Food, its use is partially implemented. Curtailment to the indiscriminate antibiotic usage in poultry feed along with better farm managerial practices leads to decreased multidrug resistant bacterial load. One health approach including multiple interventions is mandatory to enhance understanding, prevention, and control of NTS, as human health is completely related to the animal health and their environment. Adoption of different on-farm interventions strategies such as genetic selection of *Salmonella*-resistant birds, regular flock testing, use of natural antimicrobial products such as prebiotics or probiotics and egg washing on farms can reduce infection. The incidences of NTS infections have been observed more in individuals suffering from malaria and HIV because of immune-compromised health status in such individuals. So, adoption of strategies, such as malaria control, and antiretroviral therapy programmes, will not only lower the chances of primary sufferings but will greatly reduce NTS infections also.

Vaccination could be considered as a potential tool to control NTS, but currently no licensed vaccine is available for this in humans. The available typhoidal vaccine does not provide protection against NTS infections. Vaccination in animals may limit transmission of the micro-organism to humans. With this objective researchers are trying different vaccine strategies on livestock for NTS prevention. It includes live attenuated vaccines, killed vaccines, and a combination of both. Oral administration of live attenuated *S.* Gallinarum to chickens prevented not only wildtype infections by *S.* Gallinarum but also infections by *S.* Enteritidis [64]. Delivery of a killed vaccine comprising three different *Salmonella* serogroups i.e. Typhimurium, Mbandaka and Orion to chickens resulted in significant reduction in bacterial load when compared to the unvaccinated groups [65]. Administration of live attenuated *S.* Typhimurium vaccine followed by a killed *Salmonella* serovars Berta and Kentucky into chickens, showed a significant decrease in *Salmonella* sp. in the vaccinated animals when compared to the unvaccinated group [66].

Subunit vaccine development may pave a better way towards control scheme. Such vaccines come with an advantage of raising a protective immune response by using only a part of the infectious micro-organism. Common sub-cellular components of *Salmonella* used for development of vaccines are outer membrane proteins (Omps), porins, toxins and ribosomal fractions. Such vaccines have been tried in different animals and have variable success rates [67–70]. Many of the cell surface carbohydrates of pathogenic bacteria like capsular polysaccharides are important antigenic determinants as in case of Vi-based vaccines against *S.* Typhi in humans. Omps are the surface exposed proteins which play a crucial role in pathogenic processes such as motility, adherence and colonization of the host cells, injection of toxins and cellular proteases, and formation of channels for the antibiotics removal [71]. Administration of Omps of *S.* Enteritidis can elicit high antibody responses and prevent bacterial shedding in chicken challenged with virulent *Salmonella* [72].

These functions make them attractive targets for the development of vaccines. *Salmonella* is an intracellular pathogen and generation of both B-cell and T-cell immune responses are essential. Live attenuated vaccines provide both humoral and cell mediated immune response; but, they may pose a risk in immunocompromised individuals. Whereas, inactivated vaccines induce only humoral immunity. Hence, the development of subunit vaccines after B-cell and T-cell epitope prediction and assessment of peptides with high affinity for class I and II MHC proteins are a better approach, and studies focussing this [73], increase the likelihood of developing a successful vaccine. Successful induction high levels of anti-porin antibodies and enhanced cell mediated immunity against *Salmonella* also have been demonstrated [74].

## 9. Conclusion

The spread of non typhoidal salmonellosis is ubiquitous and persists in environment for a very long time duration. This poses difficulty in reducing the spread of infection. Infection from the poultry farm to fork level leads to severe complications in humans especially in immunocompromised individuals, children, and elderly. Moreover, the emergence of antimicrobial resistance in NTS is a major challenge in its effective treatment. Furthermore, till now no known vaccine is available which can control all the serotypes of NTS. Hence, in the present circumstances, implementation of one health approach could be a possible answer to prevent NTS infections.

**Author details**

Mamta Pandey[1]* and Emmagouni Sharath Kumar Goud[2]

1 Molecular Reproduction Lab, Animal Biotechnology Center, ICAR-National Dairy Research Institute, Karnal, Haryana, India

2 Department of Obstetrics and Gynaecology, University of Helsinki, Helsinki, Finland

*Address all correspondence to: drmamtavet@gmail.com

## References

[1] Sexton M. *Salmonella* and *Campylobacter* in poultry in Australia. The Australian Poultry Science Symposium; 2016; Sydney: The Poultry Research Foundation, The University of Sydney.

[2] Kirk MD, Pires SM, Black RE, Caipo M, Crump JA, Devleesschauwer B, Dopfer D, Fazil A, Fischer-Walker CL, Hald T, Hall AJ. World Health Organization estimates of the global and regional disease burden of 22 foodborne bacterial, protozoal, and viral diseases. 2010:a data synthesis. PLoS medicine. 2015;12(12):e1001940. DOI:10.1371/journal.pmed.1001940

[3] Majowicz SE, Musto J, Scallan E, Angulo FJ, Kirk M, O'Brien SJ, Jones TF, Fazil A, Hoekstra RM. The global burden of nontyphoidal *Salmonella* gastroenteritis. Clinical Infectious Diseases. 2010;50(6):882-889. DOI:10.1086/650733

[4] Glenn LM, Lindsey RL, Frank JF, Meinersmann RJ, Englen MD, Fedorka-Cray PJ, Frye JG. Analysis of antimicrobial resistance genes detected in multidrug-resistant *Salmonella enterica* serovar Typhimurium isolated from food animals. Microbial Drug Resistance. 2011;17(3):407-418. DOI:10.1089/mdr.2010.0189

[5] Crump JA, Sjolund-Karlsson M, Gordon MA, Parry CM. Epidemiology, clinical presentation, laboratory diagnosis, antimicrobial resistance, and antimicrobial management of invasive *Salmonella* infections. Clinical Microbiology Reviews. 2015;28:901-937. DOI: 10.1128/CMR.00002-15

[6] Mastroeni P, Chabalgoity JA, Dunstan SJ, Maskell DJ, Dougan G. *Salmonella*: immune responses and vaccines. The Veterinary Journal. 2001;161(2):132-164. DOI:10.1053/tvjl.2000.0502

[7] Grimont PA, Weill FX. Antigenic formulae of the Salmonella serovars. WHO collaborating centre for reference and research on Salmonella. 2007;9:1-66.

[8] Feasey NA, Dougan G, Kingsley RA, Heyderman RS, Gordon MA. Invasive non-typhoidal *Salmonella* disease: an emerging and neglected tropical disease in Africa. Lancet. 2012;379:2489-2499. DOI:10.1016/S0140-6736(11)61752-2

[9] World Health Organization. [Internet] 2021. Available from: https://www.who.int/news-room/fact-sheets/detail/*salmonella*-(non-typhoidal). Accessed 2021-01-20

[10] Centers for Disease Control and Prevention. 2011. Compendium of measures to prevent disease associated with animals in public settings. 2011. MMWR Morbidity Mortality Weekly Report 60:1-24.

[11] Liljebjelke KA, Hofacre CL, Liu T, White DG, Ayers S, Young S, Maurer JJ. Vertical and horizontal transmission of *Salmonella* within integrated broiler production system. Foodbourne Pathogens and Disease. 2005;2(1):90-102. DOI: 10.1089/fpd.2005.2.90

[12] Hugas M, Beloeil PA. Controlling *Salmonella* along the food chain in the European Union-progress over the last ten years. Eurosurveillance. 2014;19(19):20804.

[13] Dunkley KD, Callaway TR, Chalova VI, Anderson RC, Kundinger MM, Dunkley CS, Nisbet DJ, Ricke SC. Growth and genetic responses of *Salmonella* Typhimurium to pH-shifts in an anaerobic continuous culture. Anaerobe. 2008;14(1):35-42. DOI: 10.1016/j.anaerobe.2007.10.001

[14] Bavishi C, Dupont HL. Systematic review: the use of proton pump

inhibitors and increased susceptibility to enteric infection. Alimentary Pharmacology and Therapeutics. 2011;34(11-12):1269-1281. DOI:10.1111/j.1365-2036.2011.04874.x

[15] Jones TF, Ingram LA, Cieslak PR, Vugia DJ, Tobin-D'Angelo M, Hurd S, Medus C, Cronquist A, Angulo FJ. Salmonellosis outcomes differ substantially by serotype. The Journal of Infectious Diseases. 2008;198(1):109-114. DOI: 10.1086/588823

[16] MacLennan CA, Gilchrist JJ, Gordon MA, Cunningham AF, Cobbold M, Goodall M, Kingsley RA, Van Oosterhout JJ, Msefula CL, Mandala WL, Leyton DL. Dysregulated humoral immunity to nontyphoidal *Salmonella* in HIV-infected African adults. Science. 2010;328(5977):508-512. DOI: 10.1126/science.1180346

[17] Nyirenda TS, Mandala WL, Gordon MA, Mastroeni P. Immunological bases of increased susceptibility to invasive nontyphoidal *Salmonella* infection in children with malaria and anaemia. Microbes and Infection. 2018;20(9-10):589-598. DOI: 10.1016/j.micinf.2017.11.014

[18] Uche IV, MacLennan CA, Saul A. A systematic review of the incidence, risk factors and case fatality rates of invasive nontyphoidal *Salmonella* (iNTS) disease in Africa (1966 to 2014). PLoS Neglected Tropical Diseases. 2017;11(1):e0005118. DOI: 10.1371/journal.pntd.0005118

[19] CDC. Antibiotic Resistance Threats in the United States, 2019. Atlanta, GA: U.S. Department of Health and Human Services, CDC; 2019.

[20] European Food Safety Authority and European Centre for Disease Prevention and Control (EFSA and ECDC). The European Union one health 2018 zoonoses report. EFSA Journal. 2019;17(12):e05926. DOI:10.2903/j.efsa.2019.5926

[21] Reddy EA, Shaw AV, Crump JA. Community-acquired bloodstream infections in Africa: a systematic review and meta-analysis. The Lancet Infectious Diseases. 2010;10(6):417-432. DOI: 10.1016/S1473-3099(10)70072-4

[22] Marks F, Von Kalckreuth V, Aaby P, Adu-Sarkodie Y, El Tayeb MA, Ali M, Aseffa A, Baker S, Biggs HM, Bjerregaard-Andersen M, Breiman RF. Incidence of invasive *salmonella* disease in sub-Saharan Africa: a multicentre population-based surveillance study. The Lancet Global Health. 2017;5(3):e310-23. DOI: 10.1016/S2214-109X(17)30022-0

[23] Oneko M, Kariuki S, Muturi-Kioi V, Otieno K, Otieno VO, Williamson JM, Folster J, Parsons MB, Slutsker L, Mahon BE, Hamel MJ. Emergence of community-acquired, multidrug-resistant invasive nontyphoidal *Salmonella* disease in rural Western Kenya, 2009-2013. Clinical Infectious Diseases. 2015;61(suppl_4):S310-6. DOI: 10.1093/cid/civ674

[24] SCRA. [Internet]. Statistical Committee of the Republic of Armenia/Socio-Economic Situation of RA, January–December/Public Health. Available at: https://www.armstat.am/en/?nid=82. Accessed 2021-01-05.

[25] Menezes GA, Khan MA, Harish BN, Parija SC, Goessens W, Vidyalakshmi K, Baliga S, Hays JP. Molecular characterization of antimicrobial resistance in non-typhoidal *salmonellae* associated with systemic manifestations from India. Journal of Medical Microbiology. 2010;59(12):1477-1483. DOI: 10.1099/jmm.0.022319-0

[26] Chen PL, Li CY, Hsieh TH, Chang CM, Lee HC, Lee NY, Wu CJ, Lee CC, Shih HI, Ko WC. Epidemiology,

disease spectrum and economic burden of non-typhoidal *Salmonella* infections in Taiwan, 2006-2008. Epidemiology and Infection. 2012;140(12):2256-2263.

[27] Research SA. Causes and outcomes of sepsis in southeast Asia: a multinational multicentre cross-sectional study. The Lancet Global Health. 2017;5(2):e157-67. DOI: 10.1016/S2214-109X(17)30007-4

[28] Shahunja KM, Leung DT, Ahmed T, Bardhan PK, Ahmed D, Qadri F, Ryan ET, Chisti MJ. Factors associated with non-typhoidal *Salmonella* bacteremia versus typhoidal *Salmonella* bacteremia in patients presenting for care in an urban diarrheal disease hospital in Bangladesh. PLoS Neglected Tropical Diseases. 2015;9(9):e0004066. DOI: 10.1371/journal.pntd.0004066

[29] Kaur AA, Fadzilah MN, Mariam M, Abidin ASZ, Adnan, S, Nor, NSM. Community-acquired bacteremia in Paediatrics: epidemiology, aetiology and patterns of antimicrobial resistance in a tertiary care Centre, Malaysia. Medical Journal of Malaysia. 2016;71(3):117-121.

[30] Sinwat N, Angkittitrakul S, Coulson KF, Pilapil FM, Meunsene D, Chuanchuen R. High prevalence and molecular characteristics of multidrug-resistant *Salmonella* in pigs, pork and humans in Thailand and Laos provinces. Journal of Medical Microbiology. 2016;65(10):1182-1193. DOI: 10.1099/jmm.0.000339

[31] Noor Uddin GM, Larsen MH, Barco L, Minh Phu T, Dalsgaard A. Clonal occurrence of *Salmonella* Weltevreden in cultured shrimp in the Mekong Delta, Vietnam. Plos one. 2015;10(7):e0134252. DOI: 10.1371/journal.pone.0134252

[32] Chiu CH, Su LH. *Salmonella*, Non-Typhoidal Species (*S.* choleraesuis, *S.* enteritidis, *S.* hadar, *S.* typhimurium). Dostupno na http://www. antimicrobe. org/b258. asp. Zadnji pristup. 2019;21.

[33] Chen YH, Chen TP, Lu PL, Su YC, Hwang KP, Tsai JJ, Cheng HH, Peng CF. *Salmonella* choleraesuis bacteremia in southern Taiwan. The Kaohsiung Journal of Medical Sciences. 1999;15(4):202-208.

[34] Guery R, Habibi A, Arlet JB, Lionnet F, de Lastours V, Decousser JW, Mainardi JL, Razazi K, Baranes L, Bartolucci P, Godeau B. Severe, non specific symptoms in non-typhoidal *Salmonella* infections in adult patients with sickle cell disease: a retrospective multicentre study. Infectious Diseases. 2018;50(11-12):822-830. DOI: 10.1080/23744235.2018.1500706

[35] Molyneux EM, Mankhambo LA, Phiri A, Graham SM, Forsyth H, Phiri A, Walsh AL, Wilson LK, Molyneux ME. The outcome of non-typhoidal *salmonella* meningitis in Malawian children, 1997-2006. Annals of Tropical Paediatrics. 2009;29(1):13-22. DOI: 10.1179/146532809X401980

[36] Hsu RB, Tsay YG, Chen RJ, Chu SH. Risk factors for primary bacteremia and endovascular infection in patients without acquired immunodeficiency syndrome who have nontyphoid salmonellosis. Clinical Infectious Diseases. 2003;36(7):829-834. DOI:10.1086/367932

[37] Di Bonaventura G, Di Girolamo A, Catamo G, Nicoletti M, Piccolomini R. *Salmonella* typhimurium-endocarditis secondary to an acquired environmental infection: a case report. Microbiologica-Bologna. 2001;24(1):85-90.

[38] Van Sorge NM, Zialcita PA, Browne SH, Quach D, Guiney DG, Doran KS. Penetration and activation of brain endothelium by *Salmonella enterica* serovar Typhimurium. Journal of Infectious Diseases. 2011;203(3):401-405. DOI:10.1093/infdis/jiq048

[39] Genzen JR, Towle DM, Kravetz JD, Campbell SM. *Salmonella* typhimurium pulmonary infection in an immunocompetent patient. Connecticut medicine. 2008;72(3).

[40] Arda IS, Ergin F, Varan B, Demirhan B, Aslan H, Ozyaylali İ. Acute abdomen caused by *Salmonella* typhimurium infection in children. Journal of Pediatric Surgery. 2001;36(12):1849-1852. DOI:10.1053/jpsu.2001.28867

[41] Abdollahi A, Moradi-Tabriz H, Rasoulinejad M. Pneumonia due to *Salmonella* typhimuriumin an HIV-Infected Patient. Iranian Journal of Pathology. 2010;5(4):208-211.

[42] Samal B, Oommen S, Swami A, Maskey M, Shastri J. *Salmonella* brain abscess in an infant. Indian Journal of Pathology and Microbiology. 2009;52(2):269.

[43] Rybak MJ. Resistance to antimicrobial agents: an update. Pharmacotherapy: The Journal of Human Pharmacology and Drug Therapy. 2004;24(12P2):203S–15S.

[44] Alanis AJ. Resistance to antibiotics: are we in the post-antibiotic era? Archives of Medical Research. 2005;36(6):697-705. DOI:10.1016/j.arcmed.2005.06.009

[45] Croft AC, D'Antoni AV, Terzulli SL. Update on the antibacterial resistance crisis. Medical Science Monitor. 2007;13(6):RA103-18.

[46] Molbak K, Baggesen DL, Aarestrup FM, Ebbesen JM, Engberg J, Frydendahl K, Gerner-Smidt P, Petersen AM, Wegener HC. An outbreak of multidrug-resistant, quinolone-resistant *Salmonella enterica* serotype Typhimurium DT104. New England Journal of Medicine. 1999;341(19):1420-1425. DOI: 10.1056/NEJM199911043411902

[47] Su LH, Chiu CH, Chu C, Ou JT. Antimicrobial resistance in nontyphoid *Salmonella* serotypes: a global challenge. Clinical Infectious Diseases. 2004;39(4):546-551. DOI:10.1086/422726

[48] Chiu CH, Su LH, Chu C, Chia JH, Wu TL, Lin TY, Lee YS, Ou JT. Isolation of *Salmonella enterica* serotype choleraesuis resistant to ceftriaxone and ciprofloxacin. The Lancet. 2004;363(9417):1285-1286. DOI:10.1016/S0140-6736(04)16003-0

[49] Foley SL, Lynne AM. Food animal-associated *Salmonella* challenges: pathogenicity and antimicrobial resistance. Journal of Animal Science. 2008;86(suppl_14):E173-87. DOI:10.2527/jas.2007-0447

[50] Mascaretti OA. Bacteria versus antibacterial agents: an integrated approach. American Society for Microbiology (ASM); 2003.

[51] Alcaine SD, Warnick LD, Wiedmann M. Antimicrobial resistance in nontyphoidal *Salmonella*. Journal of Food Protection. 2007;70(3):780-790. DOI:10.4315/0362-028X-70.3.780

[52] Cannon M, Harford S, Davies J. A comparative study on the inhibitory actions of chloramphenicol, thiamphenicol and some fluorinated derivatives. Journal of Antimicrobial Chemotherapy. 1990;26(3):307-317. DOI:10.1093/jac/26.3.307

[53] Cloeckaert A, Chaslus-Dancla E. Mechanisms of quinolone resistance in *Salmonella*. Veterinary Research. 2001;32(3-4):291-300.

[54] Chopra I, Roberts M. Tetracycline antibiotics: mode of action, applications, molecular biology, and epidemiology of bacterial resistance. Microbiology and Molecular Biology Reviews. 2001;65(2):232-260. DOI: 10.1128/MMBR.65.2.232-260.2001

[55] Antunes P, Machado J, Sousa JC, Peixe L. Dissemination of sulfonamide resistance genes (*sul1, sul2*, and *sul3*) in Portuguese *Salmonella enterica* strains and relation with integrons. Antimicrobial agents and chemotherapy. 2005;49(2):836-839. DOI: 10.1128/AAC.49.2.836-839.2005

[56] National Academy of Sciences Committee on Drug Use in Food Animals. The use of drugs in food animals: benefits and risks. Washington, DC: National Academy Press, 1999.

[57] Al-Zenki S, Al-Nasser A, Al-Safar A, Alomirah H, Al-Haddad A, HendriksenRS, AarestrupFM. Prevalence and antibiotic resistance of Salmonella isolated from a poultry farm and processing plant environment in the State of Kuwait. Foodborne pathogens and disease. 2007;4(3):367-373.

[58] Sapkota AR, Kinney EL, George A, Hulet RM, Cruz-Cano R, Schwab KJ, Zhang G, Joseph SW. Lower prevalence of antibiotic-resistant *Salmonella* on large-scale US conventional poultry farms that transitioned to organic practices. Science of the Total Environment. 2014;476:387-392.

[59] Food and Drug Administration (FDA). 2014-2015 Retail Meat Interim Report. 2017. Available online:https://www.fda.gov/downloads/AnimalVeterinary/SafetyHealth/AntimicrobialResistance/NationalAntimicrobialResistanceMonitoringSystem/UCM498134.pdf. Accessed 2018-06-15.

[60] Gordon MA, Banda HT, Gondwe M, Gordon SB, Boeree MJ, Walsh AL, Corkill JE, Hart CA, Gilks CF, Molyneux ME. Non-typhoidal *salmonella* bacteraemia among HIV-infected Malawian adults: high mortality and frequent recrudescence. Aids. 2002;16(12):1633-1641.

[61] Tennant SM, Diallo S, Levy H, Livio S, Sow SO, Tapia M, Fields PI, Mikoleit M, Tamboura B, Kotloff KL, Nataro JP. Identification by PCR of non-typhoidal *Salmonella enterica* serovars associated with invasive infections among febrile patients in Mali. PLoS Neglected Tropical Diseases. 2010;4(3):e621. DOI:10.1371/journal.pntd.0000621

[62] Tennant SM, Zhang Y, Galen JE, Geddes CD, Levine MM. Ultra-fast and sensitive detection of non-typhoidal *Salmonella* using microwave-accelerated metal-enhanced fluorescence ("MAMEF"). PLoS One. 2011;6(4):e18700. DOI:10.1371/journal.pone.0018700

[63] Kuhn KG, Falkenhorst G, Ceper TH, Dalby T, Ethelberg S, Mølbak K, Krogfelt KA. Detecting non-typhoid *Salmonella* in humans by ELISAs: a literature review. Journal of Medical Microbiology. 2012;61(1):1-7. DOI:10.1099/jmm.0.034447-0

[64] Penha Filho RA, de Paiva JB, da Silva MD, de Almeida AM, Junior AB. Control of *Salmonella* Enteritidis and *Salmonella* Gallinarum in birds by using live vaccine candidate containing attenuated *Salmonella* Gallinarum mutant strain. Vaccine. 2010;28(16):2853-2859. DOI:10.1016/j.vaccine.2010.01.058

[65] Pavic A, Groves PJ, Cox JM. Utilization of a novel autologous killed tri-vaccine (serogroups B [Typhimurium], C [Mbandaka] and E [Orion]) for *Salmonella* control in commercial poultry breeders. Avian Pathology. 2010;39(1):31-39. DOI:10.1080/03079450903454277

[66] Dorea FC, Cole DJ, Hofacre C, Zamperini K, Mathis D, Doyle MP, Lee MD, Maurer JJ. Effect of *Salmonella* vaccination of breeder chickens on contamination of broiler chicken

carcasses in integrated poultry operations. Applied and Environmental Microbiology. 2010;76(23):7820-7825. DOI: 10.1128/AEM.01320-10

[67] Singh BR, Singh Y, Agarwal MC, Agarwal RK, Sharma VD. *Salmonella* vaccines for veterinary use: an overview. Haryana Vet. 2005;44:1-12.

[68] Vasava KA, Singh BR, Verma JC. Detection of cytotoxigenicity among strains of *Salmonella enterica* subspecies enterica serovar Abortusequi by an indirect ELISA. Indian Journal of Veterinary Research. 2004; 13:31-34.

[69] Barman TK, Sharma VD, Kumar S. Optimization of dose of *Salmonella* toxoid vaccine in poultry. Indian Journal of Experimental Biology. 2002;79:106-110.

[70] Mishra RS, Sharma VD. Comparative efficacy of various toxoids against salmonellosis in poultry. Veterinary Research Communications. 2001;25(5):337-344.

[71] Cordwell SJ. Technologies for bacterial surface proteomics. Current Opinion in Microbiology. 2006;9(3):320-329. DOI:10.1016/j.mib.2006.04.008

[72] Meenakshi M, Bakshi CS, Butchaiah G, Bansal MP, Siddiqui MZ, Singh VP. Adjuvanted outer membrane protein vaccine protects poultry against infection with *Salmonella enteritidis*. Veterinary Research Communications. 1999;23(2):81-90.

[73] Pandey M, Saxena MK. Cloning and immunopotential analysis of Omp 28 of *Salmonella* Typhimurium for the development of subunit vaccine for poultry salmonellosis. Indian Journal of Poultry Science. 2015;50(2):138-142.

[74] Lim E, Koh WH, Loh SF, Lam MS, Howe HS. Non-thyphoidal salmonellosis in patients with systemic lupus erythematosus. A study of fifty patients and a review of the literature. Lupus. 2001;10(2):87-92. DOI:10.1191/096120301675973164

Chapter 6

# *Salmonella* and the Immune System

*Weam Saad Al-Hamadany*


## Abstract

The human body has many mechanisms to resist invaders like pathogenic bacteria to avoid harm according to the living creature's law "survival for the best". On the opposite; *Salmonella* as pathogenic bacteria have many weapons that they utilize to invade the human body. The resistance mechanisms expressed by the human body are called immunity which represented by the immune system that has many different types of resistance processes, either specific (adaptive immune response) or non-specific (Innate Immune Response) against certain pathogenic invaders. As far as these processes are strong they will be enough to avoid infections occurrence, otherwise, the human body will get infected with *Salmonella*, be ill, show the disease symptoms, transmit the disease to others, and may become a carrier for the pathogen according to many circumstances. Prevention is still stood the most effective way to avoid getting infected with *Salmonella* by personal hygiene or suitable vaccination if available.


**Keywords:** immune system, *Salmonella* virulence factors, *Salmonella* infection, *Salmonella* resistance and *Salmonella* vaccination

## 1. Introduction and overview

### 1.1 Introduction to immunology and immune system

Immunology is the study of our protection against foreign macromolecules or invading organisms and our responses to them. These invaders include viruses, bacteria, protozoa or even larger parasites. Any Human body is continuously exposed to pathogenic microorganisms. The immune system is composed of two major subdivisions of immune system, the innate or nonspecific immune system and the adaptive or specific immune system [1].

The innate immune system is our first line of defense against invading organisms while the adaptive immune system acts as a second line of defense and gives protection against re-exposure to the same pathogen. Each of the major subdivisions of the immune system has both cellular and humoral components by which they carry out their protective function and help each other to do these functions. Since pathogens may replicate intracellularly (viruses and some bacteria and parasites) or extracellularly (most bacteria, fungi and parasites), different components of the immune system have evolved to protect against these different types of pathogens [2].

### 1.2 Innate or non-specific defenses

Include first line of defense which acts before invasion of pathogenic microbes. And the second line of defense which acts after invasion. The anatomical barriers that works mainly against infections with microbial invaders. This first line of defense represented by the epithelial surfaces and skin form the physical barriers that are very impermeable to most infectious agents [1].

The shedding of skin epithelium also helps remove bacteria and other infectious agents that have adhered to the epithelial surfaces. Movement due to cilia or peristalsis helps to keep air passages and the gastrointestinal tract free from microorganisms. The trapping effect of mucus that lines the respiratory and gastrointestinal tract helps protect the lungs and digestive systems from infection [2].

Chemical barriers like Lysozyme and phospholipase found in saliva and other secretions can breakdown the cell wall of bacteria and destabilize bacterial membranes. The low pH of gastric secretions prevents the growth of bacteria [3].

The microbiota of the skin and in the gastrointestinal tract can prevent the colonization of pathogenic bacteria by secreting toxic substances or by competing with pathogenic bacteria for nutrients or attachment to cell surfaces. They represent the biological barriers of the innate immunity [2].

The anatomical barriers are very effective in preventing colonization of tissues by microorganisms. However, when there is damage to tissues the anatomical barriers are breeched and infection happens. Once infectious agents have penetrated tissues, another innate defense mechanisms comes into play, namely acute inflammation as the second line of innate immune defense. Many Humoral and cellular factors play an important role in inflammation against microbial invasion, which is characterized by edema and the activation of phagocytic cells [4].

These humoral factors are found in serum or they are formed at the site of infection. They contain Complement system, Interferons and Lysozymes. The most important humoral barrier is the Complement system, since it acts as with the phagocytic cells as a bridge between specific and non-specific immune response. Complement system represents a set of glycoproteins in blood. Once they are activated after rapid cascade events that can lead to increase vascular permeability, activation of phagocytic cells, opsonization of bacteria and lysis [5].

Complement glycoproteins are synthesized by liver cells (hepatocytes) and macrophages and many other cell (e.g. gut epithelial cells). All normal individuals have complement components in their blood. This system can be activated by [1, 2]:

a. Antigen-antibody complexes containing IgG or IgM activate complement by the classical pathway that starts with C1 (complement 1).

b. Membranes and cell walls of microbial organisms (e.g. Lipopolyccharides layer [LPS] of gram –ve bacteria) and many other substances can activate complement by the alternative pathway.

c. Proteolytic enzymes released either from microbes or from host cells during immune defence mechanisims, can also activate the complement system by breaking down critical components.

The complement system takes part in both specific and non-specific resistance and generates a number of products of biological and immunological importance. The functions of the complement system are summarized in **Table 1** [3, 5]:

On the other side; the cellular factors are the main line of defense in the nonspecific immune system, they are listed in the **Table 2** [2, 5].

| No. | Function |
|---|---|
| 1. | Binding and neutralizing foreign substances that activate it. |
| 2. | Induce the ingestion of complement-coated substances by phagocytic cells (help fn the opsonization process when C3b and C4b linked with the surface of microorganisms and attach to Complement receptor on phagocytic cells then induce phagocytosis). |
| 3. | Activation of many cells including polymorphonuclear cells (PMNs) and macrophages. |
| 4. | Have roles in regulation of antibody responses. |
| 5. | Clearance of immune complexes and apoptotic cells. |
| 6. | Have roles in inflammation and tissue damage. |
| 7. | Some components (C3a, C4a and C5a), have role in Anaphylaxis (a dangerous case of type 1 hypersensitivity), hence they are called anaphylotoxins. |
| 8. | Some complement components acts as chemotactic facters e.g. C5a and MAC. |

**Table 1.**
*The functions of the complement system.*

| No. | Cell | Function |
|---|---|---|
| 1. | Neutrophils | Polymorphonuclear cells (PMNs) migrate to the site of infection where they phagocytose invading organisms and kill them intracellularly. In addition, PMNs contribute to tissue damage that occurs during inflammation. |
| 2. | Macrophages | Tissue macrophages and activated monocytes, which differentiate into macrophages, also function in phagocytosis and intracellular killing of microorganisms. In addition, macrophages are capable of extracellular killing of infected or transformed cells (self-target). Furthermore, macrophages have role in tissue repair and act as antigen presenting cells APC, which are required for the induction of specific immune responses. |
| 3. | Natural killer | NK cells can nonspecifically kill virus infected and tumor cells. These cells are not part of the inflammatory response but they are important in nonspecific immunity to viral infections and tumor surveillance. |
| 4. | Eosinophils | Eosinophils have proteins in granules that are effective in killing certain parasites. |

**Table 2.**
*Cellular factors of the nonspecific immune system and their function.*

### *1.2.1 Phagocytosis and intracellular killing*

Phagocytosis is a very important process during non-specific immune response when specialized cells engulf foreign body like bacteria or molecule like toxin or virus. The phagocytosis has four steps, **Figure 1** [2]:

1. Chemotaxis. Phagocytic cells response and migrate to the site of infection or injury by the effect of complement products and cytokines released from tissue macrophages that have encountered bacteria or any foreign body in tissue.

2. Endocytosis. Starts with pseudopodia formation then phagocytic cells bind to the foreign body by: Fc receptors–Bacteria with IgG antibody on their surface have the Fc region exposed and this part of the Ig molecule can bind to the receptor on phagocytes. Complement receptors–Phagocytic cells have a receptor for the complement C3b. Scavenger receptors mainly for invading bacteria.

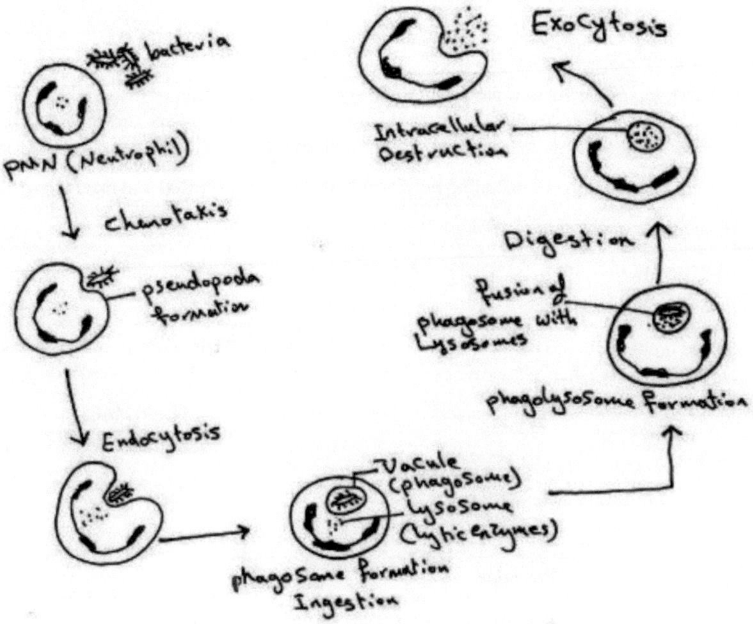


**Figure 1.**
*Phagocytosis process steps.*

3. Phagolysosome formation and degradation of foreign substances. After attachment of the bacteria the phagocyte begins to extend pseudopods around the bacteria and surround and engulf them forming the phagosome. During phagocytosis the granules or lysosomes of the phagocytes bind or fuse with the phagosome and empty their contents. The result is the foreign bodies or bacteria engulfed in the phagolysosome which have the contents of the granules or lysosomes.

Intracellular killing and Digestion (Lysis and excretion): There are three means of killing the microorganisms inside phagocytic cells; either Oxygen dependent killing by formation of NADPH using Oxygen, then production of the toxic oxygen compounds like $H_2O_2$ and hydroxyl radical (OH•). These compounds are toxic to microbes and kill them, Oxygen independent killing by production of toxic hypochlorite (OCl-) and singlet oxygen (1O2) from $H_2O_2$ using the enzyme Myeloperoxidase that released into the phagolysosome or Nitric oxide dependent killing by Toxic nitric oxide synthesis and production (NO) when microorganism binds to the macrophage because of cytokines release (TNF-$\alpha$ and IFN-$\gamma$) [3].

Oxygen -dependent killing and Oxygen -independent killing both are called the Respiratory burst. After killing, the enzymatic system of the cell will digest all the phagosome components then absorb the useful materials and excrete the residues to the environment (blood) by fusing the phagolysosome with the cell membrane.

The cells that able to do phagocytosis are (monocytes, macrophage, PMNs and dendric cells). The results of phagocytosis are either a complete destruction of foreign body and excretion (PMNs). Or a complete destruction of foreign body and some parts (polypeptides) of it will be processed and presented on the surface of the phagocytic cells (monocytes, macrophage and dendric cells) then the phagocytic cell will be antigen presenting cell (APC) [2, 3].

## 1.3 Cells involved in specific immune responses

### *1.3.1 Antigen presenting cell (APC)*

These cells are the messengers between innate (non-specific) immunity and the adaptive (specific). Specialized APC are macrophage (MØ), B-cells and Dendric cells (DC) [3].

Roles of Antigen Presenting Cell (APC) can be summarized as [5]:

1. Engulfment of foreign Ag, processing it and presenting it (or a olypeptide from it) on the surface near the Major Histocompatibility Complex MHC class I or II.

2. Communication during the immune response between immune cells especially T- cells to induce the proper immune response cellular or humoral.

3. Secretion of cytokines which are substances (glycoproteins) that regulate the immune response.

### *1.3.2 Lymphocytes*

B-cells or B-lymphocytes, T-cells or T-lymphocytes (T-helper cells including Th1 and Th2, T-Cytotoxic Tc and T-suppressor Ts) [3].

### *1.3.3 Natural killer cells (NK)*

Natural killer cells (NK) have no CD markers on the surface so they are usually called null cells [3].

It is important to know that B-cells are able to be APC by internalization of Ag inside the cell and do the processing and presenting, which will be discussed later. Also Dendric cells (DC) are cells found only in the mammalian immune system; their function is to engulf and process Ag then present it on the surface to other immune cells. Found in tissues that in contact with external environment such as skin, lung, stomach and intestine [2].

## 1.4 Mucosal immune response

In the mucosal surfaces and sites, the mucosal immune response come to play role in resistance against infection establishment. Many lymphoid tissues are associated with mucosa which are usually called mucosa-associated tissues play major role in protection since they are rich with both T-cells and B-cells, produce many types of Lymphokines that acts as signals of the immune system actions, produce IgA (sIgA Secretory IgA); the main effective immunoglobulin type in the surfaces of the body and the most important part is that mucosal surfaces have the receptors of microbiota that play as a biological barrier and support innate immunity. Many secretions are also produced by the mucosa to protect surfaces like gastric acid and continuous mucous secretion and shedding helps in renewing normal flora population and shed colonized pathogens.

Mucosa-Associated Lymphoid Tissues (MALT) Include the lymphoid tissues of the intestinal tract, genitourinary tract, tracheobronchial tree, and mammary glands. All of the mucosa-associated lymphoid tissues are unencapsulated and contain both T and B lymphocytes [2, 5].

### 1.5 Gut-associated lymphoid tissue (GALT)

It is found along the digestive tract. Three major areas of GALT that can be identified are the tonsils, the Peyer's patches, located on the submucosa of the small intestine, and the appendix. In addition, scanty lymphoid tissue is present in the lamina propria of the gastrointestinal tract [3, 5].

1. Tonsils, located in the oropharynx, are predominantly populated by B-lymphocytes and are the site of antigenic stimulation.

2. Peyer's patches (PPs), they are lymphoid structures disseminated through the submucosal space of the small intestine

Physiological roles of secondary lymphoid organs:

a. The follicles of the intestinal Peyer's patches are extremely rich in B-cells, which differentiate into IgA-producing plasma cells.

b. T-lymphocytes are also present in the intestinal mucosa, the most abundant of them expressing membrane markers that are considered typical of memory helper T-cells. This population is involved in the induction of humoral immune responses (HMI) [1, 2].

### 1.6 Antibody mediated immune response (humeral mediated immunity, HMI)

B-cells have normal Ag receptors on the surface they are natural Igs, these Igs are able to form Ag-Ab complex on the surface of B-cell. This complex will be internalized inside B-cell, then the foreign Ag will be processed within B-cell and presented (or polypeptides from it) on the surface of B-cell near MHC class II and now B-cell is APC.

T-helper (Th) cells come near the APC B-cell and by the help of TCR and CD4; Th will interact and communicate with APC B-cell and Th cell will be activated and release cytokines or lymphokines (IL-2, IL-4, IL-5 and IFN-$\gamma$), these products will induce other B-cells for dividing, proliferation and differentiation. IgM will be the first Ig produced then B- cell will switch to make IgG. This response is called T-dependent Ag immune response. The other type of response is T-independent Ag immune response, this type of Ag stimulates B-cells without need for T-helper lymphocytes interfere [1, 2].

After B-cells activation, series of events happen (proliferation, clonal expansion, division and maturation), ending with Ab and memory B-cells production. These series of events called B-cell Maturation. During the second exposure to the same Ag that started the first immune response (perhaps after year from first exposure), the B-memory cells will remember the Ag and will be activated and divide into a clone of plasma cells to start the Secondary immune response (Memory response) [3].

Antibodies or Immunoglobulins (Ig) that are produced after specific humoral response are in five types; IgG; IgM, IgA, IgD and IgE based on differences in the amino acid sequences in the constant region of the heavy chains. In addition, the classes of immunoglobulins can be divided into subclasses based on small differences in the amino acid sequences in the constant region of the heavy chains [1].

**IgG immunoglobulin:** is a Monomer, have4 subclasses (IgG1, IgG2, IgG3 and IgG4), the subclasses differ in the number of disulfide bonds and length of the hinge region. IgG is the major Ig in serum and extra vascular spaces of total serum Igs. Able to cross placenta transfer is mediated by receptor on placental cells for the Fc region of IgG, able to fix complement an binding to cells - Macrophages, monocytes, PMN's and some lymphocytes have Fc receptors for the Fc region of IgG. The term opsonin is used to describe substances that enhance phagocytosis. IgG is a good opsonin [2].

**IgM immunoglobulin:** is a pentamer, but it can also exist as a monomer. It is the third most common serum Ig and it is the first Ig to be made by the fetus and the first Ig to be made by B cells when it is stimulated by antigen. This type of Ig is a good complement fixing Ig. Thus, IgM antibodies are very efficient in leading to the lysis of microorganisms. Also a good agglutinating Ig. Thus, IgM antibodies are very good in clumping microorganisms for elimination from the body [5].

**IgA immunoglobulin: is** a monomer but IgA found in secretions is a dimer. When IgA is found in secretions is also has another protein associated with it called the secretory piece or T piece (sIgA), this secretory piece is made in epithelial cells and is added to the IgA as it passes into the secretions. The secretory piece helps IgA to be transported across mucosa and also protects it from degradation in the secretions. It is the second most common serum Ig and the major class of Ig in secretions - tears, saliva, colostrum, and mucus. Since it is found in secretions secretory IgA is important in local (mucosal) immunity but does not fix complement [5].

**IgD immunoglobulin: is** a monomer **and** found in low levels in serum; found on B cell surfaces where it functions as a receptor for antigen also does not bind complement [5].

**IgE immunoglobulin: is** a monomer and rare on serum. It has role in allergic reactions and does not fix complement [5].

### 1.7 Cell mediated immune response (cellular mediated immunity, CMI)

This response occurs against cells, which are called Target cells. During both HMI and CMI, T-helper cells recognize foreign Ag processed on the surface of APC. If this Ag was processed and presented near MHC class II, then Th cells will activate HMI by B- cells activation, but if the presented Ag on APC was near MHC class I, then Th cells will activate CMI by activation of Tc, NK and MØ. Th cells able to activate and regulate CMI and HMI by many cytokines production.

In addition, in both CMI and HMI, when Th cells recognize the foreign Ag, Th cells will start T-cells activation by series of events (expanding, clonal proliferation and differentiation), then become mature to give specific activated T-helper cells in HMI and give specific activated T-helper cells and memory T-cells in CMI [2, 5].

Role of CMI response: is the defense against Tumor cells or cancer cells, Grafts Rejection, against Intracellular parasite infected cells with foreign Ag presented near MHC class I. Target cell is the infected cell with parasite and Types 4 hypersensitivity (Delayed type of hypersensitivity) [5].

#### *1.7.1 Activation of CMI cells*

When T-helper cells recognize foreign Ag on the surface of target cell in association with (or near) MHC class I. The TCR and CD4+ play role in recognition. Then Th cell will be activated and produce cytokines (especially IL-2 and IFN- $\gamma$). These cytokines will activate Tc CD8+ cells, MØ and NK cells. This activation will increase these cells ability for killing and became more effecter.

#### *1.7.2 Mode of action for killing target cells*

After T-cytotoxic cells and NK cells activation by Th cells, T-cytotoxic cells come into close contact with target cell; they will bind to the Ag by their specific Ag receptors. While NK cells will attach to Ag (on Target cell surface) by their non-specific receptors for Ag.

T-cytotoxic cells and NK cells will kill target cells by the following mechanisms [1–5]:

a. Direct contact killing: Production of perforin, which is a protein able to form pores in target cell membrane at the point of contact between Tc cell and target cell, lead to osmotic lysis of target cell.

b. Indirect killing: By secretion of a toxin protein in the space between the two cells, which causes fragmentation of target cell nuclear DNA, then the death of target cell by Apoptosis: the programmed cell death.

c. Antibody-dependent cellular cytotoxicity (ADCC) killing: it is specific mode of killing occurs when the parasites Ags have ability to induce both HMI and CMI, target cells will be coated with specific Abs formed after HMI against some parts of intracellular parasite like virus. These Abs will bring Tc and NK cells very close to the target cell by acting like a bridge because Tc and NK have receptors to the constant region of Ab. Then Tc and NK cells will be activated and kill the target cell by extracellular products (toxins and enzymes).

This type of CMI occurs when the foreign Ag persist for long time (e.g. *Mycobacterium tuberculosis* infection is long standing intracellular infection), also, against some kinds of cancer cells [2, 5].

### 1.8 Primary immune response

It is the first exposure to the Ag resulting of forming specific Abs and memory B-cells for HMI or T-cells and memory T-cells for CMI, the phases are, **Figure 2** [1–5]:

1. Latent Phase: start after first time exposure to an Immunogen or after induction, include the followings

    - No Ab level increase (Steady titer).
    - Recognizing Ag as foreign after processing the Ag inside APC.
    - Cellular proliferation and differentiation.
    - Duration of this phase (period) is variable depending on many factors (Ag immunogenicity, Ag dose, Ag solubility, Ag route of immunization or exposure).

2. Logarithmic phase: starts when Ab titer begin to increase (active biosynthesis of Ab), last for 10-14 days till reach peak.

3. Steady phase: starts when the rates of both formation (synthesis) and catabolism are equal, then serum concentration of Ab is constant.

4. Decline phase: starts when the Ab titer starts to fall down due to increase Ab catabolism rate than synthesis.

Note: during early primary response, IgM class antibodies is predominant and first rise than IgG appears later [2, 5].

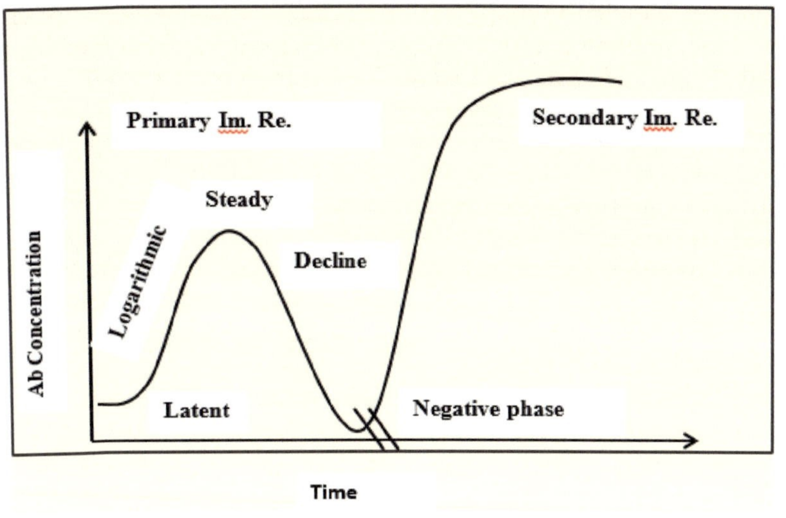


**Figure 2.**
*Primary and secondary immune response.*

### 1.9 Secondary immune response

It is the second exposure to the same immunogen that induced the first immune response (after booster dose of vaccination) may be after weeks, months, or even years later, includes [1–5]:

1. Accelerated or fast appearance of Abs.

2. Shorter latent period.

3. Rapid rate of Ab synthesis.

4. Higher peak titer of Ab.

5. More presence of memory cells.

6. Dose of immunogen needed is lower than primary.

7. Predominant Ab Class is IgG.

8. Long standing steady phase, whereas Ab titer will stay high longer time.

Negative phase: occur between primary and secondary Immune response when immunogen second dose is small and/or there is pre-existing antibodies from the first immune response (primary), then immunogen will be all consumed in Ag-Ab complex formation and phagocytosed then removed with no induction to secondary immune response [5].

## 2. *Salmonella* as a pathogenic Bacteria

*Salmonella* is a Gram-negative facultative rod-shaped bacterium in the same family as *Escherichia coli,* Enterobacteriaceae. *Salmonella* live in the intestinal tracts of warm and cold blooded animals. In humans, *Salmonella* are the cause of two diseases called salmonellosis: enteric fever (typhoid), resulting from bacterial invasion of the bloodstream, and acute gastroenteritis, resulting from a foodborne infection/ intoxication. Most common species related to human infections are *Salmonella Typhi* and *Salmonella Paratyphi* cause host-specific infections, being human the host. But other serotypes as *S.Typhimurium* and *S. Enteritidis*, mainly related to food products are also important serotypes that cause human diseases [5, 6].

*Salmonella* are found in the natural environment (water, soil, sometimes plants used as food). These bacteria are zoonotic, human or animal can excrete *Salmonella* either when they are infected with disease Salmonellosis and when they remain carriers. Also this disease is called food handling born disease due to the infected or carrier food handling workers [7].

*Salmonella* is intracellular parasite pathogen, do not multiply significantly in the natural environment (out of digestive tracts), but they can survive several weeks in water and several years in soil if conditions of temperature, humidity, and pH are favorable [5].

### 2.1 Antigenic structure and virulence factors

The bacterial antigens are the components or products of pathogens that are able to induce the immune defenses of the host to defend against, and to eliminate, the pathogen or disease. As with all *Enterobacteriaceae*, the genus *Salmonella* has three kinds of major antigens with diagnostic or identifying applications: somatic, surface, and flagellar [8–10].

#### *2.1.1 Somatic (O) or cell wall antigens*

Somatic antigens are the O side chain of LPS; they are heat stable and alcohol resistant. Cross-absorption studies individualize a large number of antigenic factors, 67 of which are used for serological identification. O factors labeled with the same number are closely related.

#### *2.1.2 Surface (envelope) antigens*

Surface antigens, commonly in enteric bacteria (e.g., *Escherichia coli and Klebsiella*), may be found in some *Salmonella*. Surface antigens in *Salmonella* may mask O antigens, and the bacteria will not be agglutinated with O antisera. One specific surface antigen is well known: the Vi antigen. The Vi antigen occurs in only three species *Salmonella Typhi*, *Salmonella Paratyphi* C, and *Salmonella dublin*.

#### *2.1.3 Flagellar (H) antigens*

Flagellar antigens are heat-labile proteins. Mixing *Salmonella* cells with flagella-specific antisera gives agglutination. Also, anti-flagellar antibodies can immobilize bacteria H antigens. Antigenic changes of the flagella known as the phase variation of H1 and H2 occurs in *Salmonella Typhimurium*.

#### *2.1.4 Exotoxins*

*Salmonella* strains may produce a thermos-labile enterotoxin that which has a limited relatedness to cholera toxin and *E. coli* (enterotoxin LT) in both structurally and antigenically characters. Additionally, a cytotoxin that inhibits protein synthesis. Both of these toxins play a role in the diarrheal symptoms of Salmonellosis.

### 2.2 *Salmonella* between pathogenicity and immunity

Innate immunity barriers play a good role in defense against *Salmonella* adhesion and colonization. But, upon infection specific immunity come to act; both humoral and cellular specific immune response will be activated to control this infection.

Primary infections with *S. Typhi* or *Salmonella ParaTyphi* usually induce a degree of immunity. Reinfection may occur but is often milder than the first infection. Circulating antibodies to O and Vi are related to resistance to infection and disease [11, 12].

*Salmonella* infections in humans vary with the bacterial species, the infectious dose upon ingested contaminated food, and the host health. The oral dose of at least $10^5$ *Salmonella Typhi* cells are the most effective dose to cause typhoid in 50% of human volunteers as agreed by many references, whereas at least $10^9$ *S. Typhimurium* cells (oral dose) are needed to cause symptoms of a toxic infection. Infants, immunosuppressed patients, and those affected with blood disease are more susceptible to *Salmonella* infection than healthy adults [12].

### 2.3 Salmonellosis (Typhoid fever)

In the **pathogenesis** of typhoid the bacteria enter the human digestive tract, penetrate the intestinal mucosa (causing no lesion), and stope in the mesenteric lymph nodes. Enteric Fever, Salmonellosis or Enterocolitis occurs after attachment to enterocytes of the ileum and colon. About 12-24 hours following ingestion of contaminated food (containing a sufficient number of *Salmonella*), the ingested *Salmonella* reach the small intestine, from which they enter the lymphatics and then the bloodstream. They are carried by the blood to many organs, including the intestine. Then *Salmonella* cells will attach to Microfold cells (M cells) and dendric cells of the jejunum. These specialized epithelial cells are found in the Peyer's patches and initiate mucosal immunity by endocytosis process for these bacteria antigens to the Macrophages and B-Lymphocytes to form APC specific for these antigens. Invasion can occur in this stage via the means of endocytosis, transfer, and exocytosis. Phagocytosis in the subserosa by macrophages and translocation into the mesenteric lymph nodes. Lymphogenous and hematogenous dissemination combined by immune cells proliferation and specific immune response is integrated. Of course the complement system upon these events is already activated, since LPS layer can activate the alternative pathway as soon as this endotoxin liberated from bacterial cells due to destruction leading to more inflammatory reactions at the site of invasion as described in the complement system roles. Moreover, the mannose residues that are found on the surface of *Salmonella* undergo lectin pathway activation of the complement system [5–12].

The organisms usually multiply in intestinal lymphoid tissue and are excreted in stools. However, in the case of *S. Typhi*, the bacteria survive ingestion by the phagocytes, and multiply within these cells. This period of time, during which the bacteria are multiplying within the phagocytes, is the 10–14 day is known as the incubation period. When huge numbers of bacteria fill an individual phagocyte, the bacteria are discharged out of the cell and into the bloodstream, where their presence begins to cause symptoms. Secondary foci in the spleen, liver, bone marrow, bile ducts, skin (roseola), and Peyer's patches then develop [5].

The presence of increasingly large numbers of bacteria in the bloodstream (called bacteremia) is responsible for an increasingly high fever, rising in stages throughout the first week to 39/40/41°C and may last throughout the four to eight weeks of the disease, in untreated individuals. Other symptoms include constipation (initially), extreme fatigue, headache and joint pain. Further symptoms:

leukopenia, bradycardia, splenic swelling, abdominal roseola, beginning in the third-week diarrhea, sometimes with intestinal bleeding due to ulceration of the Peyer's patches and inflammation of the gallbladder, severe irritation and inflammation of the lining of the abdominal cavity, called peritonitis, which is frequently a fatal outcome of typhoid fever [5, 12].

From the mesenteric lymph nodes, viable bacteria and LPS (endotoxin) will be released into the bloodstream resulting in septicemia. Moreover the effect of LPS as pyrogenic toxin, it causes activation of the complement alternative pathway which ends with membrane attack complex MAC, and that will increase LPS levels in bloodstream due to breakage of more bacterial cells leading to more harmful pyrogenic effects. The fever rises to a high plateau, and the spleen and liver become enlarged. Rose spots or rash usually on the skin of the abdomen or chest may be seen in some cases. Another scientific fact, LPS can induce both T-Dependent and T-Independent specific immune response. Specific antibodies against *Salmonella* antigens will be formed after primary infection occurrence, but, T-independent specific antibodies are with no memory B-cells formation. And that is the cause of short time specific immunity resultant after *Samonella* infections according to many scientists' opinions [7].

The complications of typhoid fever include liver and spleen enlargement (sometimes so extreme that the spleen ruptures), anemia (low red blood cell count due to blood loss from the intestinal bleeding), joint infections (especially frequent in patients with sickle cell anemia and immune system disorders), pneumonia (due to a superimposed infection, usually by *Streptococcus pneumoniae*), heart infections, meningitis, and infections of the brain (causing confusion and even coma). Untreated typhoid fever may take several months for full resolve. Spontaneous cure usually occurs [12].

### 2.4 Immune response features of *Samonella*

Due to that *Salmonella* behave as intracellular parasite inside host, these bacteria can survive inside phagocytic cells and escape the immune system meeting. Escape of destruction inside phagocytic cells like macrophage referred to the resistance of *Salmonella* to the oxidative burst used by these immune cells to kill and digest invading bacterial cells. Phagocytosis is the key process for induction of specific immune response but in the condition of contact with invading microbes and process the antigens to become APCs. Hence encountering of these bacteria sometimes can be late due to lack facing with immune cells and bacteria hide inside phagocytic cells. Also in certain circumstances as in immunocompromised patients like diabetic and old people, they usually suffer from late or weak immunological response against *Salmonella* and almost become carriers when the diagnosis and treatment are late [7, 8].

*Salmonella* induce both Th1 immunity (T-helper 1 or immunity or cellular mediated immunity) and Th2 immunity (T-helper 2 or humoral mediated immunity). When APC formed, then the immune response will be turned from innate or non-specific immunity to the specific humoral and cellular immune response, APC will present the processed Antigens of *Salmonella* to the cells of the specific immunity. Concerning Humoral mediated Immunity, specific IgG, IgM and IgA antibodies are formed against *Samonella* antigens, LPS- O antigen Vi antigen and H-antigen. Agglutinating antibodies can give positive reaction after one week post symptoms rise according to Gruber-Widal against H and O *Salmonella* antigens then the antibodies titer continue to elevate with infection time going. The white blood cell count can be found as normal or low at these stages [7, 8].

Antigenic variation can occur due to that *Salmonella* is able to generate genetic exchange and mutation abilities leading to the flagellar phase during infection course. This phenomena will cause the sero-variation and disease phases properties that is usually a characteristic of the infectious disease resultant from *Salmonella*.

Cellular mediated immunity is induced after APC formation, since *Salmonella* act as intracellular parasite and multiply inside macrophages. Then specific activated Cytotoxic T-cells will be produced and specific T-memory cells are released. Some scientists attribute joints inflammation that combined with Typhoid infections to the cellular immune response and due to accumulation of antigen-antibody complexes in patients' joints mainly during high bacterial load infections [7–12].

Cytokines of both Th1 and Th2 levels increase during Salmonellosis, Interlukins (IL-1, IL-2, IL-4, IL-6, IL-8, IL-9, IL-10, IL-13, IL-15, IL-17). Also Interferon-gamma (IFN-$\gamma$) play a great role during cellular immune response and its levels elevates in patients' blood even after cure. Another important cytokine is Tumor necrosis factor-alpha (TNF-$\alpha$), its levels raise upon infection start and stay elevated along the disease time [5, 12].

### 2.5 Carriers

After manifest or subclinical infection, some individuals continue to harbor *Salmonella* in their tissues for different times (for example convalescent carriers or healthy permanent carriers). Three percent of survivors of typhoid become permanent carriers, harboring the organisms in the gallbladder; biliary tract; or, rarely, the intestine or urinary tract. Carriers of *S. Typhi* must be treated even when asymptomatic, as they are responsible for the majority of new cases of typhoid fever. Eliminating the carrier state is actually a difficult and require two different medications for four to six weeks at least.

In the case of carriers with gall stones, surgery may need to be performed to remove the gall bladder, because the *S. Typhi* bacteria are often housed in the gall bladder, where they may survive despite antibiotic treatment [7, 8, 12].

### 2.6 Re-infections and healthy carriers

Despite of that some patients with *Salmonella* will get spontaneous cure, *Salmonella* excretion by human patients may continue long after clinical cure. About 5% of patients clinically cured from typhoid remain carriers for months or even years. Antibiotics are sometimes ineffective on *Salmonella* but can reduce mortality which may reach was 10% [9].

However, relapses may occur in 2–3 weeks after recovery despite specific antibodies titer rise. Secretory IgA antibodies may prevent attachment of *Salmonella* to intestinal epithelium during next time exposure and avoid secondary infection establishment.

Some genetic factors can make person susceptible host for re-infection easier like persons with S/S hemoglobin (sickle cell disease) are susceptible to *Salmonella* infections. Persons with A/S hemoglobin (sickle cell trait) may be more susceptible than normal individuals (those with A/A hemoglobin) [9, 10].

The incidence of human disease decreases when the level of development of a country increases (like controlled water sewage systems, improve hygiene, pasteurization of milk and dairy products). Bad ways in having food like eating raw or undercooked egg can cause illness due to these bacteria called *Salmonella* Enteritidis Infection or Egg-associated salmonellosis which is an important public health problem.

Plasmid-borne antibiotic resistance is very frequent among and can be considered as a virulence factor upon ongoing infections. *Salmonella* strains can get resistance against ampicillin, streptomycin, kanamycin, chloramphenicol, tetracycline, and sulfonamides [11–13].

### 2.7 Vaccination against Typhoid fever

Vaccination is very good health measure in eradication of *Salmonella* infections. Tourists and visitors for countries endemic with *Salmonella* must be vaccinated with *Salmonella* vaccines as a prophylactic health measure.

Early research produced two vaccines made from the entire (whole-cell) bacterium. The first one became available in the 1890s, the second in 1952. Both protected about 65% of recipients. However, the frequency and severity of the adverse effects they caused dissuaded many countries from using them. These shortcomings, combined with drug treatment failures, as a result of increasingly widespread resistance to antibiotic therapy [12–14].

Before the end of the 20th century, two new-generation typhoid vaccines had entered the scene. One, named (Ty21) and first licensed in 1983, is given in three to four oral doses and consists of a live but genetically modified *S. Typhi* strain. The second, named "Vi" and licensed in 1994, is given by injection and consists of a sugar molecule (polysaccharide) located on the surface of the bacterium. In clinical trials and early field use, the duration of efficacy of both vaccines varied to some degree. Moreover, no evidence of efficacy has been reported in children under two years of age [15]. Both vaccines are licensed, internationally available, and safe, and both are effective enough not only to reduce the incidence of typhoid fever in endemic areas but also to control outbreaks [9–11].

Meanwhile, third-generation typhoid vaccines are under trial. One is a Vi conjugate vaccine that protects about 85% of recipients, according to late-stage clinical trials, and appears to be effective in children under two years of age. A second candidate vaccine, is, like Ty21a, a live attenuated vaccine but, unlike Ty21a, can be given in a single oral dose [15].

Three types of typhoid vaccines are currently available for use nowadays:

1. Oral live-attenuated vaccine.

2. Heat-phenol-inactivated vaccine; killed bacterial vaccine.

3. The Vi capsular polysaccharide vaccine for intramuscular use.

A fourth vaccine, an acetone-inactivated parenteral vaccine, is currently available only to the armed forces. While Typhoid fever vaccinations for tourists and travelers to the endemic areas is best be done with the oral attenuated vaccine Virotif Ty 21a.

Despite of that; No typhoid vaccine is 100% effective and provide only short-term protection (sometimes for a few months), it is not a substitute for being careful and elevate hygiene [15].

## 3. Conclusions

*Salmonella*, whatever species, is dangerous microbe that is able to invade human body due to many weapons owned. Good and healthy immune system can stand against these bacteria. But with bad nutrition, low hygiene and immunosuppression;

infection with *Salmonella* will occur upon exposure and may develop to a systemic disease. *Salmonella* induce human immunity with different types of resistance processes, either specific (adaptive immune response) or non-specific (Innate Immune Response) that act against these bacteria and lead to cure during treatment course. Despite of many effective vaccines that have been produced; elevation personal hygiene is still the best way to eradicate this infectious disease. Healthy carriers of *Salmonella* are a public health problem.

## Author details

Weam Saad Al-Hamadany
Medical College, IbnSina University for Medical and Pharmaceutical Sciences, Baghdad, Iraq

*Address all correspondence to: dr.weam2016@gmail.com

## References

[1] Kuby, J. Immunology: Text Book, 3rd ed. W. H. Freeman comp. New York. 2003.

[2] Owen, J.; Puntm J. and Jones, P. Kuby Immunology. 7th ed. Freeman Comp. USA.2013.

[3] Virella, G. Introduction to Medical Immunology. 4th ed. Medical University of South Carolina. USA. 1998.

[4] Hogg, H. Essential Microbiology. Wiley and Sons Ltd.UK.2005.

[5] Brooks, G.F.; Carroll, K.C.; Butel, J.S.; Morse, S.A. and Mietzner, T.A. Jawetz, Melnick, & Adelberg's Medical Microbiology.26th ed. McGraw-Hill. 2013.

[6] Keogan, M.; Wallace, E. and O'Leary, P. Concise Clinical Immunology for Health Care Professionals. Routledge. USA.2006.

[7] Murry, P.; Rosenthal, K. and Pfaller, M. Medical Microbiology. 7th ed. Elsevier. Philadelphia.2013.

[8] Lerner, K. and Lerner, B. World of Microbiology and Immunology. Thomson. Gale group.USA.2003.

[9] Narins, B. World of Microbiology and Immunology. Vol. 2 M-Z. Thomson. Gale Comp. USA. 2003.

[10] Fritz H. Kayser; Kurt A. Bienz and Rolf M. Zinkernagel. Medical Microbiology. 1st ed. Thieme. Stuttgart. New York. 2005.

[11] Brigham Narins. World of Microbiology and Immunology. Vol. 2. Thomson Gale. New York. 2007.

[12] Samuel Baron; *et. al*. Medical Microbiology, 4th ed. University of Mexico. 2006.

[13] Guilfoile, P., Alcamo, I. and Heymann, D. Deadly Diseases and Epidemics/Antibiotic-Resistant Bacteria. Infobase Publishing. USA. 2007.

[14] Al-Hamadany, W. S. Infectious Diseases Outbreaks Management, Control and Prevention, 2nd. EuroScieCon Conference on Virology; Online Event- July 27, Journal of HIV and Retro virus. Vol.6.No.3. 2020. ISSN 2471-9676.

[15] WHO, UNICEF, World Bank. State of the world's vaccines and immunization, 3rd ed. Geneva, World Health Organization, 2009.

Chapter 7

# Pathogenicity Island in *Salmonella*

*Sarika Kombade and Navneet Kaur*

## Abstract

Considering a complex set of interplay with its host, *Salmonella* needs numerous genes for its full virulence. These genes responsible for invasion, survival, and extra intestinal spread are located on pathogenicity islands known as *Salmonella* pathogenicity islands (SPIs) that are thought to be acquired by horizontal gene transfer. A total of 17 SPIs (1–17) are recognized so far. The type III secretion system (T3SS) encoded by SPI-1 is considered as the most important virulence factor for *Salmonella* that delivers effector proteins necessary for invasion and production of enteritis. Among various SPIs, the role in virulence is well proven for SPI1 and SPI2 and further insight into the complex regulatory network of SPIs can contribute to drug investigation and prevention of infection.

**Keywords:** *Salmonella*, virulence genes, *Salmonella* pathogenicity islands, Type III secretion system

## 1. Introduction

Salmonellae are gram-negative bacteria and members of the family Enterobacteriales. They are chiefly intestinal parasites of human and a wide variety of animals including wild birds, domestic pets, rodents, chickens etc. They are also found in sewage, rivers and waters and soil. The genus *Salmonella* is divided into two species: *Salmonella enterica* that encompasses six subspecies (I, II, IIIa, IIIb, IV, and VI), and *Salmonella bongori*, which was earlier subspecies V [1]. Members of the seven *Salmonella* species can be serotyped into more than 2500 serotypes (serovars) based on somatic O and H antigens [1].

## 2. Pathophysiology of *Salmonella*

*Salmonella* is noted to cause diverse disease spectrum in humans and animals, varying from localized inflammation and gastroenteritis to typhoid fever which can lead to life-threatening systemic infection. The prime issue is that of asymptomatic healthy carriers who possibly shed bacteria in feces causing risk to community. There is diversity seen among certain *Salmonella* serovars based on host adaptation, such as *Salmonella* Typhi, *Salmonella* Paratyphi, and *Salmonella* Sendai are known to be very well adapted to only human host while *Salmonella* Typhimurium and *Salmonella* Enteritidis has a broad host range infecting animals and humans. Others produce diseases in farm animals like S. Choleraesuis in swine, S. Gallinarum in fowl. *Salmonella* Dublin (cattle) and Arizonae (reptiles) are mainly adapted to an animal species and seldom infect humans [1].



IntechOpen

## 3. Pathogenesis of *Salmonella*

For the *Salmonella* infection to commence the bacteria is ingested through contaminated food and water. The infectious dose varies considerably ranging between $10^3$–$10^6$ colony-forming units [2].

The first hurdle to *Salmonella* colonization is acidity of stomach and certain situations which either decreases stomach acidity (antacids, proton pump inhibitors, achlorhydric disease) or integrity of intestine (previous surgery of gastrointestinal tract, altered intestinal flora due to antibiotic use, inflammatory bowel disease) increases the chances of *Salmonella* infection [3].

Salmonellae exhibit an adaptive acid tolerance response on exposure to acid in vitro that possibly eases its survival in the stomach and movement to the small intestine.

When it reaches the small intestine, it attaches to the mucosal epithelial cells by fimbriae. Now, the penetration of the mucosal epithelium is achieved by *bacteria-mediated endocytosis* (BME) [4].

When the bacteria adheres to the apical epithelial surface, an extensive cytoskeletal rearrangements is followed shortly which disturbs the normal epithelial brush border prompting the configuration of membrane ruffles. These membrane ruffles reach out and encloses adherent bacteria in large vesicles. M cells (specialized cells overlying the Peyer's patches) are probably considered the primary portal of entry in case of Enteric fever and the generalized intrusion of enterocytes is thought to play a prominent role in enteritis caused by Non-Typhoidal *Salmonella* (NTS) serotypes [5].

There are several large insertions in the genome of *Salmonella* that are considered to arise from bacteriophages or plasmids, called as the *Salmonella* pathogenicity islands (SPIs). These SPIs encode genes that are crucial for survival in the host. The virulence genes are responsible for invasion, survival, and extra intestinal spread. For instance, Salmonellae encode a type III secretion system (T3SS) within *Salmonella* pathogenicity island 1 (the SPI-1 T3SS), which is necessary for bacteria-mediated endocytosis and epithelial invasion in the intestine.

## 4. Definitions

### 4.1 Genomic Island

Genomic islands (GIs) such as integrative and conjugative elements (ICEs) and integrative mobilizable elements (IMEs) are clusters of genes inside a bacterial genome which seems to be acquired by horizontal gene transfer [6]. Initially noticed in pathogenic bacteria, designated as pathogenicity islands because they carried virulence genes or other pathogenicity factors, now are also identified in various non-pathogenic bacteria. Therefore, GIs are frequently named based on the adaptive properties they bestow such as metabolic islands, antibiotic resistance islands, symbiosis islands, pathogenicity islands etc. [7]. Furthermore, GIs bless their hosts with new traits, like resistance to antimicrobials and enhanced virulence.

### 4.2 Pathogenicity island (PAI)

Pathogenicity islands are a definite class of GIs acquired by microorganisms by horizontal gene transfer. They constitute large genomic regions (10–200 kilobases in size) that are integrated in the genome of pathogenic bacteria and are not seen in non-pathogenic bacteria of the same or closely related species [6]. The concept of

| S.No. | Characteristic features of Pathogenicity Islands |
|---|---|
| 1. | Carrying of one or more genes of virulence |
| 2. | Present in pathogenic bacteria and not seen in non-pathogenic ones of the same or closely related species |
| 3. | Constitute large genomic regions (10–200 kilobases in size) |
| 4. | Possess DNA content that varies markedly from the rest of the host genome, especially percentage of G + C content and codon usage |
| 5. | Commonly situated adjacent to tRNA genes |
| 6. | Frequent association with mobile genetic elements, often flanked by direct repeats<br>Presence of integrase gene at one end of the island |
| 7. | Genetic instability that can lead to loss of the Pathogenicity islands |

**Table 1.**
*Features of Pathogenicity Islands.*

pathogenicity islands was established in the late 1980s by Jorg Hacker and his colleagues while probing the genetic grounds of virulence of uropathogenic *Escherichia coli* strains 536 and J96 [8]. The important features of PIs are summarized in **Table 1** [8].

## 5. *Salmonella* Pathogenicity Islands

Pathogenicity islands in *Salmonella* spp. are generally known as '*Salmonella* Pathogenicity Island' or SPI. They are found in large number and are the central elements for virulence in *Salmonella*. A total of 17 SPIs (1–17) are recognized so far [9].

### 5.1 SPI1

*Salmonella* species has the capability to penetrate non-phagocytic host cells. For the penetration or invasion to take place, there is requirement of several genes which were first identified for *S. typhimurium* [10]. At the later stages of research, it was established that all the genes responsible for invasion were bunched within a region at centisome 63 of the *Salmonella* chromosome [11]. After that, the second cluster of genes were identified which were required because of the ability of *Salmonella* to proliferate in different organs of the infected host, lead the researchers to designate Invasion Locus *Salmonella* Pathogenicity Island 1 and accordingly newly identified locus as *Salmonella* Pathogenicity Island 2 [12].

The size of SPI-1 is nearly 40 kilo bases in size and encodes a type III secretion system (T3SS) that is needed for BME and intestinal epithelial invasion (**Figure 1**). T3SS are considered as complex macromolecule machines that emerge to bring down the function of host cell by translocation of virulence proteins straight from the bacterial cytoplasm into the host cell. T3SS are also known as injectisomes' or 'molecular needles because of their capability to translocate proteins in a cell contact-dependent manner [13]. T3SS is also found in several species of several Gram-negative bacteria (e.g. *Salmonella*, *Yersinia*, *Shigella*, *E. coli*, *Pseudomonas*) and encompasses at least 20 different subunits that enables these bacteria to translocate specific substrates (or 'effectors') directly into the host cell cytoplasm which exerts a broad range of virulence functions [14]. The mutants of *Salmonella* not having a functional SPI-1 T3SS do not invade epithelial cells in tissue culture [15].

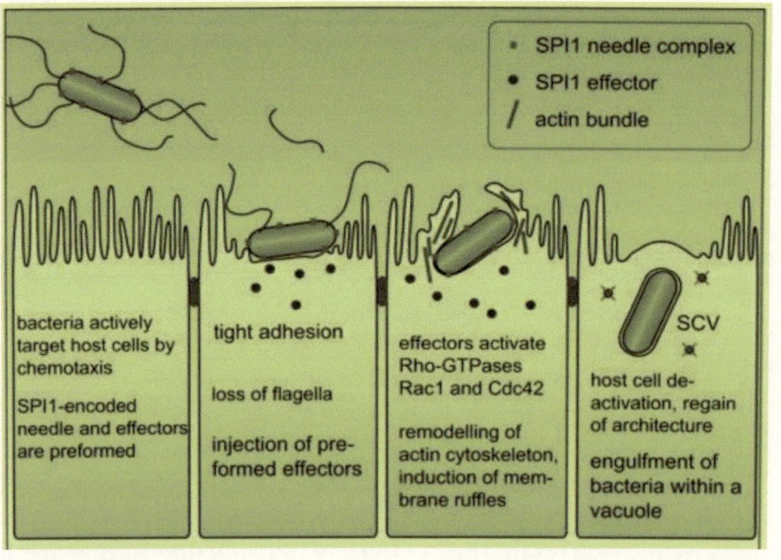


**Figure 1.**
*Invasion of* Salmonella *into non-phagocytic cells by SPI1. At the time of contact with host cell, there is injection of different effectors into cytoplasm of host cell by SPI1 encoded T3SS. This leads to stimulation of small rho GTPases that causes massive cytoskeleton rearrangements. This results in intake of bacteria by macropinocytosis. Now, bacteria live in vacuole and the host cells regain a normal architecture. (from Gerlach RG, Hensel M.* Salmonella *pathogenicity islands in host specificity, host pathogen-interactions and antibiotics resistance of* Salmonella enterica*. Berliner und Munchener tierarztliche Wochenschrift. 2007 Jul 1;120(7/8):317).*

There is requirement of at least five translocated proteins adequate invasion of cultured epithelial cells, whereas invasion is more complex and diverse in animal tissues [16]. Two subsets of effector proteins are generated by SPI-1 in which one subset mediates invasion by *Salmonella* of non-phagocytic cells through alteration of active cytoskeleton system of host cell and the other second subset is related with entero-pathogenesis and inflammation of cells of intestinal epithelium. The important effector proteins are summarized in **Table 2**.

| Effector protein | Function | Mechanism |
|---|---|---|
| SipC & SipA | Promotes membrane ruffling and *Salmonella* invasion | By direct interactions with the actin cytoskeleton |
| SopE & SopE2 | Promotes membrane ruffling ans *Salmonella* invasion | Directly activate Rac1 and Cdc42 in vitro by acting as GDP/GTP exchange factors (GEFs) and induce membrane ruffling and macropinocytosis after microinjection into epithelial cells |
| SopB (Additional SPI-1 translocated protein) | Promote membrane ruffling and *Salmonella* invasion | Targets inositol phosphate signaling within the host cell by acting as an inositol polyphosphatase. |
| SopA and SopD | Intestinal secretory and inflammatory responses | Recruitment of immune cells and secretion of fluid in intestinal lumen |
| SptP | Reverse the cytoskeletal rearrangements induced by SopE/E2 and SopB | GTPase activating protein (GAP) acts on Rac1 and Cdc42, |

**Table 2.**
*Important translocated proteins of SP-I of* Salmonella.

### 5.2 The SPI1 regulon

The expression of SPI1 is driven by central transcription factor, SPI1 encoded HilA which is a member of OmpR/ToxR family of transcriptional regulators. The expression of SPI1 regulon is directed by specific blend of environmental signals such as osmolarity, antimicrobial peptides, oxygen, pH and other unidentified signals. These signals are perceived by a set of 2 component regulatory systems: BarA/SirA [17], OmpR/EnvZ [18], PhoBR [19] and PhoPQ [20]. The phosphorylated connected response regulators can promote the expression of either HilD or HilE that in turn either stimulates or represses SPI1-expression.

1. PhoB, PhoP and FimZY can activate hilE expression that can negatively impact hilA expression [21]

2. HilD along with HilC and RtsA make up a feed-forward loop, where each factor can promote the expression of itself, rtsA, hilC and hilA, thus integrating and greatly enhancing the signal [22]

3. Nucleoid proteins HU and Fis are essential for hilA expression [23]

4. HilA binds to cis-elements present in respective promoters that activates prg/org and inv./spa operons within SPI1 which in turn leads to production of InvF, a member of the AraC family of transcriptional regulators. Now, InvF along with the chaperone SicA induces expression of a set of genes encoded within SPI1 and on various loci elsewhere in the chromosome.

5. RtsA/HilD/HilC can activate the expression of dsbA required for the function of the SPI1-encoded T3SS [23] (**Figure 2**).

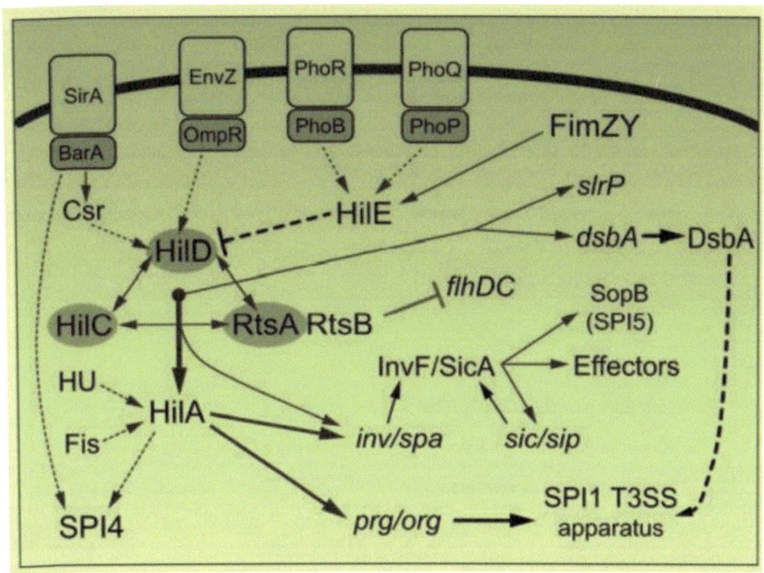


**Figure 2.**
*The SPI1 regulon. The main regulatory factor of SPI1 expression is HilA whose activity is mainly controlled on the transcriptional level and relies on a complex network of transcription factors and two-component regulatory systems. Arrows indicate activation of gene expression. Positive feedback loop are highlighted by bold arrows. Repression is noted as lines with blunt ends. Solid lines represent direct transcriptional regulation. (from Gerlach RG, Hensel M.* Salmonella *pathogenicity islands in host specificity, host pathogen-interactions and antibiotics resistance of* Salmonella enterica. *Berliner und Munchener tierarztliche Wochenschrift. 2007 Jul 1;120(7/8):317).*

## 5.3 SPI2

*Salmonella* has a second T3SS that is essential for survival within the macrophage and for establishment of systemic infection. Proteins delivered by both type III secretion systems are vital for intracellular survival. The second T3SS is encoded on *Salmonella* pathogenicity island 2 (SPI-2). The activity of SPI2 is needed to establish and maintain the *Salmonella-* containing vacuole (SCV) as an intracellular niche in which *Salmonella* can remain live and replicate.

SPI2 is of nearly 40 Kb in size and comprises of 2 distinct regions [24]:

1. The larger region relatively 25 Kb in size found exclusively in *S. enterica,* implicated in systemic pathogenesis. It encodes second T3SS.

2. The smaller region of 15 Kb in size was identified in *S. bongori. It* encodes the tetrathionate reductase (Ttr) that is involved in anaerobic respiration.

This second T3SS expressed by intracellular bacteria translocates proteins across the SCV membrane into the macrophage cytosol. With the help of these SPI2 translocated proteins, *Salmonella* escapes intracellular killing by altering the phagosome membrane to tubulate [25]. Phagosome tubulation is dynamic and rapid process and occurs to be dependent on the recruitment of microtubule motors, membrane lipid alteration and the activation of small GTPases, and membrane lipid alteration [25]. Phagosome tubulation is also correlated with the virulence by unknown mechanisms.

A total of seventeen effectors are recognized to be translocated over the SCV membrane into the host-cell cytoplasm, most of them being encoded outside the SPI2-locus [26]. Only 3 effectors are known to be encoded within SPI2 which includes SpiC, SseF and SseG. The SPI-2 translocated proteins, including SifA, SifB, SseJ, SopD2, PipB, and PipB2, localize to the surface of the SCV and either contributes to tubulation or other alterations of the phagosome [27]. The summary of important effectors has been given in **Table 3**.

## 5.4 The SPI2 regulon

The expression of SPI2 genes is controlled governed by global regulatory system: SsrAB system. It is a typical two-component system that is necessary for SPI2 regulon expression in intracellular bacteria. The main global regulatory systems that affect the expression levels of SPI2 genes are the EnvZ/OmpR and PhoPQ two-component systems, SlyA and Fis [23].

| Effectors | Functons |
|---|---|
| SpiC | Block fusion of the SCV with lysosomes |
| Sif A | *Salmonella* containing vacuole membrane integrity |
| SseJ | Cytoskeleton rearrangements |
| SsPH2 | Cytoskeleton rearrangements |
| Ttr genes | Tetrathionate respiration and outgrowth in the intestine |
| SseFG | Maintaining a juxtanuclear position of the SCV in HeLa cells |
| SpvB | *Salmonella* virulence protein that is secreted into the macrophage cytoplasm, |

**Table 3.**
*Functions of major effectors of SPI-2.*

### 5.5 SPI3

The SPI3 locus is of size 17 kilobases and is inserted at the selC tRNA gene locus. The primary known virulence determinant is Mgt CB (Magnesium transport system) operon: MisL and Mar T. This determinant is necessary for survival of *Salmonella* in the intra-phagosomal habitat in nutritionally deprived conditions. Mis L, a anti-transport protein of SPI3, is identical to the autotransported AIDA-1 adhesin of enteropathogenic *E. coli* (EPEC) while Mar T shows resemblance with Tax R (Toxin gene regulator) of *Vibrio cholerae* and it is implicated in the activation of Mis L [28].

MisL is proved to work as an adhesion [29] and it is vital for the long term persistence of *Salmonella* in the intestine as observed in animal studies. Another autotransporter, ShdA is seen to have a function in adhesion and virulence in case of *S.* Typhimurium.

A high degree of sequential variation exists in SPI3 among different serovars; however it is conserved in cases of *S.* Typhi and *S.* Typhimurium.

### 5.6 SPI4

The size of SPI4 is identified as 27 Kb. Sequencing of the *Salmonella* Typhimurium genome anticipated that the pathogenicity island constitute of not more than six genes. Hence the genes of the locus SPI4 are named as siiA-F. SiiC, SiiD and SiiF encodes components of type I secretion system which secretes SiiE. This, SiiE is huge protein (approximately 600 kDa) that is known to colonize the bovine intestine [30]. The molecular functions of SPI4 encoded proteins are not known. The role of SPI4 in *Salmonella* virulence was investigated in one of the studies using refined cell culture and infection models, there it was observed that SPI4 contributes to gastrointestinal inflammation in murine colitis model and is also required for adhesion to epithelial cells [31]. De Keersmaecker et al. suggested a role for SPI4 in intra-macrophage survival as shown for SPI2 [32].

SPI4 seems to be highly conserved among different *Salmonella* serovars [33].

### 5.7 SPI5

The size of SPI5 locus is nearly 7.6 Kb. It encodes the effector proteins for both the T3SS that is encoded by SPI-1 and SPI-2. Pip A and Pip B are also known to be encoded by SPI5 locus. Pip A is implicated in the development of systemic infection and Pip B is involved in the accumulation of lipid rafts and is a translocated effector of SPI-2 encoded T3SS which is under the control of Ssr AB two-component systems. However, PipB is neither needed for bacterium's intracellular survival nor for systemic virulence as studied in mice [34, 35].

In enteropathogenicity in a cattle infection model, significant attenuation of SPI5-deficient *Salmonella* was observed. However SPI5 mutants showed only a minor virulence defect in mouse model [36].

### 5.8 SPI6

The SPI6 locus is also known as '*Salmonella* centisome 7 genomic island' or SCI [37]. It is of size 59 Kb and it has been recognized in *S.* Typhi and *S.* Typhimurium. It is investigated to contain [35]:

1. saf gene which codes for fimbriae

2. pag N gene which encodes for invasion protein

A microarray analysis indicated the conservation of SPI6 among serovars of S. enterica subspecies I serovars was indicated by microarray analysis.

Deletion of SPI6 had no influence on the systemic pathogenesis but decreased invasiveness of the bactetia in tissue cultured cells. SPI-6 was detected to be conserved among serovars of S. *enterica as* indicated by microarray analysis. Some of the portion of SPI-6 that was also identified in subspecies III b, IV, and VII. Further, SPI-6 has shown sequential homology with the genome of *P. aeruginosa* and *Y. pestis* [38].

### 5.9 SPI7 and SPI8

The size of SPI7 and SPI8 is approximately 133 Kb and 6.8 Kb respectively. SPI7, also termed as major Pathogenicity Island is specific to *S.* Typhi, *S.* Dublin and *S.* Paratyphi. It encodes for Vi antigen and constitute pil gene cluster that encodes for putative virulence factors. Its genetic organization is very complex and composed of several horizontally acquired elements. It also constitutes few genes of conjugative plasmid-like *tra* and *sam.*. The locus is said to not stable and loss of the capsule can be seen in *S.* Typhi isolates. Additionally SPI7 also encodes a type IV fimbrial adhesin. There exists a sequential homology with few other bacteria like *Xanthomonas axonopodis* and *Pseudomonas aeruginosa* in the case of SPI7 [39].

SPI8 has been identified in *Salmonella* Typhi and the genes located here encode for putative virulence factors, whose exact function has not been reported so far.

### 5.10 SP19

The size of SPI-9 locus is nearly 16,281 basepairs. SPI9 from S. Typhi harbors three ORFs (STY2876,STY2877,STY2878) presenting 98% identity with a type 1 secretory apparatus (T1SS) and a single ORF (STY2875) that is similar to a large RTX-like protein exhibiting repeated Ig domains. It encodes for virulence factors of type I secretion system. Furthermore, as it is functional in *S.* Typhi and encodes for adhesion which is induced under conditions of high osmolarity in culture. However it does not participate in biofilm formation [40].

### 5.11 SP110

SPI10 has a size of 32.8 Kb and is defined as an insertion at the tRNA leuX gene. It appears to be hyper variable and is a point of insertion for several different DNA fragments. Sef and pef gene clusters which encodes for fimbrial adhesions have been detected in *S.* Enteritidis and cryptic bacteriophage has been seen within this locus in case of *S.* Typhi and *S.* Paratyphi A. On the other hand, *S.* Typhimurium has entirely different gene content. Because of these findings, the leuX locus represents a hot spot for the insertion of various mobile genetic elements [41].

### 5.12 SPI11 and SPI12

The SPI11 and SPI12 were identified in *Salmonella choleraesuis.* Both these islands shows properties of PAI such as association with bacteriophage genomes and tRNA genes. The low G + C content of 41.32% was seen for SPI11. The proteins encoded by these SPIs contributes to virulence of *Salmonella* but exact role is still not clear and awaits further characterization [42].

### 5.13 SPI13 and SPI14

SPI13 and SPI14 were first identified in avian adapted S. Gallinarum which is causative agent of typhoid in fowls. SPI13 is close to the tRNA pheV gene and is composed of 18 ORFs while SPI14 is not associated with a tRNA gene and constitutes 6 ORFs. Both these islands are not present in *S*. Typhi and *S*. Paratyphi A but are seen in *S*. Typhimurium and *S*. Enteritidis. This may indicate a possible role of the loci in host specificity. The role of proteins encoded by these SPIs is not clear yet and requires further molecular characterization [43].

### 5.14 SPI15, SPI16, and SPI17

SPI15, SPI16 and SPI17 were identified in S. Typhi using bioinformatics approach. All these exhibit association with tRNA genes. SPI16 and SPI17 encodes for genes that are responsible for LPS modification. There is presence of SPI15 in only S. Typhi isolate CT18 and role of its effecter proteins is not clear till date. SPI16 and SPI17 are seen in S. Typhi and most other S. enterica genome sequences [44].

**Table 4** summarizes the various SPIs.

| SPI & Insertion point | Distribution among *Salmonella* species | Variable or conserved | Function |
|---|---|---|---|
| SPI1 | | | |
| flhA-mutS | *Salmonella* spp. | Conserved | T3SS, iron uptake |
| SPI2 | | | |
| tRA val V | S.enterica | Conserved | T3SS, tetrathionate reductase |
| SPI3 | | | |
| tRNA sel C | *Salmonella* spp. | Variable | Mg2+ uptake, Misc. adhesin |
| SPI4 | | | |
| tRNA like | *Salmonella* spp. | Conserved | T1SS adhesin |
| SPI5 | | | |
| tRNA ser T | *Salmonella* spp. | variable | T3SS effectors SopB, PipB |
| SPI6 | | | |
| tRNA asp V | Subsp. I. parts in IIIB, IV, VII | Conserved In subsp. I | Saf fimbriae |
| SPI7 | | | |
| tRNA pheU | Subsp. I serovars | Instable | Vi antigen, pilus assembly, SopE |
| SPI8 | | | |
| tRNApheV | sv. Typhi | NK | NK |
| SPI9 | | | |
| prophage | Subsp.. I serovars | NK | T1SS, adhesin BapA |
| SPI10 | | | |
| tRNA leuX | subsp. I serovars | variable | Sef fimbriae |
| SPI11 | | | |
| prophage | S.Choleraesuis | NK | NK |
| SPI12 | | | |
| tRNA pro | S.Choleraesuis | NK | NK |
| SPI13 | | | |

| SPI & Insertion point | Distribution among *Salmonella* species | Variable or conserved | Function |
|---|---|---|---|
| tRNA pheV | S.Gallinarum,S. Typhimurium | NK | NK |
| SPI14 | | | |
| NK | S.Gallinarum,S. Typhimurium | NK | NK |
| SPI15 | | | |
| tRNA gly | S. Typhi | NK | NK |
| SPI16 | | | |
| tRNA arg S. Typhi, and others ? serotype conversion | S. Typhi, and others | NK | serotype conversion |
| SPI17 | | | |
| tRNA arg | S. Typhi, and others | NK | serotype conversion |
| SGI1 | | | |
| thdF–yidY | subsp. I serovars | variable | 5 antibiotic resistance genes |
| CS54 | | | |
| xseA-yfgK | subsp. I serovars | NK | adhesion |

*NK – Not Known.*

**Table 4.**
*Summary of various* Salmonella *Pathogenicity Islands.*

### 5.15 SGI1

Strains showing resistance to multiple antibiotics is a usual phenomenon seen in pathogenic bacteria and is also mostly observed in S. enterica. Resistant *Salmonella* isolates harbor resistance plasmids of variable size and composition of resistance genes. A multidrug resistance phenotype conferred by '*Salmonella* genomic island 1' or SGI1 was recognized in epidemic strain S. Typhimurium DT104 by molecular testing though It can also be present in other strains as well. The SGI1 confers resistance to the antibiotics such as ampicillin, chloramphenicol, streptomycin, sulfonamides and tetracycline [23].

## 6. Conclusion

PAI phenomenon frequently identified in pathogenic bacteria and encodes virulence genes which help pathogens to establish infections. The molecular characterization of individual virulence genes and genome sequences demonstrated large numbers of PAI in S. *enterica* serovars. Among various *Salmonella* pathogenicity islands, only SPI1 and SPI2 have well proven role in virulence while knowledge of the molecular function of the rest of the SPIs is lacking. Furthermore, molecular analysis of SPI is vital for improvement of prevention and treatment of *Salmonella* infection in human and animals. Also the varied degrees of disease severity and of bacterial pathogenesis can be explained better by understanding SPI.

*DOI: http://dx.doi.org/10.5772/intechopen.96443*

## Author details

Sarika Kombade* and Navneet Kaur
Department of Microbiology, All India Institute of Medical Sciences (AIIMS), Jodhpur, Rajasthan, India

*Address all correspondence to: spkombade@gmail.com

## References


[1] Popoff MY, Bockemuhl J, Gheesling LL. Supplement 2002 (no. 46) to the Kauffmann-White scheme. Res Microbiol.2004;155:568-570.

[2] Blaser MJ, Neuman LS. A review of human salmonellosis: Infective dose. Rev Infect Dis. 1982;4:1096.

[3] Foster JW, Hall HK. Adaptive acidification tolerance response of *Salmonella typhimurium*. J Bacteriol. 1990;172:771-778.

[4] Francis CL, Starnbach. MN, Falkow S. Morphological and cytoskeletal changes in epithelial cells occur immediately upon interaction with *Salmonella typhimurium* grown under low-oxygen conditions. Mol Microbiol. 1992;6:3077-3087.

[5] Jones BD, Falkow S. Identification and characterization of a *Salmonella typhimurium* oxygen-regulated gene required for bacterial internalization. Infect Immun. 1994;62:3745-3752.

[6] Juhas M, Van Der Meer JR, Gaillard M, Harding RM, Hood DW, Crook DW. Genomic islands: tools of bacterial horizontal gene transfer and evolution. FEMS microbiology reviews. 2009 Mar 1;33(2):376-393.

[7] da Silva Filho AC, Raittz RT, Guizelini D, De Pierri CR, Augusto DW, dos Santos-Weiss IC, Marchaukoski JN. Comparative Analysis of Genomic Island Prediction Tools. Frontiers in genetics. 2018;9:619.

[8] Hacker J, Bender L, Ott M, Wingender J, Lund B, Marre R, Goebel W. Deletions of chromosomal regions coding for fimbriae and hemolysins occur in vitro and in vivo in various extra intestinal Escherichia coli isolates. Microbial pathogenesis. 1990 Mar 1;8(3):213-225.

[9] Riquelme S, Varas M, Valenzuela C, Velozo P, Chahin N, Aguilera P, Sabag A, Labra B, Álvarez SA, Chávez FP, Santiviago CA. Relevant genes linked to virulence are required for *Salmonella* typhimurium to survive intracellularly in the social amoeba dictyostelium discoideum. Frontiers in Microbiology. 2016; 7:1305. DOI: 10.3389/fmicb.2016.01305.

[10] Galan, J.E., R.Curtiss (1989): Cloning and molecular characterization of genes whose products allow Salmonella typhimurium to penetrate tissue culture cells. Proc. Natl. Acad. Sci. USA 86, 6383-6387.

[11] Mills, D. M., V. Bajaj, C. A. Lee (1995): A 40 kb chromosomal fragment encoding Salmonella typhimurium invasion genes is absent from the corresponding region of the Escherichia coli K-12 chromosome. Mol. Microbiol. 15, 749-759.

[12] Shea, J. E., M. Hensel, C. Gleeson, D. W. Holden.Identification of a virulence locus encoding a second type III secretion system in Salmonella typhimurium. Proc. Natl. Acad. Sci. 1996;93:2593-2597.

[13] Cornelis, G. R. The type III secretion injectisome. Nat. Rev. Microbiol.2006; 4: 811-825.

[14] Ghosh, P. Process of protein transport by the type III secretion system. Microbiol. Mol. Biol. Rev.2004; 68:771-795.

[15] Watson PR, Paulin SM, Bland AP, et al. Characterization of intestinal invasion by *Salmonella typhimurium* and *Salmonella dublin* and effect of a mutation in the invH gene. Infect Immun. 1995;63:2743-2754.

[16] Haraga A, West TE, Brittnacher MJ, et al. *Burkholderia thailandensis* as

a model system for the study of the virulence-associated type III secretion system of *Burkholderia pseudomallei*. Infect Immun. 2008;76:5402-5411.

[17] Ahmer, B. M., J. van Reeuwijk, P. R. Watson, T. S. Wallis, F. Heffron. Salmonella SirA is a global regulator of genes mediating enteropathogenesis. Mol. Microbiol. 1999;31:971-982.

[18] Lucas, R. L., C. A. Lee. Roles of hilC and hilD in regulation of hilA expression in Salmonella enterica serovar Typhimurium. J. Bacteriol.2001; 183:2733-2745.

[19] Lucas, R. L., C. P. Lostroh, C. C. DiRusso, M. P. Spector, B. L. Wanner, C. A. Lee. Multiple factors independently regulate hilA and invasion gene expression in Salmonella enterica serovar Typhimurium. J. Bacteriol. 2000;182:1872-1882.

[20] Pegues, D. A., M. J. Hantman, I. Behlau, S. I. Miller. PhoP/PhoQ transcriptional repression of Salmonella typhimurium invasion genes: evidence for a role in protein secretion. Mol. Microbiol. 1995;17:169-181.

[21] Baxter, M. A.,T. F. Fahlen, R. L.Wilson, B. D. Jones. HilE interacts with HilD and negatively regulates hilA transcription and expression of the *Salmonella enterica* serovar Typhimurium invasive phenotype. Infect. Immun. 2003;71,:1295-305.

[22] Ellermeier, C. D., J. R. Ellermeier, J. M. Slauch .HilD, HilC and RtsA constitute a feed forward loop that controls expression of the SPI1 type three secretion system regulator hilA in Salmonella enterica serovar Typhimurium. Mol. Microbiol. 2005;57:691-705.

[23] Gerlach RG, Hensel M. Salmonella pathogenicity islands in host specificity, host pathogen-interactions and antibiotics resistance of Salmonella enterica. Berliner und Munchener tierarztliche Wochenschrift. 2007 Jul 1;120:317.

[24] Khule V, Hensel M. Cellular microbiology of intracellular *Salmonella enterica*: Functions of the type III secretion system of encoded by *Salmonella* pathogenicity island 2. Cellular and Molecular Life Sciences. 2004;61:2812-2826.

[25] Ramsden AE, Mota LJ, Munter S, et al. The SPI-2 type III secretion system restricts motility of *Salmonella*-containing vacuoles. Cell Microbiol. 2007;9:2517-2529.

[26] Kuhle, V., M. Hensel. Cellular microbiology of intracellular Salmonella enterica: functions of the type III secretion system encoded by Salmonella pathogenicity island 2. Cell. Mol. Life Sci. 2004;61:2812-2826.

[27] Uchiya, K., M. A. Barbieri, K. Funato, A. H. Shah, P. D. Stahl, E. A. Groisman .A Salmonella virulence protein that inhibits cellular trafficking. EMBO J. 1999;18:3924-33.

[28] Amavisit P, Lightfoot D, Browning GF, Makham PF. Variations between the pathogenic serovars within *Salmonella* pathogencity islands. Journal of Bacteriology. 2003;185:3624-3635.

[29] Dorsey, C. W., M. C. Laarakker, A. D. Humphries, E. H. Weening, A. J. Bäumler (2005): Salmonella enterica serotype Typhimurium MisL is an intestinal colonization factor that binds fibronectin. Mol. Microbiol. 57, 196-211.

[30] Gerlach RG, Jäckel D, Stecher B, Wagner C, Lupas A, Hardt WD, Hensel M. Salmonella Pathogenicity Island 4 encodes a giant non-fimbrial adhesin and the cognate type 1 secretion system. Cellular microbiology. 2007 Jul;9(7):1834-1850.

[31] Kiss T, Morgan E, Nagy G. Contribution of SPI-4 genes to the virulence of Salmonella enterica. FEMS microbiology letters. 2007 Oct 1;275(1):153-159.

[32] De Keersmaecker, S. C., K. Marchal,T. L.Verhoeven, K. Engelen, J.Vanderleyden, C. S. Detweiler. Microarray analysis and motif detection reveal new targets of the Salmonella enterica serovar Typhimurium hilA regulatory protein, including hilA itself. J. Bacteriol. 2005;187:4381-4391.

[33] Edwards, R. A., G. J. Olsen, S. R. Maloy (2002): Comparative genomics of closely related salmonellae. Trends Microbiol. 10, 94-99.

[34] Knodler LA, Celli J, Hardt WD, Vallance BA, Yip C, Finlay BB. *Salmonella* effector withina single pathogenicity island are differently expressed and translocated by separate type III secretion system. Molecular Microbiology. 2002;**43**:1664-1668.

[35] Knodler, L. A., J. Celli, W. D. Hardt, B. A. Vallance, C. Yip, B. B. Finlay. Salmonella effectors within a single pathogenicity island are differentially expressed and translocated by separate type III secretion systems. Mol. Microbiol. 2002;43;1089-1103.

[36] Wood, M.W., M. A. Jones, P. R.Watson, S. Hedges,T. S.Wallis, E. E. Galyov.Identification of a pathogenicity island required for Salmonella enteropathogenicity. Mol. Microbiol. 1998;29:883-891.

[37] Folkesson, A., S. Lofdahl, S. Normark.The Salmonella enterica subspecies I specific centisome 7 genomic island encodes novel protein families present in bacteria living in close contact with eukaryotic cells. Res. Microbiol. 2002;153:537-545.

[38] Hensel M. Evolution of pathogenicity island of *Salmonella enterica*. International Journal of Medical Microbiology. 2004;**294**:95-102.

[39] Pickard et al. Composition, acquisition and distribution of vi exo-ploysaccharide- encoading *Salmonella* pathogencity island SPI-7. Journal of Bacteriology. 2003;**185**:505-506.

[40] Velásquez JC, Hidalgo AA, Villagra N, Santiviago CA, Mora GC, Fuentes JA. SPI-9 of Salmonella enterica serovar Typhi is constituted by an operon positively regulated by RpoS and contributes to adherence to epithelial cells in culture. Microbiology. 2016 Aug 1;162(8):1367-1378.

[41] Bishop, A. L., S. Baker, S. Jenks, M. Fookes, P. O. Gaora, D. Pickard, M. Anjum, J. Farrar, T. T. Hien, A. Ivens, G. Dougan .Analysis of the hypervariable region of the Salmonella enterica genome associated with tRNA(leuX). J. Bacteriol. 2005;187:2469-2482.

[42] Chiu, C. H., P. Tang, C. Chu, S. Hu, Q. Bao, J. Yu, Y. Y. Chou, H. S. Wang, Y. S. Lee.The genome sequence of Salmonella enterica serovar Choleraesuis, a highly invasive and resistant zoonotic pathogen. Nucleic Acids Res.2005; 33:1690-1698.

[43] Shah, D. H., M. J. Lee, J. H. Park, J. H. Lee, S. K. Eo, J. T. Kwon, J. S. Chae. Identification of Salmonella gallinarum virulence genes in a chicken infection model using PCR-based signature-tagged mutagenesis. Microbiology. 2005;151:3957-3968.

[44] Vernikos, G. S., J. Parkhill. Interpolated variable order motifs for identification of horizontally acquired DNA: revisiting the Salmonella pathogenicity islands. Bioinformatics.2006; 22:2196-2203.

Chapter 8

# Tracking *Salmonella* Enteritidis in the Genomics Era: Clade Definition Using a SNP-PCR Assay and Implications for Population Structure

*Dele Ogunremi, Ruimin Gao, Rosemarie Slowey, Shu Chen, Olga Andrievskaia, Sadjia Bekal, Lawrence Goodridge and Roger C. Levesque*

## Abstract

*Salmonella enterica* serovar Enteritidis (or *Salmonella* Enteritidis, SE) is one of the oldest members of the genus *Salmonella*, based on the date of first description and has only gained prominence as a significant bacterial contaminant of food over the last three or four decades. Currently, SE is the most common *Salmonella* serovar causing foodborne illnesses. Control measures to alleviate human infections require that food isolates be characterized and this was until recently carried out using Pulsed-Field Gel Electrophoresis (PFGE) and phage typing as the main laboratory subtyping tools for use in demonstrating relatedness of isolates recovered from infected humans and the food source. The results provided by these analytical tools were presented with easy-to-understand and comprehensible nomenclature, however, the techniques were inherently poorly discriminatory, which is attributable to the clonality of SE. The tools have now given way to whole genome sequencing which provides a full and comprehensive genetic attributes of an organism and a very attractive and superior tool for defining an isolate and for inferring genetic relatedness among isolates. A comparative phylogenomic analysis of isolates of choice provides both a visual appreciation of relatedness as well as quantifiable estimates of genetic distance. Despite the considerable information provided by whole genome analysis and development of a phylogenetic tree, the approach does not lend itself to generating a useful nomenclature-based description of SE subtypes. To this end, a highly discriminatory, cost-effective, high throughput, validated single nucleotide based genotypic polymerase chain reaction assay (SNP-PCR) was developed focussing on 60 polymorphic loci. The procedure was used to identify 25 circulating clades of SE, the largest number so far described for this organism. The new subtyping test, which exploited whole genome sequencing data, displays the attributes of an ideal subtyping test: high discrimination, low cost, rapid, highly reproducible and epidemiological concordance. The procedure is useful for identifying the subtype designation of an isolate, for defining the population structure of the organism as well as for surveillance and outbreak detection.



IntechOpen

**Keywords:** *Salmonella* Enteritidis, clades, WGS, SNP-PCR, PFGE, phage typing, nomenclature, population structure

## 1. Introduction

The genus *Salmonella* contains a large number of Gram-negative bacteria primarily found in the gastrointenstinal tract of vertebrate organisms including humans, cattle, pigs, horses, companion animals, avian, reptiles and fish [1]. There are two species of *Salmonella*, namely *Salmonella enterica* and *S. bongori* [2]. *Salmonella enterica* is the species of relevance in food safety, and consists of five subspecies of varying importance in human health. *Salmonella enterica* subspecies *enterica* has received the greatest attention because of its large number of constituent organisms, now estimated at about 2,600, each defined as a serovar based on the Kauffman-White classification [1]. *Salmonella enterica* serovar Enteritidis (commonly written as *Salmonella* Enteritidis or SE) is the most prominent. The organism was originally described as a distinct species and named as *Salmonella enterica* alongside two other species namely *Salmonella choleraesuis* and *Salmonella typhi*. Since those early days, the taxonomy of *Salmonella* has changed to reflect two species and hundreds of serovars. Curiously, a limited number of *S. enterica* serovars is associated with foodborne illnesses of which SE has emerged over the last few decades as the most prevalent cause of foodborne salmonellosis in humans worldwide [3]. However, this has not always been the case and prior to the 1970s there was only the occasional report of foodborne salmonellosis attributable to SE.

The earliest reports of foodborne illnesses caused by *Salmonella* were attributed to duck egg sources as summarized by Scott [4]. Subsequently, the organisms was found in live chicks, ducks and ducklings [5, 6]. Although these early reports came from different countries, SE did not become a common cause of foodborne illnesses until the 1980s [7]. By 1994, SE was the most commonly reported *Salmonella* serotype, with an incidence of 110 laboratory-confirmed infections per 100,000 population in the Northeast of US, and shell eggs from hens were identified as the major vehicle for SE infection in humans [8], in contrast to the earlier reports incriminating duck eggs. A 2010 outbreak of egg-related SE infections in the US resulted in an estimated 1,939 illnesses and a recall of over 500 million eggs, which ranked as the largest egg recall in history and one of the most expensive food recalls ever [9]. Similar events occurred in other parts of the world and were severe enough to warrant a warning of a new pandemic [7]. Together with two other serovars namely, Typhimurium and Heidelberg, the three most common serovars alone account for 59% of *Salmonella* outbreaks in humans in Canada, while the 10 most commonly observed *Salmonella* serovars account for about 76% of the total *Salmonella* infections reported. Establishing epidemiological linkages between contaminated products and human disease for *Salmonella* serovars has been particularly difficult for a number of reasons. One of the historically important reasons has been the clonal nature of many of the dominant serovars, especially Enteritidis which makes discrimination of strains difficult and an attribution of a particular strain linked with illness to a food source particularly challenging.

One resource that has been used by researchers to study SE is the strain P125109 phage type 4 (PT4) which was isolated from an outbreak of human food poisoning in the United Kingdom, and traced back to a poultry farm. The strain is highly virulent in newly hatched chickens and is also invasive in laying hens, resulting in egg contamination [10, 11]. The complete genome sequences of the host-promiscuous SE PT4 isolate P125109 was determined by Thomson *et al.* in 2008 [12].

Next generation sequencing (NGS) and especially whole genome sequencing (WGS) has emerged in recent years and has made it possible to sequence bacterial genomes within hours, a remarkable feat that is revolutionizing the field of microbiology. With the advent of microbial WGS, new light is shed on the nature of pathogens and our understanding of the biology of *Salmonella* is steadily increasing as *Salmonella* genomes are generated increasingly at a rapid rate and are deposited in public databases. Further understanding of genome diversity and variation of bacterial pathogens has the potential to improve quantitative risk assessment and assess the evolution of *Salmonella*, relationship among strains and serovars, emergence of new strains and the role of mobile genetic elements especially plasmids and bacteriophages in *Salmonella* [13]. The recent development of the *Salmonella* SystOmics database (SalFoS https://salfos.ibis.ulaval.ca/), a rich collection of over 3000 *Salmonella* genomes and their metadata represents a milestone and an important resource for future approaches to mitigate the burden of foodborne salmonellosis [14].

Food safety which is significantly impacted by *Salmonella* has gained from the advent of microbial genomics. Subspecies characterization including serovar identification and strain differentiation can now be done using genomics approach. As will soon be evident to the reader, there is much work yet to be done as the new capacity is yet to translate to tangible benefits to the consumer. Outbreaks caused by SE have remained at a high level or even increasing and there is a need to evaluate the efficacy of procedures used to detect the organism in food as well as approaches used in tracking the organism through the entire spectrum of the food chain, from farm to fork.

## 2. Laboratory culture and identification of organism

### 2.1 Culture procedures for *Salmonella*

Culture-based methods are commonly employed to detect pathogens in food, and in clinical and environmental samples. The Compendium of Analytical Methods (https://www.canada.ca/en/health-canada/services/food-nutrition/research-programs-analytical-methods/analytical-methods/compendium-methods.html) and the Bacteriological Analytical Manual (https://www.fda.gov/food/laboratory-methods-food/bacteriological-analytical-manual-bam) are compilations of laboratory procedures developed by the food safety regulatory agencies in Canada and the United States, respectively and each contains a catalog of official and recommended methods for isolating and detecting *Salmonella*. Briefly, *Salmonella* detection in food relies on a series of culture steps in broth formulations optimized to resuscitate *Salmonella* following injury caused by food handling, processing and storage and to reduce the abundance of competing bacteria [15]. In many enrichment protocols, broth and culture plates have been described for the isolation of *Salmonella* in different types of samples and matrices [16–18]. Typically, the first step is to culture a suspect food sample in a non-selective pre-enrichment broth, examples of which are lactose broth, buffered peptone water, trypticase soy, brilliant green water, powdered milk with brilliant green and universal pre-enrichment [16]. Following an overnight incubation commonly performed at 37°C, the culture material is subsequently transferred into a selective enrichment broth which suppresses and inhibits the growth of non-salmonellae while expanding the *Salmonella* population, facilitating isolation by plating on the appropriate media plates [19, 20]. Tetrathionate (TT) and Rappaport-Vassiliadis (RV) broths and RV semi-solid medium are the most commonly used selective culture conditions, performed at 37° or 42°C overnight for several days [15, 19].

When used to detect the presence of a microorganism in a food sample, laboratory culture procedures are slow and time consuming, requiring the sequential use of non-selective and selective enrichment broths and could take a week or longer. Another disadvantage is the documented inherent bias in the performance of selective broths which results in the preferential recovery of certain *Salmonella* serovars and not others [17, 21, 22]. For instance, different *Salmonella* serotypes are recovered by culture procedures performed on non-clinical, non-human sources when compared to samples tested in hospitals and other clinical settings from patients experiencing symptoms. Experimental results show that members of some *Salmonella* serogroups are unable to effectively compete with other serovars leading to a reduced efficiency of recovery of some *Salmonella* organisms including SE, from contaminated food [21]. The use of culture-independent procedures that can lead to rapid and sensitive detection of *Salmonella* [23] may in time eclipse the routine use of culture methods for detection. Nevertheless, the recovery of *Salmonella* in food is currently required to establish risk to the consumer and in support of a regulatory action. For this reason, and for the purpose of building inventories of microbial organisms for clinical and regulatory food microbiology, culture procedures are expected to remain in use. A wide variety of selective plating media are available for the isolation of *Salmonella* and a number of them will now be examined.

#### *2.1.1 Xylose lysine desoxycholate (XLD) agar*

XLD agar is a selective growth medium originally shown to facilitate the isolation of *Shigella* but was demonstrably useful for *Salmonella* isolation and has been further modified since its first description [24, 25]. At pH 7.4, the XLD agar appears bright pink or red as a result of the phenol red indicator. *Salmonella* ferments xylose, a sugar molecule, to produce acid and the bacterial colony turns yellow. In time, xylose is consumed and lysine is in turn utilized which upon decarboxylation produces an acidic environment and colonies turn back to red. In contrast, *Shigella* cannot ferment xylose and the colony remains red. *Salmonella* is able to metabolize thiosulfate to produce hydrogen sulphide, leading to the formation of colonies with black centres, which is an important feature in differentiating *Salmonella* colonies from *Shigella*. XLD agar is capable of supporting other members of *Enterobacteriaceae* such as *Escherichia coli* however the colonies and media turns yellow because of the fermentation of lactose which is also present in the agar. *Pseudomonas aeruginosa* is also able to grow on XLD plates as pink, flat, rough colonies but will not metabolize thiosulfate nor turn black. *Proteus* organisms can grow on XLD to give rose colored colonies and can sometimes metabolize thiosulfate to render the colonies black which will be readily confused with *Salmonella*. In addition, *Salmonella* strains have been described that do not metabolize thiosulfate and will grow as pink colonies which will be readily confused with *Shigella.* Thus, XLD agar is a moderately selective medium for isolating *Salmonella* and for differentiating it from other organisms.

#### *2.1.2 Xylose lysine Tergitol-4 (XLT-4) agar*

Similar to XLD agar, XLT-4 agar is also a selective culture medium which is used to isolate and identify *Salmonella* in food and environmental samples. Compared to XLD agar, XLT-4 is supplemented with a surfactant, 7-ethyl-2-methyl-4-undecanol hydrogen sulfate commonly referred to as Tergitol 4 while lacking sodium chloride and sodium desoxycholate. The surfactant is responsible for the inhibition of *Proteus* spp. and other non-salmonellae. XLT-4 agar is clearly one of the most stringent of all selective culture plates used for isolating *Salmonella* with positive colonies growing up as red and eventually turning black starting from the centre as a result of hydrogen sulfide production. However, *Salmonella* strains that fail to produce hydrogen sulfide appear as yellow colonies on XLT-4 agar [26, 27].

### *2.1.3 XA medium - modified XLD agar by adding D-arabinose*

XA medium is an improved selective and differential medium over XLD agar following its supplementation with arabinose, a sugar that is fermented by *Citrobacter* and *Proteus* but not by *Salmonella* [28]. The sensitivity of isolation of *Salmonella* using the XA and XLD media are equally high, however, the specificity of XA medium (92.0%) is superior to that of XLD (73.0%) [28]. Many *Salmonella* organisms appear as black colonies on XA agar whereas non-salmonellae will either not grow or appear as pink colonies. The use of arabinose to differentiate *Salmonella* from other closely related organisms represents a cost-effective approach, especially when compared to chromogenic plates (see Section 2.1.7).

### *2.1.4 Hektoen enteric (HE) agar*

HE agar is a selective and differential medium for isolating and distinguishing members of the genera of *Salmonella* and *Shigella* from the other *Enterobacteriaceae*. HE agar has a blue appearance and contains indicators of lactose fermentation and hydrogen sulfide production while inhibiting the growth of Gram-positive bacteria. Species belonging to Enterobacteriaceae that are capable of fermenting one or more carbohydrates produces yellow or salmon-orange colored colonies, e.g., *Klebsiella pneumonia* which ferments lactose. Non-fermenters produce blue-green colonies. Organisms that reduce sulfur to hydrogen sulfide such as *Salmonella* will produce black colonies or blue-green colonies with a black center. In contrast, colonies of *Shigella* remain green and do not turn black because of inability to metabolize sulfur.

### *2.1.5 MacConkey agar*

MacConkey agar is used for the isolation of Gram-negative enteric bacteria which represents a large group of bacteria prominent among which includes *Salmonella, E. coli, Proteus*, *Citrobacter*, *Klebsiella*, *Pseudomonas*, *Shigella*, *Enterobacter* and *Yersinia*. These organisms grow on the agar because of the selective property conferred by crystal violet and bile salts to inhibit the growth of Gram-positive bacteria. The indicator system is the neutral red dye which turns red at a pH below 6.8 but is colorless at higher pH. Thus, lactose fermenters such as *E. coli*, *Klebsiella* and *Enterobacter* which contain the *lac* operon form red or pink colonies on McConkey agar. In contrast, the other organisms including *Salmonella* which are generally non-lactose fermenters do not change color. Because *Salmonella* produce colonies similar to other non-lactose fermenters on MacConkey, the medium does not allow for identification of *Salmonella,* an objective that has to be achieved by employing other more selective agars. At the same time, lactose fermenting *Salmonella* have historically been shown to be causes of severe infections and outbreaks in humans [29] which is attributable to the presence of the *lac* operon carried in the chromosome or on plasmids [30] and leading to colonies that appear pink or reddish on MacConkey agar. Despite its limitations, the MacConkey agar can still be a very useful addition to the collection of media needed to comprehensively isolate and identify *Salmonella* in contaminated samples.

### *2.1.6 Brilliant green sulfa (BGS) agar*

The selectivity of the BGS agar is due to the presence of brilliant green and sulfadiazine, two components that individually inhibits Gram-positive and most Gram-negative bacilli. Phenol red is the pH indicator that detects changes in pH due to the fermentation of sucrose and/or lactose. *Salmonella* colonies range from reddish or pink to nearly white in color with a red zone. Lactose or sucrose fermenters occasionally grow on this medium and appear as yellow-green colonies surrounded by a yellow-green zone. The presence of sulfadiazine in the media is effective in inhibiting the growth of *E. coli* and *Proteus* and to a large extent *Shigella* species [31]. In a latter modification of the BGS agar, the replacement of lactose with glucose and of sulfadiazine with novobiocin to create the novobiocin-brilliant green agar (NBG), led to a higher recovery of *Salmonella* but the medium could not differentiate it from hydrogen sulfide-positive *Citrobacter* organism [32].

### *2.1.7* Salmonella *chromogenic agar*

Chromogenic plates have been developed for *Salmonella* as an improved alternative to procedures that rely on the ability of the organism to produce hydrogen sulfide or their inability to ferment lactose, attributes that are not fully diagnostic of *Salmonella*. This often result in *Citrobacter* and *Proteus* species being mistakenly identified as *Salmonella* while some atypical *Salmonella* are missed entirely, using agar plates described above. There are a number of commercially available chromogenic culture media which incorporate different chromogenic substrates and result in different colors of *Salmonella* colonies. Using the *Salmonella* chromogenic agar marketed by Oxoid (United Kingdom) as an example, the medium contains the substrate, Magenta-cap (5-bromo-6-chloro-3-indolylcaprylate) which is hydrolyzed by *Salmonella* species to give magenta colonies. The second substrate, X-Gal (5-bromo-4-chloro-3-indolyl-D-galactopyranoside), is hydrolyzed by many non-*Salmonella* species including *Citrobacter* and *Proteus* to give blue colonies [33, 34]. The selection for *Salmonella* is further enhanced by the presence of bile salts which inhibit Gram-positive bacteria, and of two antibiotics namely, novobiocin and cefsulodin which inhibit *Proteus* and *Pseudomonas*, respectively.

The isolation of *Salmonella* colonies in contaminated food demonstrates the presence of live organisms that can potentially cause harm. As indicated above, the procedure requires a combination of culture conditions, and takes time. Molecular procedures that can rapidly detect *Salmonella* are often used to accelerate the process, to improve on sensitivity of detection and also to confirm colonies as *Salmonella* because of the challenges with the isolation of the bacteria as outlined above. Many molecular techniques are now available for serotype-specific identification of SE.

## 2.2 Identification of *Salmonella* Enteritidis

Many laboratory diagnostic platforms have been applied to detect and identify *Salmonella* contamination in food and these include the PCR, enzyme-linked immunosorbent assay and the lateral flow assay [35–37]. Examples are available as commercial products. Currently, the most popular platform is the PCR and the most frequently used gene target is the *invA* gene. Nevertheless, many commercial offers do not disclose their target for proprietary reasons. PCR assays have also been developed with other gene targets present either in the chromosome, e.g., *flagellin* [38], *OriC* [39] *hilA* [40], *ttr* [41] or on plasmids, e.g., *SpvR* operon [42]. Multiplex PCR assays that are able to detect and distinguish among multiple serovars have also been developed by including serovar-specific gene targets such as STM4449 (Typhimurium [43]), STM 4497 (Typhimurium [44], *fliC* (Typhimurium [45]), *sdfI* (Enteritidis [46]) and *sefA* [29]. Recent work by Nadin-Davis and colleagues showed that many of the previously identified serovar specific markers were shared by other serovars especially *sefA* and *fliC* while highlighting the limitation with the use of a plasmid encoded target [47].

A multiplex PCR method which is capable of detecting all *Salmonella* spp., while identifying and distinguishing SE from the other two most prevalent serovars namely Typhimurium [48] and Heidelberg (Ogunremi et al., unpublished) is now available. The PCR was designed to amplify DNA fragments from four *Salmonella* genes, namely, *invA* gene (211-bp fragment), *iroB* gene (309-bp fragment), Typhimurium *STM 4497* (523-bp fragment), and Enteritidis *SE147228* (612-bp fragment) and has lately incorporated a 124-bp Heidelberg-specific fragment.

The identification of members of genus *Salmonella* to the subspecies level i.e., serovar is pivotal in tracking these pathogens along the food chain and the above molecular methods are very promising replacements to replace the traditional biochemical tests because of ease of application and high specificity for identifying SE and the other serotypes.

## 3. Typing of *Salmonella* Enteritidis

### 3.1 Serotyping

Serotyping has consistently been the basis of public health surveillance of *Salmonella* and has retained this primary role, as a first-line typing method, in the era of WGS based on the development of novel bioinformatics tools (see Section 3.3). Serotypes of *Salmonella* are defined by the presence of two types of antigens, namely, a heat stable, somatic O antigen, a component of the lipopolysaccharide envelope covering the organism which is an important virulence factor, and the H antigen which is present on the flagella of the organism [49]. The antigenic properties of the O antigen are depicted as numerals, e.g., 1,9,12 for SE. In contrast, the H antigens are described using one or a few letters for the phase I antigen (e.g., g, m for SE) or as a combination of letters and numbers for antigens that are expressed should the flagella bear a phase II antigen (e.g., r and 1, 2 for Heidelberg). Agglutination assays are performed on the organisms using antibodies that are able to recognize specific antigenic molecules developed through laborious cross-absorption process against other serovars [50]. The result is an elaborate classification scheme, developed by Kauffman and White [51, 52] and which has now led to the identification of some 2,600 serotypes of *Salmonella*. The complexity has been further enhanced by the ability of plasmids and prophages to alter the expression of some of the antigens, and this had led to a frequent re-evaluation of some serovar designations. Fortunately, these alterations are fairly rare and the serotyping scheme has served well since first proposed by Schüte in 1920 [53]. Of the large number of *Salmonella* serovars identified so far, only a relatively small numbers, perhaps no more than 100 serovars are commonly associated with foodborne illnesses [54, 55].

### 3.2 Traditional subtyping procedures for *Salmonella* Enteritidis

There are two approaches for the subspecies characterization of SE. Phenotypic tests rely on the biochemical properties of the live organism and the most prominent example is phage typing. More recently, DNA based approaches or genotypic tests have dominated the field. The most widely used genotypic test being the Pulsed-Field Gel Electrophoresis. Whole genome sequencing of the DNA of SE, has over the last few years, become the dominant subtyping method in the developed world.

### *3.2.1 Pulsed-field gel electrophoresis (PFGE)*

The PFGE can been used to characterize bacteria isolates based on the pattern of distribution of restriction enzyme sites present in the organism's DNA. For *Salmonella*, the electrophoretic mobility of DNA fragments digested by the restriction enzyme *Xba*I or *Bln*I produces a characteristic fingerprinting pattern that is used to subtype the isolate. During the period between 2009 and 2019, the Canadian Food Inspection Agency used the PFGE for outbreak investigations as one of the two subtyping tests for SE, the other being the phage type. Despite the presence of hundreds of different PFGE types among field isolates of SE only two PFGE types predominated and each consisted of thousands of isolates in the Canadian PulseNet database. The two commonest Canadian primary PFGE types, namely SEN.XAI 0003 and SEN.XAI 0006, were responsible for 33.8 and 19.2% of Canadian SE isolates documented in the PulseNet database between 2012 and 2017 (Ogunremi, Allain and Nadon, unpublished). The predominance of only a few PFGE SE types was long recognized as a consequence of the poor discriminatory ability of the technique for analyzing the relatedness of SE isolates (**Table 1**) rather than a reflection of an evolutionary dominance of a few circulating strains [56]. These observations led to the pursuance of WGS as an alternative approach [57].

### *3.2.2 Phage typing*

In contrast to the PFGE, phage typing is a phenotypic test that exploits the ability of certain bacteriophages, i.e., viruses that infect bacteria, to differentially attach and gain entrance into strains of bacteria. Phage typing of SE is the outcome of the pattern of susceptibility of different strains to a bacteriophage or a combination of bacteriophages, resulting in lysis of the bacterial cell [58]. A large number of phage types of SE have been described in Canada and elsewhere, however phage types 8, 13 and 13a were observed to predominate in Canada [59]. This observation may not reflect the presence of a few, circulating dominant strains of SE in Canada, but instead may be a consequence of the inadequacy of phage typing as a discriminatory tool that can accurately delineate the population structure of SE in Canada, similar to the PFGE as discussed above (see Section 3.2.2 and **Table 1**). The plasticity of phage types also diminishes its use as a subtyping tool. Factors such as the restriction system within the bacteria, ability of lipopolysaccharides and outer membranes to adsorb the bacteriophage, and the immune system of the vertebrate host infected by the bacteria can alter the phage type of an organism [60]. The reagents used for phage typing require very rigorous quality control and yet, test performance can be remarkably different among laboratories [61]. Changes occurring within an organism such as the acquisition or loss of IncN plasmid [62, 63], transfer of IncX plasmid [64] or loss of the lipopolysaccharide layer [65] have been shown to lead to poor test reproducibility. Thus, two isolates with the same phage type may in fact be unrelated and conversely, two isolates that show distinct phage types may be closely related. As a result of these factors, phage typing shows inadequate discriminatory power, partial typeability and poor reproducibility [66].

| Clade | Strain identification | Source description | Phage type | PFGE type SENXAI, SENBNI | EnteroBase | |
|---|---|---|---|---|---|---|
| | | | | | MLST (7 gene) | cgMLST_v2 + HierCC_v1 |
| 1 | 2007-MI-0187-0006 | Poultry environment | Atypical | 0214, 0225 | 814 | 259062 |
| 2 | 08OTH012 6–4 | Poultry environment | 9b | 0214, 0225 | 814 | 259068 |
| 3 | 06-1472 | Animal feed | 13a | 0006, N/A | 639 | 273915 |
| 4 | OLF 10012–1 | Sea food, clams | 13, 1b | 0009, 0013 | 11 | 5485 |
| 5 | ID094888 | Clinical case | 6a | N/A, 0011 | 11 | 259098 |
| 6 | dart-1997-742-B2 | Cheese lunchables | 8 | 0003, 0003 | 11 | 259481 |
| 7 | S-MBS4754A | Chicken ceacum | 51 | N/A | 8471 | 259064 |
| 8 | SE974-OLF-2015-NSub | Bovine, heifer | N/A | N/A | 11 | 260728 |
| 9 | S-MBS1982A | Chicken thigh | N/A | N/A | 11 | 259069 |
| 10 | 10OTH025 7–14 | Poultry environment | 13 | 0038, 0016 | 11 | 259063 |
| 11 | S-MBS0737R | Chicken carcass | 13a | N/A | 11 | 259067 |
| 12 | 05–3936 | Chicken breast | 13a | 0068, N/A | 11 | 259480 |
| 13 | 07–1474 | Chicken nuggets | 8 | 0003, N/A | 11 | 30959 |
| 14 | S-MBS3492A | Chicken breast | N/A | N/A | 11 | 259071 |
| 15 | S-MBS7608A | Chicken carcass | 8 | N/A | 11 | 259072 |
| 16 | 10SU010 19–1 | Poultry environment | 8 | 0003, 0003 | 11 | 5490 |
| 17 | 07–1485 | Chicken nuggets | 14b | 0003, 0003 | 11 | 30959 |
| 18 | S-MBS3006A | Chicken ceacum | 8 | N/A | 11 | 259070 |
| 19 | 11OTH025 11-5 | Poultry environment | 8 | 0003, 0003 | 11 | 273916 |
| 20 | S-MBS8825A | Chicken ceacum | 8 | N/A | 11 | 259066 |
| 21 | SA20100239 | Bovine liver | 2 | N/A | 11 | 14029 |
| 22 | 00D989 83–4 | Poultry environment | 23 | 0003, 0009 | 11 | 5498 |
| 23 | SE972-OLF-2015-NSub112-S19 | Water treatment plant | 8 | N/A | 11 | 259100 |
| 24 | ID112184 | Human | 8 | 0007, 0212 | 11 | 259479 |
| 25 | EN1811 | Food processing equipment | 13 | 0076, 0003 | 11 | 233056 |

*The single nucleotide-polymorphism chain reaction (SNP-PCR) was used to test* Salmonella *Enteritidis (SE) isolates and a representative strain for each designated clade (from 1 to 25) is shown in comparison to traditional and whole genome sequence based subtyping results. Only the SNP-PCR and EnteroBase core-genome multi-locus sequence typing (cg-MLST) supplemented with Hierarchical level analysis (HierCC) showed distinct resolution of the representative strains. All other methods including 7 gene MLST, phage typing and pulsed-field gel electrophoresis (PFGE) did not provide adequate discriminatory ability relevant for strain differentiation, outbreak investigation or tracking SE from farm to fork. N/A: Not available.*

**Table 1.**
*Clade designation of* Salmonella *Enteritidis organisms depicting a representative strain for each clade and comparison with the results of traditional and new subtyping assays.*

### *3.2.3 Multiple locus variable-number tandem repeat analysis (MLVA) assay*

MLVA is a molecular typing method that is based on PCR amplification of polymorphic regions of the DNA containing variable numbers of tandemly repeated sequences. The method has been standardized by PulseNet International and applied to the epidemiological investigations of SE either as a supplement or substitute for PFGE subtyping [67, 68]. An advantage of the MLVA is the designation of the typing results with a numeric sequence of tandem repeats. This represents a simple, easy-to-understand nomenclature which facilitated the reporting and exchange of test results between laboratories, and translated to a reliable tracking of an organism during epidemiological investigations. The discriminative ability of the MLVA has been variously shown to be superior [69], equivalent [70] or poorer than the PFGE [71].

Detailed genetic studies of SE have consistently shown the underlying causes of the poor discriminatory abilities of available subtyping tools, namely: isolates of SE are extremely similar (i.e., are highly clonal) and this poses a difficulty in finding a definitive, distinguishing trait that could be used to track lineages [70, 72, 73]. The timely arrival and increasing adoption of WGS has altered the analytical landscape.

## 3.3 Application of whole genome sequencing (WGS) in *Salmonella* Enteritidis: identification and characterization

The development of WGS procedure has heralded the application of a powerful technology for the identification and characterization of SE [57] which has been used for outbreak investigations [74], trace back procedures [75] and surveillance [76]. Furthermore, WGS analysis of SE has provided insights into phylogenetic relatedness of isolates, presence and prevalence antimicrobial resistance genes, novel mobile elements, virulence markers and bacteriophages in strains of the organism isolated from humans, food animals, production facilities and environmental sources [77–79]. Relevant to developing long term control and intervention strategies are the insights to be gained from the increasing application of WGS to the understanding of transmission dynamics of SE as was done in Chile to infer possible transmission of SE between gulls, poultry, and humans [80]. Bioinformatics approaches that allow useful information to be mined from genome sequences will now be discussed.

### *3.3.1 Whole genome-based serotyping*

Serovar prediction can now be done on *Salmonella* isolates if the whole genome sequence is available by replacing the laborious agglutination assay (see Section 3.1) with an *in silico* analysis of the nucleotide sequence of the organism. Effectively, the traditional gold standard of traditional serology based on the Kauffmann-White Scheme has been replaced in the developed economies with *in silico* approaches [81]. Two of the mostly widely tools for this purpose are the *Salmonella In Silico* Typing Resource (SISTR) software and the SeqSero2 software [82, 83].

SISTR is an open, web-based bioinformatics platform capable of rapid *in silico* analyses of minimally processed draft assemblies of *Salmonella* genomes to generate accurate serovar designations. A collection of markers previously developed for the various *Salmonella* serovars formed the basis of the new tool [84]. The performance of SISTR is enhanced by the integration of additional multilocus sequence typing tools (see Section 3.3.2) which as a separate platform has been suggested as a replacement for the use of serotypes to define taxonomic as well as evolutionary groups of *Salmonella* [55]. SeqSero, which was launched in 2015 was developed to employ the use of the *rfb* cluster, *fliC* and *flijB* to categorize *Salmonella* according to serovar using draft genome assemblies [83]. A subsequent improvement of the software, released as SeqSero2 included addition of markers at the level of the genus, species, subspecies as well as certain serotypes. Furthermore, a kmer-based algorithm was included that ensured a genome can be analyzed and the result available within seconds [85].

#### *3.3.2 Multilocus sequence typing*

Multilocus sequence typing (MLST) evaluates the nucleotide sequences of multiple housekeeping genes of an organism as a means of establishing similarities or differences among isolates [86]. Based on the sequences, each housekeeping gene is assigned an allele which can be stringed together in a nomenclature that defines the organism. Although the MLST scheme was developed using the bacterium *Neisseria meningitidis* [86], the advantage of electronic portability of sequence data and ease of incorporation of additional genes found a good synergy in the advent of WGS and has gained application in food safety. This has birthed the widely used EnteroBase (https://enterobase.warwick.ac.uk/) [87], an integrated web-based platform that permits the upload and analysis of short read Illumina sequences. This has allowed the expansion of the MLST scheme which was based on the initial six housekeeping genes [86] to a series of flexible applications and expansions for *Salmonella* including seven genes (legacy MLST), 3002 genes identified as the core genome of *Salmonella,* to produce core genome MLST (cgMLST) and 21,065 orthologous genes detected in a set of 537 *Salmonella* genomes, regarded as whole genome MLST (wgMLST). Despite the adoption of the wgMLST by PulseNet International [88], an influential international body which overlooks regulatory subtyping procedures for foodborne bacteria, EnteroBase's Sequence Type, ST, of *Salmonella* became a widely adopted subtype descriptor for *Salmonella*. However, ST does not provide adequate resolution for epidemiological concordance and outbreak level discrimination [89], and in addressing the challenge EnteroBase has additionally provided the core genome ST, cgSTs, complemented with a newly described 11 levels of genetic resolution hierarchies or HierCC for *Salmonella* (**Table 1**) [87, 90]. The result is a tool that appears to provide the needed resolution for strain differentiation in the context of disease outbreaks.

#### *3.3.3 Single nucleotide polymorphism (SNP) pipelines*

Single base substitutions represent one of the commonest variation in genomes and the resulting polymorphism can form the basis for the characterization of a microbe including SE. SNPs are detected as nucleotide changes at a specific location in a genome after aligning or comparing it to a designated reference genome. Bioinformatics pipelines have been developed to automate the aligning and identification of the variants. A number of SNP pipelines are in common use and will now be described. SNVPhyl which was developed at the Public Health Agency of Canada identifies high quality SNPs among a set of selected isolates and is useful for generating phylogenetic trees from these SNPs [91]. Public Health England developed SnapperDB, also a high-quality SNP pipeline which analyzes microbial genomes, evaluates genetic distances among the genomes and infers relatedness of strains [92]. Parsnp detects core genome SNP in bacterial genomes and with the aid of adjunct interactive tool Gingr can be used to display informative overviews for specific sub-clades and genomic regions [93]. The kSNP tool detects SNPs in the pan genome but is uniquely able to carry out comparisons among genomes without a requirement for genome alignment nor the use a reference genome [94].

### 3.4 Rationale for developing a new reliable, rapid, robust, cost-effective, epidemiologically concordant, easily implementable subtyping tool

A strategy aimed at developing a tool capable of differentiating lineages in the highly clonal *S.* Enteritidis lineages will likely require interrogating a significant amount of the bacterial DNA information. The opportunities provided by the massively parallel sequencing technology [95], which deduces the entire nucleotide sequence of an organism appeared at the onset to be the most viable option in charting a course to address the need. Use of genome sequence for taxonomy including strain differentiation could conceivably work well with strains showing significant genetic diversity, e.g., >5% differences among unrelated strains. However, this may be very difficult for a clonal organism such as SE where diversity between unrelated strains could be as little as 1% and the similar regions of the genome would have to be ignored before focusing on the dissimilar portions to demonstrate an accurate quantitative estimate of relatedness. This may explain the failure to use whole genome sequence to develop a reliable estimation of genetic distance by means of a phylogenetic tree for a group of SE isolates (Ogunremi et al., unpublished data) using a method shown to work for other bacteria [96].

Consequently, this led to an effort to develop, analyze and characterize the genomes of SE. During the early phase of this endeavor involving a select number of SE isolates from Canada, 669 SNPs were detected in the genome of SE [57]. Subsequent analysis of 135 SE genomes present in the GenBank in 2014 led to the identification of a total of 1440 SNPs providing a robust resource that was exploited for a SNP-based strain differentiation and clustering of foodborne SE isolates [57]. Thus, despite the universal acceptance of the usefulness of whole genome sequences for microbes, individual organisms such as the highly clonal SE may pose a unique challenge that might require a more focused analysis on carefully selected targets of the entire genome.

## 4. Single nucleotide polymorphism-polymerase chain reaction test (SNP-PCR) as a new, nomenclature friendly procedure

### 4.1 History and development of *Salmonella* Enteritidis lineages/clades and SNP-PCR

The existing molecular methods investigate only very small portions or attributes of the entire bacterial genome. The PFGE, as an example, identifies enzyme restriction patterns in the genome whereas WGS-based procedures have available for analysis detailed information on the entire genome to exploit as a basis for comparison and discrimination. To that end, extremely small differences, such as single nucleotide polymorphisms (SNPs), can be identified and used for subtyping as long as these attributes are consistently preserved in a particular bacterial lineage. Notably, Allard and colleagues [97] carried out bioinformatics analysis of a total of 104 SE genomes belonging, for the most part, to the predominant PFGE pattern (JEGX01.0004). They described a total of 9 clades and found 366 genes that showed variation, i.e., presence or absence, in the SE genome. This observation complemented and expanded on an earlier study by another laboratory which showed that two isolates of SE with the same phage type, PT 13a, were differentiated by a relatively large number of loci, i.e., 250 SNPs [73]. Similarly, by using a specific reference genome, for instance SE strain P125109, the WGS-based sequence reads were mapped to the reference to find SNPs which were used to build maximum-likelihood phylogenetic trees.

Another study involving 55 SE strains selected from clinical and environmental samples in Minnesota and Ohio from 2001 to 2014 showed the existence of only two major groups [98]. Furthermore, WGS based SNPs analysis of 675 SE isolates from 45 countries formed a global epidemic clade and two new clades that were found to be geographically restricted to distinct regions of Africa [99]. Using a closely related serovar - *S.* Gallinarum - as an outgroup, a maximum-likelihood phylogenetic tree was constructed based on the alignment of a total of 42,373 SNPs [99]. In addition, a SNP-based phylogenetic structure of 401 European SE isolates implicated outbreaks correlating with national and international egg distribution network [75].

Thus, genetic variation that could allow the development of a routine subtyping tool for tracking purposes is present and demonstrable within the SE genome but was apparently not fully exploited given the few number of subgroupings in each of the reported, sampled populations, and this presented a need to properly mine the SE genome and develop a very discriminatory subtyping procedure. In exploring this need, our hypothesis was that the use of a large number of SNPs may not necessarily improve the power of discrimination. More is not necessarily better. A large number of uninformative loci may be counterproductive and undesirable for strain differentiation. As a first step to address this need, whole-genome sequences of 11 SE isolates obtained in Canada were developed and compared to SE P125109 reference strain phage type 4 which led to the identification of 1361 loci where the SE genome showed SNP [100]. Subsequent selection of 60 SNPs spread throughout the genome and distributed among different gene types and in intergenic locations led to the development of a rapid, inexpensive fluorescence-based real time PCR subtyping assay [55].

### 4.2 The SNP-PCR subtyping procedure

The SNP-PCR genotype assay is an allele-specific, single amplification procedure based on the specific binding of one of two, competing forward primers, 18–20 nucleotides long, which differ by one single nucleotide at the locus of interest. The use of a single reverse primer completes the amplification process leading to the accumulation of an amplicon bearing the SNP of interest. Each primer is designed with a specific tail that allows a complementary binding with a commercially provided, customized sequence labeled with a fluorescent dye, FAM or HEX for allele 1 or 2 respectively (LGC Genomics, Beverly, MA). Thus, the first cycle of amplification ensures that the specific forward oligonucleotide present in the primer mix binds to the sequence containing the SNP and excludes the other primer. The reverse primer, also 18–20 nucleotides long, binds and elongates the fragment during amplification ensuring that the tail sequence is present, which then allows the accumulating fragment to contain either the FAM or HEX fluorescent label depending on the initial binding of one of the bi-allelic primers, which is dictated by which of the SNP corresponds to allele 1 or allele 2. Thus, detection is based on the use of fluorescent labeled sequence that assigns the allele number to either of the two nucleotides that may occupy the SNP position. The SNP alleles are compiled for all SE strains at the 60 loci and used as input to carry out evolutionary history analyses using Maximum Parsimony method, which was conducted using Molecular Evolutionary Genetics Analysis on the MEGA-X computing platform [101]. The distinct grouping of the SE isolates are identified as clades and each given a specific numerical description starting from 1.

Following the development of the SNP-PCR procedure, our initial application of the assay to a group of 55 SE isolates obtained in Canada led to the recognition of 12 clades of SE [57].

### 4.3 Twenty five circulating clades of *Salmonella* Enteritidis

Recently, the laboratory validation of the SNP-PCR assay was completed using 1,127 SE isolates obtained from food, animal, humans, and environmental sources in Canada and Europe and we observed a total of 25 circulating clades of SE (**Table 1**, Ogunremi et al., manuscript under preparation). In addition, 13 other globally distributed isolates identified from published papers [98, 99] as well as the widely used reference SE strain P125109 phage type 4 were also included in a phylogenetic comparison using the Maximum Parsimony method. These strains were distributed across the generated phylogenetic tree and homed to distinct SE clades providing further validation of the SNP-PCR tool to appropriately cluster strains and at the same time, distinguish among different strains (Ogunremi et al., manuscript under preparation). The validation procedure unambiguously demonstrated the robustness of the assay while displaying its prowess in estimating genetic distances and relatedness among and between clades, and its relevance in constructing an evolutionary map of SE following the testing of a large number of isolates.

### 4.4 Advantages of SNP-PCR: nomenclature and population structure

Previous studies aimed at evaluating the population structure of the highly clonal SE have reported fewer lineages and clades among isolates tested. For instance, a study of 675 very diverse isolates collected over many decades (1948–2013) in 45 countries and 6 continents revealed the presence of only 3 clades; a subgroup of 58 isolates was identified but could not be clustered by the method used by the authors [97]. Yet another study demonstrated 9 clades among a large but PFGE-uniform group of isolates [99]. These studies, which showed a limited diversity among SE populations, served to underscore our contrasting observations, and reinforced the excellent discrimination observed for SE using the validated SNP-PCR assay. The SNP-PCR compares well with cgMLST-HierCC function in EnteroBase in discriminating among strains chosen to represent SE clades from a very diverse SE population from a variety of sources and different continents (**Table 1**; Ogunremi et al., under preparation).

Apart from being a highly discriminatory and robust assay, the SNP-PCR is very cost-effective. Reagents cost are estimated at Can$0.25 per SNP per isolate and testing 60 SNPs is cheaper than the traditional, less discriminatory subtyping assays (Can$26 for phage typing and Can$36 for two-enzyme PFGE analysis in reagent costs) or for WGS (Can$100). The SNP-PCR validation procedure (described above) showed that only 17 SNP loci needed to be tested to assign an isolate to a clade and the test performed excellently well on crude, boiled bacterial extract, obviating the need for DNA purification and further creating an increased savings of reagents, labour and time.

Another important attribute of the SNP-PCR is its equal adaptability to few samples or a large number of samples. When compared to Illumina WGS which requires a prescribed number of samples per run (e.g., 20 *Salmonella* strains using MiSeq version 3 library kit over 600 cycle sequencing which runs for 65 hours), the SNP-PCR can be used to test one or a few samples with the appropriate controls without any cost implication on the volume of analysis. At the other end, a single PCR sample can handle a 384-well plate loaded with hundreds of samples and machine run completed in 2 hours. The labor costs of running the SE SNP-PCR test (2 h PCR time) and analyzing the results are at least an order of magnitude lower than those of any subtyping approach including traditional molecular tests or WGS. The SNP-PCR test shows very good reproducibility (95%) in tests conducted in six laboratories.

The SNP-PCR impressively satisfies all the seven criteria expected of an ideal subtyping test which includes cost effectiveness, rapid performance, robust results, typeability, high discrimination, reproducibility, and epidemiological concordance [66].

## 5. Conclusions

The bacterial pathogen, *Salmonella* Enteritidis is one of the most prevalent causes of foodborne illness in humans worldwide, yet tracking a strain of the organism through the food safety system is challenging because of its clonal nature, evident at the genomic level, which historically has resulted in poorly discriminating laboratory typing methods. The current application of genomics has led to the development of comprehensive and highly discriminatory tools however there are still challenges with the interpretation of the outputs and the application of the methods to differentiate between outbreaks and sporadic infections. The effect is a poorly understood population structure of SE.

This chapter illustrates the existence of 25 clades of SE, which should be useful for defining the population structure and tracking the pathogen from farm to fork. The phylogenetic relationships among the 25 clades of SE was obtained using a population of 1127 isolates obtained from a variety of sources in Canada and Europe. The validated SNP-PCR assay displayed the attributes of an ideal subtyping test and can be implemented in resource deprived countries where routine genome sequencing remains unaffordable, as well as in resource rich countries when characterizing a few isolates may not justify the expense of a genome sequencing run or for surveillance where interest in characterizing a large number of lower priority, non-clinical but valuable isolates is a very desirable goal.

## Acknowledgements

This work was funded by the Canadian Food Inspection Agency, Ontario Ministry of Agriculture, Food and Rural Affairs, and Genome Research and Development Initiative of the Government of Canada.

## Conflict of interest

The authors declare no conflict of interest.

## Acronyms and abbreviations

| | |
|---|---|
| PFGE | Pulsed-field gel electrophoresis |
| XLD | Xylose lysine desoxycholate |
| XLT4 | Xylose lysine tergitol-4 |
| HE | Hektoen enteric |
| BGS | Brilliant green sulfa |
| NTS | Non-typhoidal *Salmonella* |
| NGS | Next generation sequencing |
| SalFoS | *Salmonella* Foodborne Syst-OMICS database |

## Author details

Dele Ogunremi[1*], Ruimin Gao[1*], Rosemarie Slowey[2], Shu Chen[3], Olga Andrievskaia[1], Sadjia Bekal[4], Lawrence Goodridge[5] and Roger C. Levesque[6]

1 Ottawa Laboratory Fallowfield, Canadian Food Inspection Agency, Ottawa, Ontario, Canada

2 Central Veterinary Research Laboratory, Department of Agriculture Food and the Marine, Co. Kildare, Ireland

3 Laboratory Services, University of Guelph, Guelph, Ontario, Canada

4 National Institute of Public Health of Quebec, Quebec, Canada

5 Department of Food Science, University of Guelph, Guelph, Canada

6 Institute for Integrative Systems Biology (IBIS), Université Laval, Quebec City, Canada

*Address all correspondence to: dele.ogunremi@inspection.gc.ca and ruimin.gao@inspection.gc.ca

## References

[1] Özkalp B: **Isolation and identification of** ***Salmonellas*** **from different samples**. In: *A Dangerous Foodborne Pathogen* Edited by Mahmoud BSM: InTech; 2012. pp. 123-156.

[2] Brenner FW, Villar RG, Angulo FJ, Tauxe R, Swaminathan B: *Salmonella* **nomenclature**. *J Clin Microbiol* 2000, **38**(7):2465-2467.

[3] Vieira A JA, Pires SM, Karlsmose S, Wegener HC, Wong DLF: **WHO Global Foodborne Infections Network Country Databank – a resource to link human and non-human sources of** *Salmonella*. *In Proceeding of the 12th Symposium of the International Society for Veterinary Epidemiology and Economics, Durban, South Africa* 2009.

[4] Scott WM: **Food poisoning due to eggs**. *Br Med J* 1930, **2**:56-58.

[5] Garside JS, Gordon RF: *Salmonella* **infections of ducks and ducklings**. *J Comp Pathol* 1943, **53**:80-89.

[6] Anonymous: **Eggs and** ***Salmonella*** **infections**. *Br J Med* 1944:760-761.

[7] Rodrigue DC, Tauxe RV, Rowe B: **International increase in** ***Salmonella*** **Enteritidis: a new pandemic?** *Epidemiol Infect* 1990, **105**(1):21-27.

[8] Braden CR: *Salmonella enterica* **serotype Enteritidis and eggs: a national epidemic in the United States**. *Clin Infect Dis* 2006, **43**(4):512-517.

[9] **Multistate outbreak of human** ***Salmonella*** **Enteritidis infections associated with shell eggs**. In.: Centers for Disease Control and Prevention; 2010. https://www.cdc.gov/salmonella/2010/shell-eggs-12-2-10.html.

[10] Barrow PA: **Experimental infection of chickens with** ***Salmonella*** **enteritidis**. *Avian Pathol* 1991, **20**(1):145-153.

[11] Barrow PA, Lovell MA: **Experimental infection of egg-laying hens with** ***Salmonella*** **Enteritidis phage type 4**. *Avian Pathol* 1991, **20**(2):335-348.

[12] Thomson NR, Clayton DJ, Windhorst D, Vernikos G, Davidson S, Churcher C, Quail MA, Stevens M, Jones MA, Watson M *et al*: **Comparative genome analysis of** ***Salmonella*** **Enteritidis PT4 and** ***Salmonella*** **Gallinarum 287/91 provides insights into evolutionary and host adaptation pathways**. *Genome Res* 2008, **18**(10):1624-1637.

[13] Branchu P, Bawn M, Kingsley RA: **Genome variation and molecular epidemiology of** ***Salmonella enterica*** **serovar Typhimurium pathovariants**. *Infect Immun* 2018, **86**(8):e00079-18.

[14] Emond-Rheault JG, Jeukens J, Freschi L, Kukavica-Ibrulj I, Boyle B, Dupont MJ, Colavecchio A, Barrere V, Cadieux B, Arya G *et al*: **A Syst-OMICS approach to ensuring food safety and reducing the economic burden of Salmonellosis**. *Front Microbiol* 2017, **8**:996.

[15] Daquigan N, Grim CJ, White JR, Hanes DE, Jarvis KG: **Early recovery of** ***Salmonella*** **from tood using a 6-hour non-selective pre-enrichment and reformulation of Tetrathionate broth**. *Front Microbiol* 2016, 7:2103.

[16] D'Aoust JY: **Update on preenrichment and selective enrichment conditions for detection of** ***Salmonella*** **in Foods**. *J Food Prot* 1981, **44**(5):369-374.

[17] Gorski L: **Selective enrichment media bias the types of** ***Salmonella enterica*** **strains isolated from mixed strain cultures and complex enrichment broths**. *PLoS One* 2012, 7(4):e34722.

[18] Bailey JS, Cox NA: **Universal preenrichment broth for the simultaneous detection of** *Salmonella* **and** *Listeria* **in Foods.** *J Food Prot* 1992, **55**(4):256-259.

[19] Vassiliadis P: **The Rappaport-Vassiliadis (RV) enrichment medium for the isolation of** *Salmonellas*: **an overview.** *J Appl Bacteriol* 1983, **54**(1):69-76.

[20] Fagerberg DJ, Avens JS: **Enrichment and plating methodology for** *Salmonella* **detection in food.** *Journal Milk Food Technology* 1976, **39**:628-646.

[21] Singer RS, Mayer AE, Hanson TE, Isaacson RE: **Do microbial interactions and cultivation media decrease the accuracy of** *Salmonella* **surveillance systems and outbreak investigations?** *J Food Prot* 2009, **72**(4):707-713.

[22] Harvey RW, Price TH: **The examination of samples infected with multiple salmonella serotypes.** *J Hyg (Lond)* 1967, **65**(3):423-434.

[23] Ogunremi D, Dupras AA, Naushad S, Gao R, Duceppe MO, Omidi K, Marquez IG, Huang H, Goodridge L, Levesque RC *et al*: **A new whole genome culture-independent diagnostic Test (WG-CIDT) for rapid detection of** *Salmonella* **in lettuce.** *Front Microbiol* 2020, **11**:602.

[24] Nye KJ, Fallon D, Frodsham D, Gee B, Graham C, Howe S, Messer S, Turner T, Warren RE: **An evaluation of the performance of XLD, DCA, MLCB, and ABC agars as direct plating media for the isolation of** *Salmonella enterica* **from faeces.** *J Clin Pathol* 2002, **55**(4):286-288.

[25] Taylor WI: **Isolation of shigellae. I. Xylose lysine agars; new media for isolation of enteric pathogens.** *Am J Clin Pathol* 1965, **44**(4):471-475.

[26] Miller RG, Tate CR, Mallinson ET, Scherrer JA: **Xylose-lysine-tergitol 4: an improved selective agar medium for the isolation of** *Salmonella*. *Poult Sci* 1991, **70**(12):2429-2432.

[27] Albert MJ, Al Obaid K, Alfouzan W, Sheikh AR, Udo E, Izumiya H, Bulach DM, Seemann T: **Isolation of** *Salmonella enterica* **serovar Kentucky strain ST 198 and its H2S-negative variant from a patient: implications for diagnosis.** *J Clin Microbiol* 2014, **52**(11):4090-4093.

[28] Park SH, Ryu S, Kang DH: **Development of an improved selective and differential medium for isolation of** *Salmonella* **spp.** *J Clin Microbiol* 2012, **50**(10):3222-3226.

[29] Pickett G, Agate GH: **Outbreak of salmonellosis due to lactose-fermenting variant of** *Salmonella* **newington.** In. Salmonella surveillance: Public Health Service, National Communicable Disease Center, Atlanta; MMWR 1967, **16**:18.

[30] Leonard SR, Lacher DW, Lampel KA: **Acquisition of the lac operon by** *Salmonella enterica*. *BMC Microbiol* 2015, **15**:173.

[31] Moats WA, Kinner JA: **Factors affecting selectivity of brilliant green-phenol red agar for salmonellae.** *Appl Microbiol* 1974, **27**(1):118-123.

[32] Devenish JA, Ciebin BW, Brodsky MH: **Novobiocin-brilliant green-glucose agar: new medium for isolation of salmonellae.** *Appl Environ Microbiol* 1986, **52**(3):539-545.

[33] Rambach A: **New plate medium for facilitated differentiation of** *Salmonella* **spp. from** *Proteus* **spp. and other enteric bacteria.** *Appl Environ Microbiol* 1990, **56**(1):301-303.

[34] Cassar R, Cuschieri P: **Comparison of** *Salmonella* **chromogenic medium with DCLS agar for isolation of** *Salmonella* **species from stool**

**specimens.** *J Clin Microbiol* 2003, **41**(7):3229-3232.

[35] Blais BW, Martinez-Perez A: **Detection of group D salmonellae including** *Salmonella* **Enteritidis in eggs by polymyxin-based enzyme-linked immunosorbent assay.** *J Food Prot* 2008, **71**(2):392-396.

[36] Mozola MA, Peng X, Wendorf M: **Evaluation of the GeneQuence DNA hybridization method in conjunction with 24-hour enrichment protocols for detection of** *Salmonella* **spp. in select foods: collaborative study.** *J AOAC Int* 2007, **90**(3):738-755.

[37] Seo KH, Valentin-Bon IE, Brackett RE, Holt PS: **Rapid, specific detection of** *Salmonella* **Enteritidis in pooled eggs by real-time PCR.** *J Food Prot* 2004, **67**(5):864-869.

[38] Song JH, Cho H, Park MY, Na DS, Moon HB, Pai CH: **Detection of** *Salmonella* **typhi in the blood of patients with typhoid fever by polymerase chain reaction.** *J Clin Microbiol* 1993, **31**(6):1439-1443.

[39] Malorny B, Hoorfar J, Bunge C, Helmuth R: **Multicenter validation of the analytical accuracy of** *Salmonella* **PCR: towards an international standard.** *Appl Environ Microbiol* 2003, **69**(1):290-296.

[40] Cardona-Castro N, Restrepo-Pineda E, Correa-Ochoa M: **Detection of** *hilA* **gene sequences in serovars of** *Salmonella enterica* **subspecies** *enterica*. *Mem Inst Oswaldo Cruz* 2002, **97**(8): 1153-1156.

[41] Gwida MM, Al-Ashmawy MA: **Culture versus PCR for** *Salmonella* **species identification in some dairy products and dairy handlers with special concern to its zoonotic importance.** *Vet Med Int* 2014, **2014**:502370.

[42] Jenikova G, Pazlarova J, Demnerova K: **Detection of** *Salmonella* **in food samples by the combination of immunomagnetic separation and PCR assay.** *Int Microbiol* 2000, **3**(4): 225-229.

[43] Saeki EK, J. A, Bonfante RC, Hirooka EY, Moreira de Oliveira TCR: **Multiplex PCR (mPCR) for the detection of** *Salmonella* **spp. and the differentiation of the Typhimurium and Enteritidis serovars in chicken meat.** *J Food Saf* 2013, **33**:25-29.

[44] Shanmugasundaram M, Radhika M, Murali HS, Batra HV: **Detection of** *Salmonella enterica* **serovar Typhimurium by selective amplification of** *fliC, fljB, iroB, invA, rfbJ, STM2755, STM4497* **genes by polymerase chain reaction in a monoplex and multiplex format.** *World J Microbiol Biotechnol* 2009, **25**:1385-1394.

[45] Rodpai E, Moongkarndi P, Tungrugsasut W, Phosannoradej R, Kanarat S: **Comparison of multiplex polymerase chain reaction and immunoassay to detect** *Salmonella* **spp.,** *S.* **Typhimurium, and** *S.* **Enteritidis in Thai chicken meat.** *Science Asia* 2013, **39**:150-159.

[46] Agron PG, Walker RL, Kinde H, Sawyer SJ, Hayes DC, Wollard J, Andersen GL: **Identification by subtractive hybridization of sequences specific for** *Salmonella enterica* **serovar enteritidis.** *Appl Environ Microbiol* 2001, **67**(11): 4984-4991.

[47] Nadin-Davis S, Pope L, Ogunremi D, Brooks B, Devenish J: **A real-time PCR regimen for testing environmental samples for** *Salmonella enterica subsp. enterica* **serovars of concern to the poultry industry, with special focus on** *Salmonella* **Enteritidis.** *Can J Microbiol* 2019, **65**(2):162-173.

[48] Ogunremi D, Nadin-Davis S, Dupras AA, Marquez IG, Omidi K, Pope L, Devenish J, Burke T, Allain R, Leclair D: **Evaluation of a multiplex PCR assay for the identification of** *Salmonella* **serovars Enteritidis and Typhimurium using retail and abattoir samples.** *J Food Prot* 2017, **80**(2):295-301.

[49] Grimont PAD, Weill FX: **Antigenic formula of the** *Salmonella serovars*. In., 9 edn: World Health Organzation Collaborating Centre for Reference and Research on *Salmonella*; 2007: 1-166.

[50] Le Minor L, Rohde R: **Guideline for the preparation of** *Salmonella* **antisera,** 4 edn: World Health Organization Collaborating Centre for Reference and Research on *Salmonella*; 1989.

[51] Salmonella Subcommittee of the Nomenclature Committee of the International Society for Microbiology: **The Genus Salmonella Lignieres, 1900.** *J Hyg (Lond)* 1934, **34**(3):333-350.

[52] Kauffman F: **Űber einen neuen serologischen Formenweschsel der Typhusbacillen.** *Z Hyg Infeckt Kr* 1935, **116**:617.

[53] Schüte H: **The paratyphoid B group** *Lancet* 1920, **1**:93.

[54] Guibourdenche M, Roggentin P, Mikoleit M, Fields PI, Bockemuhl J, Grimont PA, Weill FX: **Supplement 2003-2007 (No. 47) to the White-Kauffmann-Le Minor scheme.** *Res Microbiol* 2010, **161**(1):26-29.

[55] Jackson BR, Griffin PM, Cole D, Walsh KA, Chai SJ: **Outbreak-associated** *Salmonella enterica* **serotypes and food Commodities, United States, 1998-2008.** *Emerg Infect Dis* 2013, **19**(8):1239-1244.

[56] Zheng J, Keys CE, Zhao S, Meng J, Brown EW: **Enhanced subtyping scheme for** *Salmonella* **Enteritidis.** *Emerg Infect Dis* 2007, **13**(12):1932-1935.

[57] Ogunremi D, Kelly H, Dupras AA, Belanger S, Devenish J: **Development of a new molecular subtyping tool for** *Salmonella enterica* **serovar Enteritidis based on single nucleotide polymorphism genotyping using PCR.** *J Clin Microbiol* 2014, **52**(12):4275-4285.

[58] Ward LR, de Sa JD, Rowe B: **A phage-typing scheme for** *Salmonella* **Enteritidis.** *Epidemiol Infect* 1987, **99**(2):291-294.

[59] Public Health Agency of Canada. **The National Enteric Surveillance Program (NESP) Annual summary for 2013.** National Microbiology Laboratory, Public Health Agency of Canada, Winnipeg, Manitoba, Canada; 2015.

[60] Rabsch W: *Salmonella* **Typhimurium phage typing for pathogens.** *Methods Mol Biol* 2007, **394**:177-211.

[61] Baggesen DL, Sorensen G, Nielsen EM, Wegener HC: **Phage typing of** *Salmonella* **Typhimurium - is it still a useful tool for surveillance and outbreak investigation?** *Euro Surveill* 2010, **15**(4):19471.

[62] Frost JA, Ward LR, Rowe B: **Acquisition of a drug resistance plasmid converts** *Salmonella* **Enteritidis phage type 4 to phage type 24.** *Epidemiol Infect* 1989, **103**(2):243-248.

[63] Threlfall EJ, Chart H, Ward LR, de Sa JD, Rowe B: **Interrelationships between strains of** *Salmonella* **Enteritidis belonging to phage types 4, 7, 7a, 8, 13, 13a, 23, 24 and 30.** *J Appl Bacteriol* 1993, **75**(1):43-48.

[64] Brown DJ, Baggesen DL, Platt DJ, Olsen JE: **Phage type conversion in** *Salmonella enterica* **serotype Enteritidis caused by the introduction of a resistance plasmid of incompatibility**

**group X (IncX)**. *Epidemiol Infect* 1999, **122**(1):19-22.

[65] Chart H, Row B, Threlfall EJ, Ward LR: **Conversion of** *Salmonella enteritidis* **phage type 4 to phage type 7 involves loss of lipopolysaccharide with concomitant loss of virulence**. *FEMS Microbiol Lett* 1989, **51**(1):37-40.

[66] van Belkum A, Tassios PT, Dijkshoorn L, Haeggman S, Cookson B, Fry NK, Fussing V, Green J, Feil E, Gerner-Smidt P *et al*: **Guidelines for the validation and application of typing methods for use in bacterial epidemiology**. *Clin Microbiol Infect* 2007, **13 Suppl 3**:1-46.

[67] Nadon CA, Trees E, Ng LK, Moller Nielsen E, Reimer A, Maxwell N, Kubota KA, Gerner-Smidt P, Group MHW: **Development and application of MLVA methods as a tool for inter-laboratory surveillance**. *Euro Surveill* 2013, **18**(35):20565.

[68] Bertrand S, De Lamine de Bex G, Wildemauwe C, Lunguya O, Phoba MF, Ley B, Jacobs J, Vanhoof R, Mattheus W: **Multi locus variable-number tandem repeat (MLVA) typing tools improved the surveillance of** *Salmonella* **Enteritidis: a 6 years retrospective study**. *PLoS One* 2015, **10**(2):e0117950.

[69] Boxrud D, Pederson-Gulrud K, Wotton J, Medus C, Lyszkowicz E, Besser J, Bartkus JM: **Comparison of multiple-locus variable-number tandem repeat analysis, pulsed-field gel electrophoresis, and phage typing for subtype analysis of** *Salmonella enterica* **serotype Enteritidis**. *J Clin Microbiol* 2007, **45**(2):536-543.

[70] Ziebell K, Chui L, King R, Johnson S, Boerlin P, Johnson RP: **Subtyping of Canadian isolates of** *Salmonella* **Enteritidis using Multiple Locus Variable Number Tandem Repeat Analysis (MLVA) alone and in combination with Pulsed-Field Gel Electrophoresis (PFGE) and phage typing**. *J Microbiol Methods* 2017, **139**:29-36.

[71] Nadin-Davis S, Pope L, Chmara J, Duceppe MO, Burke T, Devenish J, Andrievskaia O, Allain R, Ogunremi D: **An unusual** *Salmonella* **Enteritidis strain carrying a modified virulence plasmid lacking the** *prot6e* **gene represents a geographically widely distributed lineage**. *Front Microbiol* 2020, **11**:1322.

[72] Olson AB, Andrysiak AK, Tracz DM, Guard-Bouldin J, Demczuk W, Ng LK, Maki A, Jamieson F, Gilmour MW: **Limited genetic diversity in** *Salmonella enterica* **serovar Enteritidis PT13**. *BMC Microbiol* 2007, **7**:87.

[73] Guard J, Morales CA, Fedorka-Cray P, Gast RK: **Single nucleotide polymorphisms that differentiate two subpopulations of** *Salmonella enteritidis* **within phage type**. *BMC Res Notes* 2011, **4**:369.

[74] Leekitcharoenphon P, Nielsen EM, Kaas RS, Lund O, Aarestrup FM: **Evaluation of whole genome sequencing for outbreak detection of** *Salmonella enterica*. *PLoS One* 2014, **9**(2):e87991.

[75] Dallman T, Inns T, Jombart T, Ashton P, Loman N, Chatt C, Messelhaeusser U, Rabsch W, Simon S, Nikisins S *et al*: **Phylogenetic structure of European** *Salmonella* **Enteritidis outbreak correlates with national and international egg distribution network**. *Microb Genom* 2016, **2**(8):e000070.

[76] den Bakker HC, Allard MW, Bopp D, Brown EW, Fontana J, Iqbal Z, Kinney A, Limberger R, Musser KA, Shudt M *et al*: **Rapid whole-genome sequencing for surveillance of** *Salmonella enterica* **serovar Enteritidis**. *Emerg Infect Dis* 2014, **20**(8):1306-1314.

[77] Pornsukarom S, van Vliet AHM, Thakur S: **Whole genome sequencing analysis of multiple *Salmonella* serovars provides insights into phylogenetic relatedness, antimicrobial resistance, and virulence markers across humans, food animals and agriculture environmental sources.** *BMC Genomics* 2018, **19**(1):801.

[78] van den Berg RR, Dissel S, Rapallini M, van der Weijden CC, Wit B, Heymans R: **Characterization and whole genome sequencing of closely related multidrug-resistant *Salmonella enterica* serovar Heidelberg isolates from imported poultry meat in the Netherlands.** *PLoS One* 2019, **14**(7):e0219795.

[79] Moreno Switt AI, den Bakker HC, Cummings CA, Rodriguez-Rivera LD, Govoni G, Raneiri ML, Degoricija L, Brown S, Hoelzer K, Peters JE *et al*: **Identification and characterization of novel *Salmonella* mobile elements involved in the dissemination of genes linked to virulence and transmission.** *PLoS One* 2012, 7(7):e41247.

[80] Toro M, Retamal P, Ayers S, Barreto M, Allard M, Brown EW, Gonzalez-Escalona N: **Whole-genome sequencing analysis of *Salmonella enterica* serovar Enteritidis isolates in Chile provides insights into possible transmission between gulls, poultry, and humans.** *Appl Environ Microbiol* 2016, **82**(20):6223-6232.

[81] Ibrahim GM, Morin PM: ***Salmonella* serotyping using whole genome sequencing.** *Front Microbiol* 2018, **9**:2993.

[82] Yoshida CE, Kruczkiewicz P, Laing CR, Lingohr EJ, Gannon VP, Nash JH, Taboada EN: **The *Salmonella* In Silico Typing Resource (SISTR): An open web-accessible tool for rapidly typing and subtyping draft *Salmonella* genome assemblies.** *PLoS One* 2016, **11**(1):e0147101.

[83] Zhang S, Yin Y, Jones MB, Zhang Z, Deatherage Kaiser BL, Dinsmore BA, Fitzgerald C, Fields PI, Deng X: ***Salmonella* serotype determination utilizing high-throughput genome sequencing data.** *J Clin Microbiol* 2015, **53**(5):1685-1692.

[84] Franklin K, Lingohr EJ, Yoshida C, Anjum M, Bodrossy L, Clark CG, Kropinski AM, Karmali MA: **Rapid genoserotyping tool for classification of *Salmonella* serovars.** *J Clin Microbiol* 2011, **49**(8):2954-2965.

[85] Zhang S, den Bakker HC, Li S, Chen J, Dinsmore BA, Lane C, Lauer AC, Fields PI, Deng X: **SeqSero2: Rapid and improved *Salmonella* serotype determination using whole-genome sequencing data.** *Appl Environ Microbiol* 2019, **85**(23).

[86] Maiden MC, Bygraves JA, Feil E, Morelli G, Russell JE, Urwin R, Zhang Q, Zhou J, Zurth K, Caugant DA *et al*: **Multilocus sequence typing: a portable approach to the identification of clones within populations of pathogenic microorganisms.** *Proc Natl Acad Sci U S A* 1998, **95**(6):3140-3145.

[87] Zhou Z, Alikhan NF, Mohamed K, Fan Y, Agama Study G, Achtman M: **The EnteroBase user's guide, with case studies on *Salmonella* transmissions, *Yersinia pestis* phylogeny, and *Escherichia* core genomic diversity.** *Genome Res* 2020, **30**(1):138-152.

[88] Nadon C, Van Walle I, Gerner-Smidt P, Campos J, Chinen I, Concepcion-Acevedo J, Gilpin B, Smith AM, Man Kam K, Perez E *et al*: **PulseNet International: Vision for the implementation of whole genome sequencing (WGS) for global food-borne disease surveillance.** *Euro Surveill* 2017, **22**(23):30544.

[89] Burnett E, Ishida M, de Janon S, Naushad S, Duceppe MO, Gao R,

Jardim A, Chen JC, Tagg KA, Ogunremi D *et al*: **Whole-Genome Sequencing reveals the presence of the** *bla***CTX-M-65 gene in extended-spectrum β-Lactamase-producing and multi-drug-resistant clones of** *Salmonella* **serovar Infantis isolated from broiler chicken environments in the Galapagos islands**. *Antibiotics* 2021, **10**(3):267.

[90] Achtman M, Zhou Z, Alikhan NF, Tyne W, Parkhill J, Cormican M, Chiou CS, Torpdahl M, Litrup E, Prendergast DM *et al*: **Genomic diversity of** *Salmonella enterica* **- The UoWUCC 10K genomes project**. *Wellcome Open Res* 2020, **5**:223.

[91] Petkau A, Mabon P, Sieffert C, Knox NC, Cabral J, Iskander M, Iskander M, Weedmark K, Zaheer R, Katz LS *et al*: **SNVPhyl: a single nucleotide variant phylogenomics pipeline for microbial genomic epidemiology**. *Microb Genom* 2017, **3**(6):e000116.

[92] Dallman T, Ashton P, Schafer U, Jironkin A, Painset A, Shaaban S, Hartman H, Myers R, Underwood A, Jenkins C *et al*: **SnapperDB: a database solution for routine sequencing analysis of bacterial isolates**. *Bioinformatics* 2018, **34**(17):3028-3029.

[93] Treangen TJ, Ondov BD, Koren S, Phillippy AM: **The Harvest suite for rapid core-genome alignment and visualization of thousands of intraspecific microbial genomes**. *Genome Biol* 2014, **15**(11):524.

[94] Gardner SN, Slezak T, Hall BG: **kSNP3.0: SNP detection and phylogenetic analysis of genomes without genome alignment or reference genome**. *Bioinformatics* 2015, **31**(17):2877-2878.

[95] Kircher M, Kelso J: **High-throughput DNA sequencing-concepts and limitations**. *Bioessays* 2010, **32**(6):524-536.

[96] Xu Z, Hao B: **CVTree update: a newly designed phylogenetic study platform using composition vectors and whole genomes**. *Nucleic Acids Res* 2009, **37**(Web Server issue):W174-178.

[97] Allard MW, Luo Y, Strain E, Pettengill J, Timme R, Wang C, Li C, Keys CE, Zheng J, Stones R *et al*: **On the evolutionary history, population genetics and diversity among isolates of** *Salmonella* **Enteritidis PFGE pattern JEGX01.0004**. *PLoS One* 2013, **8**(1):e55254.

[98] Taylor AJ, Lappi V, Wolfgang WJ, Lapierre P, Palumbo MJ, Medus C, Boxrud D: **Characterization of foodborne outbreaks of** *Salmonella enterica* **serovar Enteritidis with Whole-Genome Sequencing single nucleotide polymorphism-based analysis for surveillance and outbreak detection**. *J Clin Microbiol* 2015, **53**(10):3334-3340.

[99] Feasey NA, Hadfield J, Keddy KH, Dallman TJ, Jacobs J, Deng X, Wigley P, Barquist L, Langridge GC, Feltwell T *et al*: **Distinct** *Salmonella* **Enteritidis lineages associated with enterocolitis in high-income settings and invasive disease in low-income settings**. *Nat Genet* 2016, **48**(10):1211-1217.

[100] Ogunremi D, Devenish J, Amoako K, Kelly H, Dupras AA, Belanger S, Wang LR: **High resolution assembly and characterization of genomes of Canadian isolates of** *Salmonella* **Enteritidis**. *BMC Genomics* 2014, **15**:713.

[101] Kumar S, Stecher G, Li M, Knyaz C, Tamura K: **MEGA X: Molecular Evolutionary Genetics Analysis across computing platforms**. *Mol Biol Evol* 2018, **35**(6):1547-1549.

Chapter 9

# *Salmonella enterica* subsp. *diarizonae* Serotype 61:k:1:5:(7) a Host Adapted to Sheep

*Inés Rubira, Luis Pedro Figueras, José Calasanz Jiménez, Marta Ruiz de Arcaute, Héctor Ruiz, José Antonio Ventura and Delia Lacasta*

## Abstract

*Salmonella* genus is widely distributed in nature and causes a spectrum of diseases in man and animals. *Salmonella enterica subsp. diarizonae* serotype 61: k: 1,5, (7) (SED) is a host adapted to sheep and its presence as saprophytic bacteria in sheep has been described in different countries. Several studies performed in abattoirs reported the presence of SED in healthy sheep in the intestinal content and also in the respiratory tract. In addition, this microorganism has also been isolated from nostril and faecal samples in healthy live animals. For this reason, this microorganism is considered well adapted to sheep, behaving normally these animals as asymptomatic carriers. However, SED has also been reported causing health disorders such as chronic proliferative rhinitis in adult sheep, abortions, testicular lesions in rams or alimentary tract disorders in young animals. The zoonotic potential of this microorganism is also discussed.

**Keywords:** sheep, *Salmonella enterica subsp. diarizonae* serotype 61: k: 1,5, (7), respiratory tract, host-adapted, chronic proliferative rhinitis, zoonosis

## 1. Introduction

*Salmonella* spp. is a gram-negative, facultative intracellular anaerobe bacterium. It is a rod-shaped bacterium belonging to the family *Enterobacteriaceae* [1, 2]. It is also a ubiquitous and hardy bacterium that can survive several weeks in a dry environment and several months in water. Most serotypes are present in a wide range of hosts [3]. *Salmonella* spp. is a primary pathogen that is distributed worldwide. It can be found in several locations and is responsible for important disorders in both animals and humans. Certain serotypes are particularly relevant due to their zoonotic potential or because they cause relevant economic losses [4].

In animals, salmonellosis is presented in four major forms, such as enteritis, septicaemia, abortion and asymptomatic carriage. In humans, salmonellosis includes several syndromes such as enteric fever, gastroenteritis, septicaemia, focal infections and, in the case of some typhoidal strains, an asymptomatic carrier state [2]. Salmonellosis is endemic in most countries and causes substantial economic losses [5, 6]. *Salmonella* infections in farm animals, and their transmission to humans, have a substantial economic and social impact [7].

Three hundred serovars have been described in *Salmonella enterica subsp. diarizonae* (IIIb) group that was firstly isolated from reptiles [7]. Most of these serovars are isolated from coldblooded animals, and some of them can also affect humans. Reptiles carry *Salmonella* spp. as part of their intestinal flora, and shed the microorganism intermittently through their faeces. Infected faeces can contaminate, directly or indirectly, humans [8, 9].

*Salmonella arizona* serotype 61:K:1,5,7 (currently *Salmonella enterica subsp. diarizonae* serotype 61:k:1,5, (7): SED) was firstly identified in sheep in 1952 from carcasses of newborn lambs [10] and *Salmonella arizona* 61:k:1,5,7 was isolated for the first time in abortifacient material from sheep in England and Wales in 1976 [11, 12]. In 1999, in the UK, this specific serovar was responsible for all the identified incidents in sheep [13]. In addition, from 1998, SED became the most common serovar isolated from sheep in England, and in 1999 represented 45.7% of the total *Salmonella* incidents [14, 15]. Likewise, SED, along with *Salmonella abortusovis,* are the *salmonella* microorganisms more often isolated from ovine in Spain [4]. Moreover, the detection of SED has been increased in recent years.

SED is considered host-adapted to sheep, and it displays a very different epidemiological pattern than does the sheep-restricted *Salmonella enterica* serovar *abortusovis*. SED is able to produce both intestinal and extraintestinal infections with faecal, vaginal, and nasal colonisation, but mostly without clinical disease. These properties deviate from the classical characteristic of ubiquitous serovars. Therefore, the term "sheep-associated serovar" appears to be more appropriate for characterising it [16]. Interestingly, a bacterium apparently "host-adapted" to the digestive tract of some reptiles jumped between species and found the respiratory tract of sheep as a location for its saprophytic existence, mainly when it is not a common pathogen of the respiratory system neither in humans or animals.

## 2. *Salmonella enterica subsp. diarizonae* serotype 61:k:1:5:(7) host-adapted to sheep

Although *Salmonella* spp. has been mostly related to digestive and reproductive disorders, SED is a microorganism well adapted to the respiratory tract of sheep. The traditional association of bacteria of the genus *Salmonella* with digestive disorders has meant that, for many years, SED was mainly sought in the digestive system and its possible location in the respiratory tract was neglected [17, 18].

Several works reported the presence of SED in the intestinal content of healthy sheep. Thus, SED has been isolated from this location in the United Kingdom [12, 14, 19], Norway [20, 21], Switzerland [22–24], Iceland [25], Sweden [26], Canada [27], the United States [28, 29] or Spain [18]. All these studies suggest that sheep are a reservoir for this microorganism, thus being considered as a saprophyte microorganism of this specie. As mentioned above, in almost all these studies only intestinal content was analysed, then the percentages of isolation found were normally low. It is described a 1% in the UK, 2% in Iceland, 17.6% in Sweden and 11–43% in Switzerland [26]. The samples collected and analysed in these studies were either stool from live animals or intestinal content or gut sections in studies carried out in abattoirs. However, for the past two decades, it has been proven that SED is a common microorganism of the respiratory tract of sheep. Bonke et al. analysed the presence of SED in tonsils and faeces of healthy sheep at the abattoir, and they found 43% of positive adult animals in tonsils, while only a 2% of the faeces samples were positive and only in young animals [23].

A recent study was performed by our research group to investigate the prevalence of SED in nostrils and stool of healthy live sheep in Spain [18]. The data collected in this study were analysed at two levels, animals and farms. The results showed that 45.3% of the animals were SED positive in nostrils or faeces, being the number of positive samples in nostrils higher than in faeces (38.5% vs. 22.5%). These data differ from those reported by Bonke et al. [23], that despite showing a high prevalence in tonsils all the adult animals analysed were negative in faeces. This was justified by the authors with poor conservation, and a small number of intestinal samples analysed. In our study, at farm level, nine of the ten analysed farms had at least one positive isolation of SED in one of the locations (nostrils or faeces). Further, all positive farms except one had SED isolations in both locations, nostrils and faeces, and in almost all positive farms sheep belonging to the youngest age ranges (0–2 and 2–4 years) accounted for more than 50% of positive isolates. The collective and individual prevalence in the studied region (Aragon, Spain) was estimated at 90% and 45.3%, respectively [18].

## 3. *Salmonella enterica subsp. diarizonae* serotype 61:k:1:5:(7) causing disease

The importance of this bacterium as a pathogen causing disease in sheep seems to have been increased in recent years, with the number of reports in international publications and conferences growing.

The first report of this microorganism as pathogen was related to abortifacient material from sheep of England and Wales in 1976 [12]. Several authors consider SED as a relevant abortive agent in sheep; however, a detailed analysis of different studies indicate that these bacteria appear along with other abortive microorganisms, suggesting a secondary role in these processes. Thus, Sojka et al. reported the presence of SED in nine abortion incidents, however in eight of them other abortifacient agents were also isolated, and on the ninth, SED was isolated in small numbers from placental cotyledons of only one of the two aborted lambs [12].

SED has also been associated with testicular lesions in rams [30, 31]. In both clinical descriptions, severe enlargement of the scrotal contents, fibrous adhesions between testicular layers, the coexistence of epididymal abscesses and testicular atrophy were described. This bacterium was isolated from the suppurative exudate in both cases, and the authors highlighted the importance of including this microorganism in the differential diagnosis of ovine genital infections.

Recently, SED was also associated with an outbreak of diarrhoea in lambs in Greece [32], where the presence of this microorganism was suggested as the cause of the digestive clinical signs in the lambs. Although SED is regularly isolated from faeces and tissues of the gastrointestinal tract from apparently healthy lambs and adult sheep [33], the pathogen has also been detected in faeces and tissues of lambs that had died from diarrhoea, thus gaining attention as a potential causative agent.

Although the previously mentioned incidents are infrequent health disorders associated with SED, chronic proliferative rhinitis (CPR) is a common disease that has been clearly related to this specific serovar of *Salmonella*. This disorder is precisely located in the upper tract of sheep where this microorganism is frequently isolated in healthy animals. CPR is a slow and progressive condition with an irreparable and poor prognosis for the untreated affected animals. It causes an inflammation of the ventral nasal turbinates causing very specific clinical signs that start with uni or bilateral thick seromucous nasal discharge together with snoring. These signs persist for several weeks or months and worsen, with almost complete nasal obstruction caused by the severe proliferation of the nasal mucosa of the turbinates in association with severe chronic inflammation, often visible at the nares [24, 34]. At this point, animals develop severe respiratory distress with striking mouth breathing. The inadequate flow of air provides a better situation for opportunistic bacteria, and secondary pulmonary diseases can also be found. The affected animals are early removed from the flocks either because of their death or because of their premature health condition deterioration [17]. At *post-mortem* examination, the ventral turbinates are shown swollen and have a roughened surface [17]. The section of the turbinate shows a proliferative tissue and affected animals frequently have nasal deformation and deviation of the nasal septum [24]. Histopathological evaluation reveals a thickened nasal mucosa with multiple polypoid projections. These polyps are covered by hyperplasic respiratory epithelium. Gram staining reveals the presence of numerous gram-negative bacilli within many epithelial cells, and *Samonella* immunohistochemistry reveals intracellular dot or rod formations inside proliferating epithelial cells and macrophages [17, 24, 29]. Although CPR generally produces a proliferative inflammation of the ventral nasal turbinates, recently it was described for the first time the affection of the dorsal turbinate and ethmoidal areas in an adult sheep [35].

In the prevalence study of SED carried out in Spain by our research group [18], a significantly higher percentage of isolates of SED was found in the flocks with previous cases of CPR than in those in which the disease had never been diagnosed. This could suggest that the infection pressure in the farm might favour the occurrence of clinical cases of the disease, since, as concluded in the experimental infection carried out in 2017 [36], the simple presence of the bacteria in the nasal secretions is not enough to trigger clinical signs of the disease. It seems that other factors, yet to be discovered, are necessary for SED to pass through the epithelial cells of the nostrils and elicit the inflammatory reaction. Further studies will be necessary to unravel why this saprophytic bacterium of the high respiratory tract is able to cross the epithelial barrier causing severe inflammation in some animals.

## 4. Zoonotic potential of *Salmonella enterica subsp. diarizonae* serotype 61:k:1:5:(7)

Salmonellosis represents one of the most important zoonosis [16]. Salmonellosis in humans is generally contracted through the consumption of contaminated food of animal origin (mainly eggs, meat, poultry, and milk), although other foods, including green vegetables contaminated by manure, have been implicated in its transmission. *Salmonella* bacteria are prevalent in food animals such as poultry, pigs, and cattle and can pass through the entire food chain from animal feed, primary production, and all the way to households or food-service establishments and institutions [3]. Person-to-person transmission can also happen through the faecal-oral route. Human cases also occur where individuals have contact with infected animals, including pets. These infected animals often do not show signs of disease [3].

*Salmonella enterica subsp. diarizonae* is frequently isolated from the environment, cold-blooded animals, sheep and humans. However, only a few studies describe the isolation of this serovar from invasive human infections [37]. The rising popularity of exotic reptile as pets has led to an increase in the number of reptile-associated salmonelloses (RAS), considering it as an emerging zoonosis in humans [38]. All the zoonotic cases that have been described associated with SED were in persons that had some contact with reptiles, mostly as pets. *Salmonella enterica subsp. diarizonae* is the serovar most commonly isolated in patients with RAS. Young children and immunocompromised people seem to be especially prone to infections with reptile-associated *Salmonella* and often experience severe clinical courses, as it was described in different studies. Gastroenteritis in a neonate was presented, in which a regular contact of her mother with several pet reptiles was confirmed. The isolated serovar isolated in this case was *S. enterica* subsp. *diarizonae* serotype 47:k:z35 [8]. Also, this serovar was reported on reptile-associated maxillary sinusitis in a Snake Handler in 2016 [38]. In the United States, more than one million cases of human *Salmonella* infection occur every year, and a great amount of these cases result from exposure to reptiles or amphibians [39]. In order to prevent RAS, the Centres for Disease Control and Prevention have recommended handwashing with soap and water after handling reptiles or reptile cages; these recommendations also stipulate that reptiles should not be kept near children and immunocompromised persons, and should not be allowed to roam freely throughout the home or living area [40, 41].

The zoonotic potential of *Salmonella enterica supsp. Diarizonae* serotype 61:k:1:5:(7) and the role of sheep in the transmission has been widely discussed [17, 26, 42], even though there have been no confirmed human cases associated with this specific serotype. In France, in 2008, a pseudo-outbreak in humans associated with SED was reported [43]. After a large number of SED positive samples from humans, trace-back investigations incriminated culture media containing contaminated sheep blood agar. None of the positive patients had suggestive symptoms of *Salmonella* infection. All samples had been taken during routine screening and SED was isolated from different body sites, including nine from usually sterile sites. The unusual clinical presentation and unusual serotype of *Salmonella* led to the suspicion that the origin of the contamination might be linked to the laboratory processing of the samples. After some investigations, they revealed that nine of the ten isolates had grown on sheep blood agar from the same manufacturer and that the batch number was similar for three cases. The manufacturer confirmed that the samples of the blood agar were positive to SED, owing to contaminated sheep blood [43].

On the other hand, and to emphasise that the discussion of zoonotic potential of this bacterium is still on the table, despite the high prevalence of SED in countries where other zoonotic salmonellas are under control, such as Sweden, Norway or Finland [18, 26], there have been no cases of human salmonellosis associated with this microorganism. As scientific opinions and evaluation of on-farm control measures performed in Sweden concluded that the impact of sheep associated *S. enterica subsp. diarizonae* on human health was very low, Swedish authorities decided to make an exemption for *S. enterica subsp. diarizonae* in sheep in the current *Salmonella* control measures and, in Norway, it was concluded that the impact of *S. enterica subsp. diarizonae* on human health appeared to be marginal [18, 26].

## 5. Conclusions

*Salmonella enterica supsp. Diarizonae* serotype 61:k:1:5:(7) is a host-adapted to sheep, being commonly isolated from upper respiratory tract of healthy sheep. This microorganism has been probably under-reported because traditionally it has been sought in the digestive tract of sheep and lambs when its more frequent isolation is from the upper respiratory tract. Some recent studies clarified the prevalence of SED in the nasal mucosa of healthy animals in different flocks.

Chronic proliferative rhinitis is an upper respiratory tract disorder clearly related to SED. However, the kinetics of the infection is not entirely understood, and further studies will be necessary to uncover why this saprophytic bacterium of the high respiratory tract is able to cross the epithelial barrier causing severe inflammation in some animals. In recent years, the number of CPR reports has been increased, what could mean that the knowledge and a proper description of the disease can lead to identifying new cases. Finally, it is essential to highlight that there are still some concerns about the zoonotic potential of this bacterium and the relevance of sheep as a reservoir of the infection.

## Author details

Inés Rubira[1], Luis Pedro Figueras[1,2], José Calasanz Jiménez[1], Marta Ruiz de Arcaute[2], Héctor Ruiz[2], José Antonio Ventura[2] and Delia Lacasta[2]*

1 Gabinete Técnico Veterinario S.L., C/ Isla Conejera sn, 50013 Zaragoza, Spain

2 Animal Pathology Department, Veterinary Faculty of Zaragoza, Instituto Agroalimentario de Aragón-IA2 (Universidad de Zaragoza-CITA), C/Miguel Servet 177, 50013 Zaragoza, Spain

*Address all correspondence to: dlacasta@unizar.es

## References

[1] Coburn B., Grassl G.A., Finlay B.B. 2007. *Salmonella*, the host and disease. Immunology & Cell Biology. 85, 112-118. DOI: https://doi.org/10.1038/sj.icb.7100007

[2] Agbaje M., Begum R.H., Oyecunle M.A., Ojo O.E., Adenubi O.T., 2011. Evolution of *Salmonella* nomenclature: a critical note. Folia Microbiologica (Praha) 56, 497-503. DOI: 10.1007/s12223-011-0075-4.

[3] World Health Organization. *Salmonella* (non-typhoidal) [Internet]. 2018. Available from: https://www.who.int/news-room/fact-sheets/detail/salmonella-(non-typhoidal) [Accessed: 2020/11/07]

[4] Figueras Ara L.P. Estudio de la prevalencia y transmisión de *Salmonella enterica* subsp. *diarizonae* serotipo 61:k:1,5,(7) en el ganado ovino y su implicación como agente etiológico de la rinitis crónica proliferativa. Estudios de campo e infección experimental en corderos. Zaragoza: Spain. doctoral thesis. Veterinary Faculty. University of Zaragoza, Veterinary Faculty. Universiti of Zaragoza. Spain. 2017.

[5] Lillehoj H, Okamura M. 2003. Host immunity and vaccine development to *Coccidia* and *Salmonella* infections in chickens. The Journal of Poultry Science. 40:151-193. DOI: 10.2141/jpsa.40.151.

[6] Center of Infectious Disease Research and Policy (CIDRAP). Salmonellosis. [Internet]. 2006. Available from: www.cidrap.umn.edu/cidrap/content/fs/food-disease/ causes/salmoview.html

[7] Old D.C. 1992. Nomenclature of *Salmonella*. Journal of Medical Microbiology. 37, 361-363.

[8] Hervás J. A., Rosell A., Hervás D., Rubio R., Dueñas J., Mena A. 2012. Reptile-pet *Salmonella enterica* Subspecies *diarizonae* Gastroenteritis in a Neonate. The Pediatric Infectious Disease Journal. Volume 31, Number 10. DOI: 10.1097/INF.0b013e318260530c.

[9] Braun S., Spalloni W., Ferreccio F., Postigo J., Fernández A., Porte L., Saldivia A., Wigant W., Triantafilo V. 2015. *Salmonella* spp. Gastroenteritis associated to pet turtles in three infants. Revista chilena de infectologia: organo oficial de la Sociedad Chilena de Infectologia 32(3):334-338. DOI: 10.4067/S0716-10182015000400013.

[10] Ryff J.F., Browne J. 1952. Paracolon abortion in ewes. Journal of the American Veterinary Medical Association 121, 266-267.

[11] Long J.R., Finley G.G., Clark M.H., Rehmtulla A.J. 1978. Ovine fetal infection due to *Salmonella arizonae*. The Canadian Veterinary Journal 19, 260-263. PMCID: PMC1789459.

[12] Sojka W.J., Wray C., Shreeve J.E., Bell J.C. 1983. The incidence of *Salmonella* infection in sheep in England and Wales, 1975 to 1981. British Veterinary Journal 139, 386-392. DOI: https://doi.org/10.1016/S0007-1935(17)30383-4.

[13] Anon. 1999. *Salmonella* Bacteria. Detection in Foods, No. 71. In 5th ed. Nordic Committee on Food Analysis (NMKL), Oslo 9.

[14] Davies R.H., Evans S.J., Preece B.E., Chappell S., Kidd S., Jones Y.E. 2001. Increase in *Salmonella enterica* subspecies *diarizonae* serovar 61:k:1,5,(7) in sheep. Veterinary Record 149, 555-557. DOI: 10.1136/vr.149.18.555.

[15] Milnes A.S., Stewart I., Clifton-Hadley F.A., Davies R.H., Newell D.G., Sayers A.R., Cheasty T., Cassar C., Ridley A., Cook A.J.C., Evans S.J.,

Teale C.J., Smith R.P., McNally A., Toszeghy M., Futter R., Kay A., Paiba G.A. 2008. Intestinal carriage of verocytotoxigenic *Escherichia coli* O157, *Salmonella*, thermophilic *Campylobacter* and *Yersinia enterocolitica*, in cattle, sheep and pigs at slaughter in Great Britain during 2003. Epidemiology & Infection 136, 739-751. DOI: 10.1017/S0950268807009223.

[16] Busch A., Tomaso H., Methner U. 2019. Whole-Genome Sequence of *Salmonella enterica subsp. diarizonae* Serovar 61:k:1,5,(7) Strain 1569 (14PM0011), Isolated from German Sheep. Microbiology Resource Announcements. American Society for Microbiology. 8(38): e00539-19. DOI: 10.1128/MRA.00539-19.

[17] Lacasta D., Ferrer L.M., Ramos J.J., Bueso J.P., Borobia M., Ruiz de Arcaute M., Figueras L., Gonzalez-Sainz J.M., De las Heras M. 2012. Chronic proliferative rhinitis associated with *Salmonella enterica* subspecies *diarizonae* serovar 61:k:1,5,(7) in sheep in Spain. Journal of Comparative Pathology 147, 406-409. DOI: 10.1016/j.jcpa.2012.03.004.

[18] Figueras L., Ferrer L.M., González J.M., Bueso J.P., Ramos J.J., Rubira I., Burian E., Lacasta D. 2020. Prevalence of *Salmonella enterica* subsp. *diarizonae* serotype 61:k:1:5:(7) in nasal secretions and stool of sheep flocks with and without cases of chronic proliferative rhinitis. Veterinary Microbiology, 247 (2020) 108767. DOI: 10.1016/j.vetmic.2020.108767.

[19] Hall M.L., Rowe B. 1992. *Salmonella arizonae* in the UK from 1966 to 1990. Epidemiology & Infection 108, 59-65. DOI: 10.1017/s0950268800049505.

[20] Alvseike O., Skjerve E. 2002. Prevalence of *Salmonella* subspecies *diarizonae* in Norwegian sheep herds. Preventive. Veterinary Medicine Journal 52, 277-285. DOI: 10.1016/S0167-5877(01)00252-5.

[21] Sandberg M., Alvseike O., Skjerve E. 2002. The prevalence and dynamics of *Salmonella enterica* IIIb 61:k:1,5,(7) in sheep flocks in Norway. Preventive Veterinary Medicine Journal 52, 267-275. DOI: 10.1016/S0167-5877(01)00250-1.

[22] Zweifel C., Zychowska M.A., Stephan R. 2004. Prevalence and characteristics of Shiga toxin-producing *Escherichia coli*, *Salmonella* spp. and *Campylobacter* spp. isolated from slaughtered sheep in Switzerland. International Journal of Food Microbiology 92, 45-53. DOI: 10.1016/j.ijfoodmicro.2003.07.005.

[23] Bonke R., Wacheck S., Bumann C., Thum C., Stueber E., Koenig M., Stephan R., Fredriksson-Ahomaa M. 2012. High prevalence of *Salmonella enterica* subsp *diarizonae* in tonsils of sheep at slaughter. Food Research International 45, 880-884. DOI: 10.1016/j.foodres.2011.01.050.

[24] Stokar-Regenscheit N., Overesch G., Giezendanner R., Roos S., Gurtner C. 2017. *Salmonella enterica* subsp. *diarizonae* serotype 61:k:1,5,(7) associated with chronic proliferative rhinitis and high nasal colonization rates in a flock of Texel sheep in Switzerland. Preventive Veterinary Medicine Journal 145, 78-82. DOI: 10.1016/j.prevetmed.2017.07.003.

[25] Hjartardottir S., Gunnarsson E., Sigvaldadottir J. 2002. *Salmonella* in sheep in Iceland. Acta Veterinaria Scandinavica 43, 43-48. DOI: 10.1186/1751-0147-43-43.

[26] Sörén K., Lindblad M., Jernberg C., Eriksson E., Melin L., Wahlstrom H., Lundh M. 2015. Changes in the risk management of *Salmonella enterica* subspecies *diarizonae* serovar 61:(k):1,5,(7) in Swedish sheep herds

and sheep meat due to the results of a prevalence study 2012. Acta Veterinaria Scandinavica 3 (57), 6. DOI: 10.1186/s13028-015-0096-0.

[27] Pritchard J., 1990. *Salmonella arizonae* in sheep. Can. Vet. J. 31 42-42.

[28] Dargatz D.A., Marshall K.L., Fedorka-Cray P.J., Erdman M.M., Kopral C.A. 2015. *Salmonella* prevalence and antimicrobial susceptibility from the national animal health monitoring system sheep 2011 study. Foodborne Pathogens and Disease 12, 953-957. DOI: 10.1089/fpd.2015.2016.

[29] Wolf C., Schefers J. 2017. Challenges Posed by a Flock Problem of *Salmonella diarizonae* Induced Proliferative Rhinitis. In: 9th International Sheep Veterinary Congress. Harrogate, United Kingdom.

[30] Ferreras Md.C., Muñoz M., Perez V., Benavides J., García-Pariente C., Fuertes M., Aduriz G., García-Marin J.F. 2007. Unilateral orchitis and epididymitis caused by *Salmonella enterica* subspecies *diarizonae* infection in a ram. Journal of Veterinary Diagnostic Investigation 19, 194-197. DOI: 10.1177/104063870701900211.

[31] Celeghini E.C.C., Gregory L., Pinheiro E.S., Piva F.M., Carneiro P.A.B., Parapinski-Santos B., Bianchi M., Benesi F.J. 2013. Orchiepididymitis in ram by *Salmonella enterica* sub *diarizonae*: first case in South America. Arquivo Brasileiro de Medicina Veterinária Zootecnia 65, 139-144. DOI: 10.1590/S0102-09352013000100021.

[32] Chatzopoulos D.C., Sarrou S., Vasileiou N.G.C., Ioannidi K.S., Peteinaki E., Valiakos G., Tsokana C.N., Papadopoulos E., Spyrou V., Mavrogianni V.S., Giannakopoulos A., Sbiraki A., Lacasta D., Bueso J.P., Athanasiou L.V., Billinis C., Fthenakis G.C. 2016. Dissemination of intestinal pathogens between lambs and puppies in sheep farms. Small Ruminant Research 141, 5-10. DOI: 10.1016/j.smallrumres.2016.06.006.

[33] Harp J.A., Myers L.L., Rich J.E., Gates N.L. 1981. Role of *Salmonella arizonae* and other infective agents in enteric disease of lambs. American Journal of Veterinary Research 42, 596-599.

[34] Rubira I., Figueras L., De las Heras M., Bueso J.P., Castells E., Climent M., Lacasta D. 2019. Chronic proliferative rhinitis in sheep: An update. Small Ruminant Research 179 (2019) 21-25. DOI: 10.1016/j.smallrumres.2019.09.001.

[35] López-Tamayo S., Rubira I., De las Heras M., Castells E., Lacasta D. 2020. Use of thermography for the diagnosis of chronic proliferative rhinitis in sheep and its application in the differential diagnosis of the rst case affecting the dorsal turbinate. Veterinary Record Case Report 2020;**8**:e001070. DOI: 10.1136/vetreccr-2020-001070

[36] Lacasta D., Figueras L., Bueso J.P., De las Heras M., Ramos J.J., Ferrer L.M., González J.M., Ruíz de Arcaute M., Ortín A., Marteles D., Navarro T., Fernandez A. 2017. Experimental infection with *Salmonella enterica* subsp. *diarizonae* serotype 61:k:1,5,(7) in sheep: study of cell mediated immune response. Small Ruminant Research 149, 28-33. DOI: https://doi.org/10.1016/j.smallrumres.2017.01.011.

[37] Giner-Lamia J., Vinuesa P., Betancor L., Silva C., Bisio J., Soleto L., Chabalgoity J.A., Puente J.L., The *Salmonella* CYTED Network, García-del Portillo F. 2019. Genome analysis of *Salmonella enterica* subsp. *diarizonae* isolates from invasive human infections reveals enrichment of virulence-related functions in lineage ST1256. BMC Genomics (2019) 20:99. DOI: https://doi.org/10.1186/s12864-018-5352-z.

[38] Horvath, L., Kraft, M., Fostiropaulos, K., Falkowski, A., Tarr, P.E., 2016. *Salmonella enterica* subspecies *diarizonae* maxilary sinusitis in a snake handler: first report. Open Forum Infectious Disease 3 (2). DOI: 10.1093/ofid/ofw066.

[39] Mermin J., Hoar B., Angulo F.J. 1997. Iguanas and *Salmonella marina* infection in children: a reflection of the increasing incidence of reptile-associated salmonellosis in the United States. Pediatrics 99:399-402. DOI: 10.1542/peds.99.3.399.

[40] U.S. Department of Health and Human Services. Centers for Disease Control and Prevention. *Salmonella* surveillance annual tabulation summary 1998. [Internet]. 1998. Available from: https://search.cdc.gov/search/?query=Salmonella%20surveillance%20annual%20tabulation%20summary%201998&sitelimit=www.cdc.gov%2Fmmwr%2F&dpage=2#results.

[41] Schröter M., Roggentin P., Hofmann J., Speicher A., Laufs R., Mack D. 2004. Pet Snakes as a Reservoir for *Salmonella enterica* subsp. *diarizonae* (Serogroup IIIb): a Prospective Study. APPLIED AND ENVIRONMENTAL MICROBIOLOGY, Jan. 2004, p. 613-615. DOI: 10.1128/AEM.70.1.613-615.2004

[42] Methner U., Moog U. 2018. Occurrence and characterisation of *Salmonella enterica* subspecies *diarizonae* serovar 61:k:1,5,(7) in sheep in the federal state of Thuringia, Germany. BMC Veterinary Research 14:401. DOI: 10.1186/s12917-018-1741-4.

[43] Thiolet J.M., Jourdan-Da Silva N., Reggiani A., De Valk H., Coignard B., Weill F.X. 2010. Nationwide pseudo-outbreak of *Salmonella enterica ssp. diarizonae*, France. Clinical Microbiology and Infection 17 (6) 915-918. DOI: https://doi.org/10.1111/j.1469-0691.2010.03343.x.

Section 3

# Alternative Methods for the Control of *Salmonella* in the Food Production Chain

Chapter 10

# Potential Roles for Bacteriophages in Reducing *Salmonella* from Poultry and Swine

*Anisha M. Thanki, Steve Hooton, Adriano M. Gigante, Robert J. Atterbury and Martha R.J. Clokie*

## Abstract

This chapter discusses application of natural parasites of bacteria, bacteriophages (phages), as a promising biological control for *Salmonella* in poultry and swine. Many studies have shown phages can be applied at different points from farm-to-fork, from pre to post slaughter, to control the spread of *Salmonella* in the food chain. Pre-slaughter applications include administering phages via oral gavage, in drinking water and in feed. Post slaughter applications include adding phages to carcasses and during packaging of meat products. The research discussed in this chapter demonstrate a set of promising data that relate to the ability of phages to reduce *Salmonella* colonisation and abundance. Collectively the studies support the viability of phage as antimicrobial prophylactics and therapeutics to prevent and control *Salmonella* in the food chain.

**Keywords:** Bacteriophages, phages, swine, poultry, delivery

## 1. Introduction

The global problem of antimicrobial resistance (AMR) is driving the search for novel treatments to control multidrug-resistant (MDR) pathogenic bacteria. Infections caused by MDR pathogens impose a significant burden on healthcare systems and economic productivity and are a major cause of mortality. Globally, AMR is associated with 700,000 deaths annually, with the prospect of this reaching 10,000,000 by 2050 if no resolution is found [1].

A One Health approach, that considers the intrinsic associations between antibiotic use in livestock and agriculture, the emergence of MDR pathogens, and the societal impact of AMR in developed and developing nations is required [2, 3]. However, integrating these approaches is challenging as antibiotic use in agriculture is generally widespread [4, 5]. For example, prophylactic administration of antibiotics to pigs during the weaning process is a standard technique employed in many countries [6]. Over recent years, efforts to limit antibiotic use other than specifically to control active bacterial infections have been implemented. Consequently, the use of antibiotics as growth promoters in food production animals was banned in the European Union (EU) in 2006 and in the United States of America (USA) in 2017 [7].

Gram negative *Enterobacteriaceae* are an important component of human, animal, and environmental microbiomes and can be associated with both health and disease. While the family contains several notorious pathogens (e.g. certain *E. coli*, *Klebsiella* spp., *Shigella* spp. *etc*.), the genus *Salmonella* presents a problem for AMR due to its ubiquitous distribution in food production environments and MDR phenotypes [8]. Worryingly, clinically important antibiotics are becoming ineffective, including colistin, which is a human critical antibiotic [9]. As such, alternative strategies to control/eliminate MDR *Salmonella* that may replace or complement antibiotics are needed.

Globally, dominant *Salmonella* serovars display a distribution pattern in pigs and poultry reflective of each industry. In pigs, *S*. Typhimurium (e.g. U288, U302, DT193, DT104), monophasic 4,[5],12:i:- and other variants such as 4,12:i:- are the dominant strains at both farm and slaughterhouse facilities in the UK and EU [10–13]. Other serovars such as *S*. Derby, *S*. Enteritidis, *S*. Bovismorbificans, *S*. Kedougou, *S*. Rissen, and *S*. Brandenburg are also reported [13, 14]. In the USA and China the dominant *Salmonella* serovars include *S*. Typhimurium, monophasic 4,[5],12:i:- *S*. Infantis, and *S*. Brandenburg [15].

For poultry, and in parallel with the global emergence of strains such as 4,[5],12:i:- the most prevalent serovar in UK production facilities is an *S*. Typhimurium derivative 13,23:i:- that accounted for almost a quarter of all isolations in 2019 [13]. Across the EU, the USA and China monophasic strains continue to expand throughout poultry production facilities. Other serovars such as *S*. Enteritidis, *S*. Berta, *S*. Typhimurium, *S*. Infantis, *S*. Hadar, *S*. Kentucky, and *S*. Heidelberg have all been isolated and/or linked to outbreaks [16–18]. The global diversity of *Salmonella* spp. within pig and poultry production constitutes a significant source of disease for humans and animals alike.

Controlling *Salmonella* requires intervention strategies capable of implementation at the national/international level. One such strategy is the targeted application of natural bacterial predators, bacteriophages (phages). Over the last decade, a robust body of evidence has demonstrated that phages can be applied at various points from farm-to-fork for pathogen control [19, 20]. Phage application could be implemented at the stage of rearing [21, 22], slaughter and processing [23], or at pre-retail/packaging [24, 25].

## 2. Phages

Phages are viruses that specifically infect and kill bacteria and with few reported side-effects in humans and animals. Phages are the most abundant biological entity on Earth, with estimated numbers ten times greater than bacterial cells [26]. Phages were independently discovered by Frederick Twort and Felix d'Herelle in 1915 and 1917 respectively. D'Herelle was the first to test phage efficacy in animals and showed phage treatment increased the survival of chickens suffering from fowl typhoid by 95–100% compared with 0–25% in untreated birds [27]. Despite this, phage therapy research slowed markedly following the discovery of antibiotics. However, research into phage therapy has been renewed since the emergence of AMR as it offers a promising alternative to antibiotics. Studies have shown phages are able to lyse MDR strains [28, 29] and there are multiple examples of successful phage therapy in humans [30] and animals [31]. Furthermore, phages can be applied to food to reduce bacterial loads and globally are being used commercially to improve food safety [32].

### 2.1 Phage morphology and infection cycle

Phages are characterised based on their virion morphology, genome type and sequence, and the infection cycle they follow. Phages are approximately a hundred times smaller than bacterial cells by volume, and generally only infect a subset of strains within a host species. Over 5,000 phages have been viewed under the transmission electron microscope (TEM) [33] and over 96% of phages studied are tailed phages and belong to the order *Caudovirales*. Siphoviruses, myoviruses and podoviruses are the most common phage types and constitute 61, 25 and 14% of all isolated tailed phages respectively (**Figure 1**) [34].

Phages are obligate parasites of bacteria as they lack the capacity to replicate independently. Phage replication occurs through either a lytic or lysogenic cycle (**Figure 2**). Phages following the lytic cycle attach to receptor(s) on the host cell surface using tail fibres, after which they inject their DNA and sequester the host's metabolic processes to produce more phage, eventually leading to cell lysis and release of the virions for further cycles of infection [35]. In comparison, during the lysogenic cycle phage DNA is incorporated into the bacterial cell and is replicated along with the host. Under certain conditions, e.g. stress and DNA damage, the phage can enter a lytic cycle as above. The lifestyle of the phage is determined via sequencing where the absence of recognisable integrases and other genes involved in the process of integration can be taken as indicative of a strictly lytic life cycle [36]. As lytic phages kill their target cells directly, they are preferred for therapeutic applications.

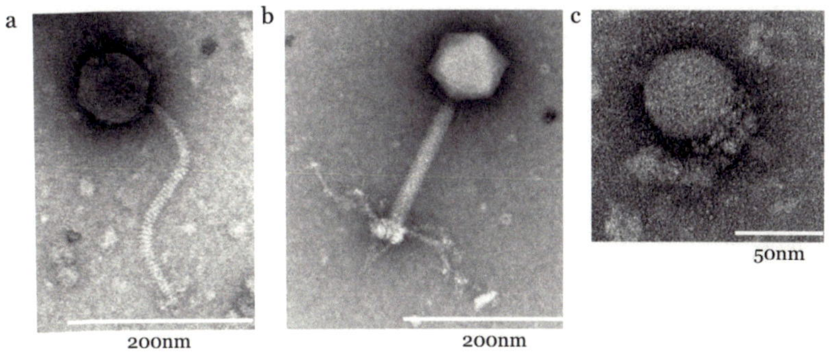


**Figure 1.**
*Morphology of tailed phages viewed under TEM. The images show the typical structure of a (a) siphovirus, (b) myovirus and (c) podovirus. TEM images were taken by the Electron Microscopy Facility at the University of Leicester.*

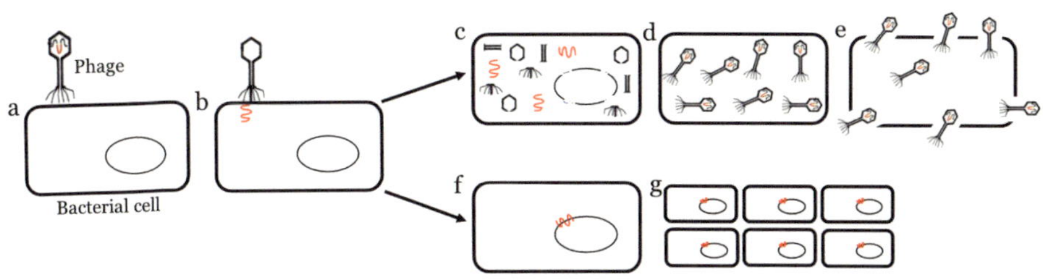


**Figure 2.**
*Phage lytic and lysogenic infection cycle. (a) phages attach to a receptor on the bacterial cell, after which (b) they inject their DNA (red line) into the cytoplasm of the cell. Phages can then go on to follow the lytic cycle (c-d) or the lysogenic cycle (f-g). In the lytic cycle (c) phages take over the host cells machinery to replicate their nucleic acids and proteins (d) to form new phage progeny. This (e) leads to lysis of the bacterial cell to release the phage progeny and the phages go on to infect more target bacterial cells. In the lysogenic cycle (f) phage DNA is integrated into the bacterial genome and (e) as the bacterial cells are replicated the prophage is replicated simultaneously.*

### 2.2 Phage isolation, host range and resistance

Phages can be isolated from any environment their hosts inhabit. *Salmonella*-specific phages have been isolated from faecal material obtained from pig and chicken farms, food processing plants, wild boar reserve [29], slurry lagoons [37], and sewage [22, 23]. Consequently, as phages are found in nature, humans and animals are continuously exposed to them, which is a major advantage in using them, as new entities would not be introduced into biological systems when phages are applied therapeutically [38].

The lytic spectrum (host range) of a phage is determined by screening against multiple strains of the target pathogen. Both narrow and broad host range phages have potential uses as therapeutics [39], for example a highly-specific, narrow host range phage can be applied with minimal perturbation to other residual microbial populations. Broad host-range phages provide a better scope of lysis and are therefore the desired components of most phage therapeutic applications. Multiple phages can be combined as a cocktail to improve phage coverage of the target species [40].

Emergence of resistance against therapeutic phages is a possibility as both phages and bacteria are in a continuous arms race. The mechanisms of phage resistance include altering the phage receptor, blocking phage DNA injection or inhibiting phage replication. This resistance can be countered by using cocktails of phage which bind to different receptors, as its unlikely resistance to all phage in the cocktail will emerge concurrently. Moreover, phage resistance can lead to a fitness cost for the bacterial cells [41]. Different multiplicities of infections (MOI's), which is ratio of phages to bacterial cells can also be trialed to limit resistance [42].

## 3. Experimental phage studies in chickens and pigs

In this section, the application of phages pre- and post-slaughter to reduce *Salmonella* numbers in chickens and pigs is discussed. Studies have varying levels of success in reducing *Salmonella* in challenge models, but with each study, valuable information is gained on phage dose, route of administration and resistance.

### 3.1 Experimental phage studies in chickens

#### *3.1.1* In vivo *phage studies in chickens at farm level*

One of the first studies that investigated phage therapy against *Salmonella* challenged chickens dates back to 1991 [43]. The authors orally challenged one day old Rhode Island Red chickens with *S.* Typhimurium ($10^8$ Colony Forming Units (CFU)) and 10 minutes later administered a single phage orally at dose $10^{12}$ Plaque Forming Units (PFU)/mL. The mortality of untreated chickens was 56% 21 days post-challenge but in chickens treated with phage mortality was reduced to 20%. The authors demonstrated phage transition and replication in the gut at sites of *Salmonella* colonization such as the crop, intestine and caecum. Similarly, Atterbury et al. [21] showed in two different broiler chicken studies phage treatment ($10^{11}$ PFU/mL) administered two days after challenge ($10^8$ CFU/mL), reduced ceacal colonisation by 4.2 and 2.2 $\log_{10}$ CFU/mL in birds challenged with Enteritidis P125109 or Typhimurium 4/74 respectively after 48 hours.

Goncalves and colleagues [44] compared the efficacy of three different phage cocktails in 45-day-old broiler chickens. The phages were administered at a dose of $10^9$ PFU/mL via oral gavage, 1 hour post challenge with *S.* Enteritidis at

$10^7$ CFU/mL. Two of the three phage cocktails reduced caecal *Salmonella* counts by ~2 $log_{10}$ CFU/mL in 12 hours, and *Salmonella* counts were below the detectable limit in the crop.

Toro et al. [45] designed a cocktail of three phages which could infect the top seven serotypes commonly associated with chickens. This cocktail was administered orally on days 4, 5, 6, 18, 19 and 20 at a dose of $5.4 \times 10^6$ PFU/bird and birds were challenged on day 7 with *S.* Typhimurium ($10^5$ CFU/mL). The phage treatment reduced *Salmonella* colonisation in the caeca by ten-fold, 4 days post-challenge, and 48 hours after treatment phages were isolated in the birds' faeces. Interestingly, the authors found phage treatment had a beneficial effect and chickens given the treatment gained more weight in comparison to challenged birds.

Delivering phages to chickens individually, via oral gavage, would be impractical commercially, however they could be administered easily through drinking water. Clavijo and colleagues [46] added a six-phage cocktail (named SalmoFREE®) at dose of $10^8$ PFU/mL to drinking water on days 18, 26 and 34 (chickens were slaughtered on day 35), which was sufficient to reduce ceacal *Salmonella* counts to below the detectable limit (below 100 CFU/mL). The trial was conducted at a commercial farm where there was a record of *Salmonella* outbreaks and included 34,680 broiler chickens. This is the biggest and the only trial to date evaluating phage efficacy against *Salmonella* in a commercial setting. There was no difference in mortality or productivity measurements between untreated control birds and those treated with phage only, suggesting the cocktail was safe. Furthermore, the authors conducted a microbiome study and showed phage treatment had no detrimental effect on the chicken's microbiota [47]. Their studies provide further valuable evidence into the effectiveness and safety of phage treatment.

Delivering phages as feed additives has been investigated. Sklar and Joerger. [48] added a single phage dose (A) and a three-phage cocktail (B) to starter broiler feed at a dose of $10^7$ PFU/g. The treated feed was available throughout the trial and chickens were challenged with *S.* Enteritidis at $10^4$ CFU on day 1. After 14 days phage A reduced caecal colonisation by 1.9 $log_{10}$ CFU/g and cocktail B by 0.6 $log_{10}$ CFU/g. The authors found that the process of mixing phage with feed and storing feed in bird rearing conditions over 14 days caused a 2 $log_{10}$ PFU/g reduction in phage numbers. Phage stability in feed could be a limitation and further research is needed to determine the impact storage conditions have on phage stability, such as factors as humidity and temperature.

### *3.1.2 Experimental post-slaughter phage studies in chickens*

Following processing and packaging, meat is refrigerated to avoid bacterial growth, but *Salmonella* can survive under these conditions and phages could be used to reduce surface contamination of *Salmonella*. Goode et al. [38] applied a single phage to chicken skin artificially contaminated with *S.* Enteritidis at $10^3$ CFU/cm$^3$. Phage applied at doses above $10^5$ PFU/mL reduced bacterial numbers by over 98% and phages amplified on the surface of the infected skin by three-fold over 48 hours. In comparison, in the uninfected samples the phage titre reduced by 1 $log_{10}$ PFU/cm$^3$, which suggests phages don't linger in absence of their target pathogen.

Atterbury et al. [49] showed phage treatment at dose $10^9$ PFU/mL reduced levels of *S.* Enteritidis and *S.* Typhimurium by 72.2% and 38.9% respectively on spiked chicken skin samples ($10^6$ CFU/ml). The authors confirmed phage infection was occurring on the surface of the chicken skin by spreading a bioluminescent *S.* Typhimurium strain on the surface of chicken skin and then monitored its growth using photon counting. Further studies have shown the efficacy of phage treatment to reduce *Salmonella* numbers on chicken skins are comparable to the typical chemical agents used by the food industry [50]. In addition, combining phage and chemical treatment was able to further decrease *Salmonella* counts to below detection levels [51].

To date only one study has investigated phage application on whole carcasses. Higgins et al. [52] spiked chicken carcasses with *S*. Enteritidis at 20 CFU, after which carcasses were sprayed with phage at different doses. The authors found only high phage doses of $10^8$ and $10^{10}$ PFU/ml were effective and after 24 hours, *Salmonella* was only isolated from one out of fifteen carcasses. The phage counts were not monitored in the study, therefore it's unclear if there was phage amplification.

Phage treatment of raw meat samples has been shown to be effective at reducing bacterial load and consequently reducing its presence in the final consumer product. Duc et al. [53] tested the lytic activity of a five-phage cocktail at dose $10^9$ PFU on chicken breasts inoculated with either *S*. Enteritidis or *S*. Typhimurium at $10^5$ CFU. The phage cocktail reduced counts of both strains by ~1.6 $\log_{10}$ CFU/piece of chicken breast, when stored at 8°C, over 24 hours. However, when the meat was stored at 25°C phage treatment was more effective and reduced *S*. Enteritidis or *S*. Typhimurium by 3.1 and 2.2 $\log_{10}$ CFU/piece respectively over 24 hours. This could suggest phage activity is temperature dependent. However, another study showed phage activity was unaltered when spiked chicken breasts ($10^5$ CFU/ml) were treated with phage at doses $10^6$ and $10^7$ PFU/mL and stored at 4°C and 25°C. Under both conditions, phage treatment reduced bacterial counts to undetectable levels after just 12 hours [54]. The studies suggest phage temperature stability can vary between phages and its stability needs to be tested to determine which are more effective at food storage temperatures.

## 3.2 Experimental phage studies in pigs

### *3.2.1 Phage therapy in pre-market and market-weight pigs*

Very few studies have examined the efficacy of phage treatments to control *Salmonella* in live pigs and this is largely due to the inherent difficulties of performing longitudinal studies from piglets to finished pigs. One pioneering study did exactly that and the efficacy of a fifteen-phage cocktail were tested in challenged piglets and market-weight pigs [22]. In the first study, the phage cocktail ($10^9$ PFU/mL) and challenge strain *S*. Typhimurium γ4232 ($5 \times 10^8$ CFU/pig) were co-administered via oral gavage to piglets. Piglets were euthanised 6 hours post-inoculation in order to mimic the amount of time spent in a holding pen. Overall, the activity of the phage cocktail was sufficient to achieve 2–3 $\log_{10}$ CFU (~99%) reductions in the ileum, tonsils and caecum. In collected ileum and caecal samples, in five out of six phage-treated pigs *S*. Typhimurium counts were reduced to below the limits of detection (~100 CFU/mL).

The authors next assessed the efficacy of the phage cocktail in market-weight pigs. Four pigs (in three replicates) were inoculated via oral gavage with $5 \times 10^9$ CFU *S*. Typhimurium and allowed to contaminate a holding pen for a period of 48 hours. Following this, sixteen naïve pigs (non-*Salmonella* infected – eight phage-treated/eight mock treatments controls) were introduced to the holding pens and allowed to co-mingle with the seeder pigs for 6 hours. Phage cocktail administration involved an initial oral gavage of $10^9$ PFU/mL followed by further identical doses every 2 hours for a total of 6 hours. After 6 hours of co-mingling between *S*. Typhimurium γ4232-infected, phage cocktail-treated, and mock control-treated pigs, each cohort was euthanised. In phage treated pigs there was 1 to 1.5 $\log_{10}$ CFU/mL reductions in *Salmonella* colonisation in ceacal and ileal samples. The role phages can play in controlling *Salmonella* infection in pigs at a critical stage of the production process is evident from the work performed by Wall et al. [22].

A similar degree of efficacy was observed when applying a microencapsulated phage cocktail treatment to control shedding of *S.* Typhimurium during a holding period of 6 hours [55]. Saez et al. found that shedding of *Salmonella* from pigs in the phage-treated group (PT) was less common than non-phage treated pigs (nPT) at 2 hours (% pigs shedding PT-38.1%, nPT-71.4%) and 4 hours (PT-42,9% - nPT-81.1%). Sampling of caecal and ileal contents 6 hours post-infection showed that phage-treated pigs had significantly less *S.* Typhimurium levels at both anatomical sites by 1 $\log_{10}$ CFU/mL. Another study produced some promising results by showing how dietary supplementation with probiotics (*Saccharomyces cerevisiae*, *Lactobacillus acidophilus*, and *Bacillus subtilis*) and phages can positively influence growth performance of pigs. A phage cocktail (~$10^9$ PFU/g) designed to target a diverse selection of bacteria (*S.* Typhimurium, *S.* Enteritidis, *S.* Choleraesuis, *S.* Derby, *Staphylococcus aureus*, *Escherichia coli*, and *Clostridium perfringens* types A and C) was administered as part of a feed supplement. Interestingly, the addition of phage was found to be more effective than probiotics. Phages may therefore offer an attractive alternative to replace the use of antibiotics as growth promoters in pigs [56].

#### *3.2.2 Phage decontamination of pigskin*

Post-slaughter application of phages has the potential to reduce risks associated with pork contaminated with *Salmonella* prior to general retail. An investigation into the stability of phages at retail temperatures (fresh 4°C and frozen −20°C) and also their ability to control the endemic UK pig pathogen *S.* Typhimurium U288 was examined [23]. Hooton et al. tested killing activity of *Salmonella*-specific phages against a diverse panel of *Salmonella* serovars prior to formulation as a four phage cocktail (PC1). PC1 consisted of three novel *Salmonella* phage isolates (ΦSH17, ΦSH18, and ΦSH19) combined with the broad-host range *Salmonella* phage Felix 01 in equal volumes/titres for a final concentration of $10^8$ PFU/mL. Initially it was shown that both *S.* Typhimurium U288 and the phage components of PC1 are both stable on experimentally-contaminated pigskin pieces stored at temperatures reflective of those at retail. The efficacy of PC1 was subsequently tested on spiked pigskin over a five-day trial under fresh conditions (4°C). A 3 × 3 matrix of CFU ($10^6$, $10^4$, and $10^3$) versus PC1 PFU ($10^7$, $10^5$, and $10^4$) was used to examine a range of MOIs (0.01–10,000) to determine the most effective combination.

The phage cocktail applied at MOI's of 1000 ($10^7$ PC1 V $10^4$ U288) and 10 ($10^5$ PC1 v $10^4$ U288) reduced *S.* Typhimurium U288 levels by ~92% after 1 hour post challenge. After 48 hours *Salmonella* counts were significantly reduced by ~1.4 $\log_{10}$ CFU/4 $cm^2$. The first reductions of *S.* Typhimurium U288 below the limits of detection were also reported at the 48 hour timepoint, specifically when an MOI of 10 was employed against low level contamination. At 96 hours post-inoculation it was evident that MOIs in excess of the target bacterium could reduce low-level bacterial contamination to below the limits of detection. The results reported here indicate that phages may provide useful tools for the post-harvest reduction of *S.* Typhimurium U288 on pork products [23].

## 4. Commercial phage products

A handful of phage products that target *Salmonella* in pre- and post- slaughter stages of the food chain are commercially available and summarized in **Table 1**.

| Product name and developers | Phages | Approval | Notes | Reference(s) |
|---|---|---|---|---|
| GPI Biotech VAM-S Gum Products International, Inc. (Newmarket, Canada) | Three lytic phages: Phi_16, Phi_78, Phi_87 | FDA approved, GRAS (GRN000917 Sep 2010) | • For use on poultry, red meat, eggs, fruits, vegetables, fish and shellfish | https://www.fda.gov/grasnoticeinventory |
| Bafasal Proteon Pharmaceuticals (Poland) | Four lytic phages: 3sent1, 8sent65, 8sent1748 and 5sent1 | EURL approved (FAD-2017-0039 - CRL/170007 Oct 2018) | • Approved as feed additive<br>• Awaiting approval in the EU | [57] https://ec.europa.eu/jrc/sites/jrcsh/files/finrep-fad-2017-0039-bafasal.pdf |
| Biotector® SCJ CheilJedang Research Institute of Biotechnology (South Korea) | Non-disclosed | Patented | • On feed to control *Salmonella* in poultry | [20] |
| SalmoFresh™ Intralytix Inc. (USA) | Six lytic phages | FDA-approved, granted GRAS status (GRN000435) Feb, 2013<br>Approved by Israel Ministry of Health; Health Canada | • Food treatment<br>• Effective against over 900 *Salmonella* strains representing more than 50 serotypes. | https://www.cfsanappsexternal.fda.gov/scripts/fdcc/?set=GRASNotices&id=435&sort=GRN_No&order=DESC&startrow=1&type=basic&search=INTRALYTIX<br>[58] |
| SalmoPro® Phagelux (Canada) | Two lytic phages: BP-63, BP-12 | FDA-approved, granted GRAS status (752) Jul 2018 | • For use as an antimicrobial processing aid to control *Salmonella* on food, when applied onto food surfaces | https://www.fda.gov/grasnoticeinventory [58] |
| Salmonelex™ (Former PhageGuard) Micreos Food Safety BV (The Netherland) | Two lytic phages: Fo1a and S16 | FDA-approved, granted GRAS status (GRN000630) | • For use as an antimicrobial on foodstuffs to control *Salmonella*<br>• Can be sprayed topically or added to chill tank water | Micreos Food Safety BV. Salmonelex™. Available online: https://www.fda.gov/media/98485/download<br>[58] |
| SalmoFREE® Sciphage (Colombia) | Six lytic phages | Patented only (patent number WO2017089947A2) | • For therapy and control *Salmonella* in poultry frm<br>• Phages are added to water | [46, 47] |

**Table 1.**
*Patented and approved* Salmonella *phage products.*

Some products (SalmoFresh® and SalmoPro®) have already obtained clearance from specific regulatory agencies, such as FDA, and are available to purchase, while others are patented but not approved by any regulatory authority, at the time of writing. However relevant scientific data about the product has been published, such as for SalmoFree® [20].

## 5. Phage/antibiotic synergy (PAS)

The efficacy of combinatorial medicinal treatments is well-documented and have proven successful in treating a range of human diseases such as cancer, HIV, and malaria [59]. Similarly, the use of phages and antibiotics synergistically (PAS) has been explored and experimental studies have shown using phages and antibiotics in combination could enhance bacterial suppression and lower emergence of bacterial resistance. Furthermore, a combined approach can lead to re-establishment of antibiotic sensitivity, for example in cases where phages bind to bacterial drug efflux pumps [60]. *In vitro* studies have investigated PAS activity for the control of *S.* Typhimurium with the well-studied phage P22 and antibiotics ceftriaxone and ciprofloxacin. The study found pre-treatment of *S.* Typhimurium with phage P22 prior to antibiotic addition was the most effective approach in comparison to treating with phages 6 hours after antibiotic treatment [61]. The timing and order of phage and antibiotics needs to be considered as it can influence PAS activity. It was also reported that the presence of antibiotics did not negatively influence phage binding to *Salmonella* cells, and a significant increase in phage lytic activity was observed [62].

To date, no *in vivo* PAS studies have been conducted in *Salmonella* challenged chickens and pigs. Therefore, further *in vivo* work is required for the underlying dynamics of PAS to be understood and developed into useful combinatorial therapies. Within the context of phage therapeutics in agricultural settings (and potentially in the clinic) PAS may well provide an exciting route of research for development into a parallel treatment with antibiotics. The emergence of resistance from the target bacterium to both antibiotics and phage treatment, choice of antibiotics and phage combinations, and potential efficacy-improving interactions with immunological responses will be important factors for consideration [59].

## 6. Potential challenges of using phages in poultry and pigs

The use of phages against *Salmonella* in farming, either pre or post-slaughter, have some challenges. Some of those difficulties are common to phage therapy in general and fall in to four categories initial phage selection; phage delivery; resistance development; and regulatory approval.

### 6.1 Phage selection

On the initial phage selection, potential phage candidates need to be virulent and propagate via the lytic cycle as opposed to temperate (can propagate via lysogenic or lytic cycle), which need to be confirmed by sequencing. This is to ensure the phage will not integrate on host genome avoiding transduction and horizontal gene transfer [21, 63, 64].

### 6.2 Phage delivery

The topic of phage therapy pharmacodynamics and pharmacokinetics is complex and more specific reviews have recently been published on this topic [65, 66]. In brief, phages need to reach the site of bacterial colonisation, and in poultry and pigs, *Salmonella* initially colonises the gut. Many studies have been designed to establish if phages can be delivered to the gastrointestinal tract and beyond via oral administration, either in feed or drinking water. For post-slaughter application, phage preparations can be applied by directly applying to carcasses, meat, skin, packaging materials as well as surfaces in the abattoir or meat processing facilities. Both pre- and post-slaughter applications present challenges to phage delivery [36].

A particular challenge to phages delivered orally to control *Salmonella* is to ensure they will be active in the gut pH, despite the fact that they are sensitive biological entities and will encounter changes in pH (**Figure 3**) and temperature. Phages are typically stable between pH 4 and pH 10 [67]. However, the studies discussed in Section 3 highlight natural phages retain lytic activity through the passage of the gut and do reach the focal point of infection.

Alternative solutions have been developed to protect phages from the acidic conditions by using dry or liquid formulation solutions. For example, it was shown that Felix O1 microencapsulation in chitosan-alginate microspheres could fully preserve phage viability upon 1 hour exposure to simulated gastric fluid (pH = 2.4 with 3.2 mg $ml^{-1}$ pepsin) and 3 hour exposure to 2% (wt/vol) porcine bile extract [68]. Other studies have shown that liposome-encapsulated phages (UAB_Phi20, UAB_Phi78, and UAB_Phi87) were significantly more stable in simulated gastric fluid (pH = 2.8 with 3.0 mg ml – 1 pepsin) when compared to free phages *in vitro* while the preparation was stable at 4°C for at least 3 months [69]. These data show that the challenge of gut pH range that the phage has to endure, when administered orally, can be overcome by selecting phages that remain viable and withstand wide pH variations or, in addition or as an alternative, shield the phages by means of pH resistant pharmaceutical formulation development.

### 6.3 Overcoming phage resistance

Phages are no different from other antimicrobials that are used to kill bacteria, and can become resistant to them following exposure. Often, the use of phage cocktails and rotation schedules is used to limit or avoid the development of resistant mutants. When phages are used post-slaughter as disinfectants, they can be deployed at a high titer, to reduce the build up of phage-resistant bacteria [70]. In order to reduce the accumulation of phages on surfaces after their intended use in

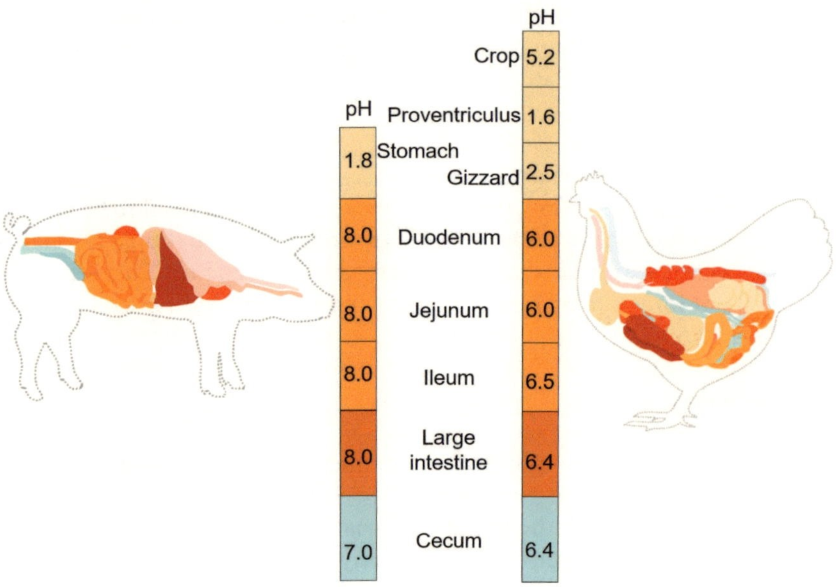


**Figure 3.**
*The gastrointestinal pH changes in the gut of pigs and chickens.*

the food industry, some disinfectants were tested and proved to be successful at neutralizing phages, such as peracetic acid [71].

When using phages within farm settings the challenge of phage persistence, spread and resistance development is more significant because successive animals will be housed in the same facilities and disinfection must be thorough. A recent study using a patented six phage cocktail against *Salmonella* (SalmoFREE) showed that after the first trial, SalmoFREE phages were detected from the beginning of the second trial in treated and control groups houses, showing that even after the cleaning and disinfection process, phages persisted in the environment and survived between trials. As a consequence, birds in the second trial (control and treatment group) showed unexpected reduction of *Salmonella* counts even before treatment/placebo administration at day 17 [46].

Effective disinfection practices and phage cocktail rotation may be the solution to the phage persistence challenge, however the effect of disinfectants should be tested on a case-by-case basis to determine their efficacy at neutralizing the respective phage(s) cocktail [72].

### 6.4 Regulating phage products

As discussed above, phages may be used as a feed additive to prevent or treat infection, as a medicine to treat infection or as a post product treatment for carcasses or meat. Phages could also be used to decontaminate either the environment that the animals are living in, or facilities regarding production of the final product. Phages go through specific regulatory pathways depending on which of these intervention points that they are used in, and on the level of claims associated with their use. To take a product to market requires the developer to know which regulatory route they will take, in order to gather appropriate data on safety and efficacy [73].

In recent years there has been a significant amount of engagement from regulating bodies, who are also acutely aware of the need to find novel antimicrobials. They are also aware that this is often seen as a major hurdle to developing the technology and are keen to help. It is important to state that by working with regulators there is an opportunity to impose a regulatory system that will allow the exploration of this technology whilst hopefully mitigating against many of the mistakes that we have previously made in terms of overusing antibiotics from the outset. Antibiotic stewardship was largely implemented after extensive bacterial resistance to antibiotics had already been achieved however sensible regulation could work hand in hand with a stewardship program to maintain effective phage use for future generations [74].

Establishing how phages fit into traditional drug/veterinary medical product regulatory systems is not always trivial. In the USA, phages are regulated by the Food and Drug Administration (FDA), regardless as to whether they are to be used in humans or animals, although they go through different parts of this depending on exactly how they are being used. Interestingly the FDA regulates phages in the same way regardless of whether they are 'natural' or engineered [73].

In Europe phages are currently regulated by the European Commission through the European Medicines Agency. Unlike the system in the USA, if phages are genetically altered, they are regulated differently. In the UK if phages are to be used within animals, they are regulated by the Veterinary Medical Directorate but if their end use is in humans they are regulated by the Medicines and Healthcare products Regulatory Agency and in food they are regulated by the Food Standards Agency. The different regulatory authorities do communicate with each other to identify commonalities and routes forward. Clearly there are parallels with other biologicals such as monoclonal antibodies, which will inform how phages are effectively regulated [73].

## 7. Future work: machine learning tools

Phage characterisation based on host range analysis, studying phage host interactions, phage infection kinetics and designing phage cocktails is resource-intensive. Machine learning (ML) based tools can be developed to predict these interactions, and the application of computational biology, artificial intelligence (AI) and modelling in phage research is rapidly developing [75]. The combination of these techniques with high-throughput Next Generation Sequencing promises greater insights into phage biology alongside the development of new tools to address previously intractable problems in phage therapy [76]. Computational tools applied to phage research are based on:

1. Homology-based methods: comparing the features (e.g. DNA/RNA/protein sequences) of an unknown phage with comparable information from databases of known phage. Examples include HostPhinder [77], VirHostMatcher [78] and ILMF-VH [79].

2. Machine Learning (ML) methods: these use combinations of algorithms and statistical techniques such as logistic regression and support vectors to find patterns in large datasets which are then used to make predictions [80].

3. Deep Learning (DL) methods: a subset of machine learning in which the key features used for pattern recognition and classification are identified by the computer algorithm directly and do not require human input [81].

A key aim of these approaches, as applied to phage therapy, is to facilitate or automate the matching of phages to target bacterial pathogens. This would revolutionize the field as it would reduce or eliminate the need for extensive host range profiling in the laboratory and would allow the rapid countering of resistance.

Homology-based methods have been used more extensively than ML so far, but more for the identification and annotation of phage DNA from metagenomic data than for phage host matching. Homology-based approaches have used genomic similarity (e.g. HostPhinder [77]), oligonucleotide frequency (e.g. VirHostMatcher [77]), and phage abundance profiling [82]. However, the success of these methods varies widely, with correct identification of the host to genus or species level only occurring between 9.5% and 75% of the time.

Phage host matching using ML has also met with varied success. Approaches include using chemical parameters of all phage and host proteins [80], or focusing on a subset of these, such as receptor binding proteins [75], which have accurately predicted phage hosts 30 to 90% of the time. Relatively few studies have used DL methods. As with the homology-based methods above, some studies have focused on the use of DL to identify and separate phage sequences from metagenomic data. DL was used by Li et al. [79] to accurately to match phage and host species 81% of the time using 27 features of phage and host proteins.

A disadvantage of ML and DL is that large datasets are required, and these are often skewed heavily towards phage which infect a small number of well-studied bacteria. For example, in one study approximately 86% of phage used in the ML model infected a single species (*M. smegmatis*). Moreover, DL methods are not readily interpretable and regarded as 'black boxes' due to the lack of human involvement in feature selection and application. Additionally, even the best performing ML and DL models are currently unable to predict phage hosts at the strain level, which will a necessary step in real-world therapeutic applications.

Phage host matching is likely to be more useful when using phage therapy for highly diverse pathogens, such as *Salmonella* and *E. coli*, than for more homogenous bacteria such as *Staphylococcus aureus*. ML and DL have the potential to automate the process of phage selection of their predictions are shown to be reliable, and potentially in the future could help design personalised phage therapeutics for human and agricultural use.

## 8. Conclusions

Phages could provide a natural alternative to traditional antimicrobial therapies in pig and poultry production. Multiple intervention points exist from farm-to-fork allowing for the development of targeted phage therapeutic strategies. The promising results obtained from diverse experimental approaches demonstrate the potential of phages to reduce *Salmonella* in live animals, as well as in finished retail products. With correct stewardship, phages may well become an integrated solution in livestock production especially within the remit of controlling significant pathogens such as *Salmonella*. While some products have made it to market, current legislation needs further development prior to widespread acceptance of phage therapeutics in animals and on retail products. The next generation of phage research is set to take advantage of developments in the fields of machine-based learning and other computationally oriented approaches. Such exciting techniques may offer a more refined approach towards the application of phages for elimination of *Salmonella* from pig and poultry production.

## Acknowledgements

Dr. Steve Hooton and Dr. Adriano M. Gigante are funded by the Biotechnology and Biological Sciences Research Council (BBSRC – BB/T006196/1) and we thank BBSRC for providing the funding for this work.

## Conflict of interest

The authors declare no conflict of interest.

## Author details

Anisha M. Thanki[1], Steve Hooton[1], Adriano M. Gigante[2], Robert J. Atterbury[2] and Martha R.J. Clokie[1*]

1 Department of Genetics and Genome Biology, University of Leicester, Leicester, UK

2 School of Veterinary Medicine and Science, University of Nottingham, Leicestershire, UK

*Address all correspondence to: mrjc1@leicester.ac.uk

## References

[1] Neill J. Tackling drug-resistant infections globally: final report and recommendations. 2016. Avaliable from: https://amr-review.org/sites/default/files/160518_Final%20paper_with%20cover.pdf

[2] Walsh TR. A one-health approach to antimicrobial resistance. Nature Microbiology. 2018;3:8:854-855. DOI:10.1038/s41564-018-0208-5

[3] Hernando-Amado S, Coque TM, Baquero F, and Martínez JL. Defining and combating antibiotic resistance from One Health and Global Health perspectives. Nature Microbiology. 2019;4:1432-1442. DOI:10.1038/s41564-019-0503-9

[4] Robinson TP, Bu DP, Carrique-Mas J, Fevre, EM, Gilbert M, Grace D, Hay SI, Jiwakanon J, Makkar M, Kariuki S, Laxminarayan R, Lubroth J, Magnusson U, Thi Ngoc P, Van Boeckel TP, Woolhouse MEJ. Antibiotic resistance: Mitigation opportunities in livestock sector development. Animal. 2017;11:1-3. DOI: 10.1017/S1751731116001828

[5] Patel SJ, Wellington M, Shah RM, Ferreira MJ. Antibiotic stewardship in food-producing animals: challenges, progress, and opportunities. Clinical Theraupectics. 2020;42:1649-1658. DOI: 10.1016/j.clinthera.2020.07.004.

[6] Diana A, Boyle LA, Leomard FC, Carroll C, Sheehan E, Murphy D, Manzanilla EG. Removing prophylactic antibiotics from pig feed: how does it affect their performance and health? BMC Veterinary Research. 2019;15,:1-8. DOI: 10.1186/s12917-019-1808-x

[7] Helm ET, Curry S, Trachsel JM, Schroyen M, and Gabler NK. Evaluating nursery pig responses to in-feed sub-therapeutic antibiotics. PLoS One. 2019;14:1-16. DOI: 10.1371/journal.pone.0216070

[8] Emond-Rheault JG, Hamel J, Jeukens J, Freschi L, Kukavica-Ibrulj I, Boyle B, Tamber S, Malo D, FRanz E, Burnett E, Daigle F, Arya G, Sanderson K, Wiedmann M, Slawson RM, Weadge JT, Stephan R, Bekal S, Gruenheid S, Goodridge LD, Levesque RC. The *Salmonella enterica* plasmidome as a reservoir of antibiotic resistance. Microorganisms. 2020;8:1-18. DOI: 10.3390/microorganisms8071016

[9] Lima T, Domingues S, and Da Silva GJ. Plasmid-mediated colistin resistance in *Salmonella enterica*: a review. Microorganisms. 2019;7:1-17. DOI: 10.3390/microorganisms7020055

[10] Hooton SPT, Timms AR, Moreton J, Wilson R, Connerton IF. Complete genome sequence of *Salmonella enterica* serovar Typhimurium U288. Genome Announcoument. 2013;1:2010-2011. DOI: 10.1128/genomeA.00467-13

[11] Hooton SPT, Timms AR, Cummings NJ, Moreton J, Wilson R, Connerton IF. The complete plasmid sequences of *Salmonella enterica* serovar Typhimurium U288. Plasmid. 2014;76:32-39. DOI: 10.1016/j.plasmid.2014.08.002

[12] Gosling RJ, Mueller-Doblies D, MartelliF, Nunez-GarciaJ, KellN, RabieA, Wales AD, Davies RH. Observations on the distribution and persistence of monophasic *Salmonella* Typhimurium on infected pig and cattle farms. Veterinary Microbiology. 2018;227:90-96. DOI: 10.1016/j.vetmic.2018.10.032

[13] APHA. *Salmonella* in livestock production in GB. Animal Plant Health Agency. 3-240. 2020

[14] Bonardi S. *Salmonella* in the pork production chain and its impact

on human health in the European Union. Epidemiology & Infection. 2017;145:1513-1526. DOI: https://doi.org/10.1017/S095026881700036X

[15] Broadway PR, Brooks JC, Mollenkopf DF, Calle MA, Loneragen GH, Miller MF, Carroll JA, Sanchez NCB, Wittum TE. Prevalence and antimicrobial susceptibility of *Salmonella* serovars isolated from U.S. retail ground pork. Foodborne Pathogens and Disease. 2021. DOI: 10.1089/fpd.2020.2853

[16] Koutsoumanis K, Allende A, Alvarez-Ordonez A, Bolton D, Bover-Cid S, Chemaly M, De Cesare A, Herman L, Hilbert F, Lindqvist R, Nauta M, Peixe L, Ru G, Simmons M, Skandamis P, Suffredini E, Dewulf J, Hald T, Michel V, Niskanen T, Ricci A, Snary E, Boelaert F, Messens W, Davies R. *Salmonella* control in poultry flocks and its public health impact. EFSA Journal. 2019. Avaliable at: https://efsa.onlinelibrary.wiley.com/doi/epdf/10.2903/j.efsa.2019.5596

[17] Velasquez CG, Macklin KS, Kumar S, Bailey M, Ebmer PE, Oliver HF, Martin-Gonzalez FS, Singh M. Prevalence and antimicrobial resistance patterns of *Salmonella* isolated from poultry farms in southeastern United States. Poultry Science. 2018;97:2144-2152. DOI: 10.3382/ps/pex449

[18] Yang X, Huang J, Zhang Y, Liu S, Chen L, Xiao C, Zeng H, Wei X, Gu Q, Li Y, Wang J, Ding Y, Zhang J, Wu Q. Prevalence, abundance, serovars and antimicrobial resistance of *Salmonella* isolated from retail raw poultry meat in China. Science of Total Environment. 2020;713:136385. DOI: 10.1016/j.scitotenv.2019.136385

[19] Hussain MA, Liu H, Wang Q, Zhong F, Guo Q, Balamurugan S. Use of encapsulated bacteriophages to enhance farm to fork food safety. Critical Reviews in Food Science and Nutrition. 2017;57:2801-2810. DOI: 10.1080/10408398.2015.1069729

[20] Sillankorva SM, Oliveira H, Azeredo J. Bacteriophages and their role in food safety. International Journal of Microbiology. 2012;2012. DOI: 10.1155/2012/863945

[21] Atterbury RJ, Van Bergen MAP, Ortiz F, Lovell A, Harris JA, De Boer A, Wagenaar JA, Allen VM, Barrow PA. Bacteriophage therapy to reduce *Salmonella* colonization of broiler chickens. Applied and Environmental Microbiology. 2007;73:4543-4549. DOI: 10.1128/AEM.00049-07

[22] Wall SK, Zhang J, Rostagno MH, Ebner PD. Phage therapy to reduce preprocessing *Salmonella* infections in market-weight swine. Applied and Environmental Microbiology. 2010;76:48-53. DOI: 10.1128/AEM.00785-09

[23] Hooton SPT, Atterbury RJ, Connerton IF. Application of a bacteriophage cocktail to reduce *Salmonella* Typhimurium U288 contamination on pig skin. International Journal of Food Microbiology. 2011;151:157-163. DOI: 10.1016/j.ijfoodmicro.2011.08.015

[24] Gouvêa DM, Mendonça RCS, Soto ML, Cruz RS. Acetate cellulose film with bacteriophages for potential antimicrobial use in food packaging. LWT - Food Science and Technology. 2015;63:85-91. DOI: https://doi.org/10.1016/j.lwt.2015.03.014

[25] Gouvêa DM, Mendonça RCS, Lopez MES, Batalha LS. Absorbent food pads containing bacteriophages for potential antimicrobial use in refrigerated food products. LWT - Food Science and Technology. 2016;67:159-166. DOI: https://doi.org/10.1016/j.lwt.2015.11.043

[26] Clokie MRJ, Millard AD, Letarov AV, Heaphy S. Phages in nature. Bacteriophage. 2011;1:31-45. DOI: 10.4161/bact.1.1.14942

[27] Duckworth DH. Who discovered bacteriophage? Bacteriology Reviews. 1976;40:793-802. Avaliable at: https://www.ncbi.nlm.nih.gov/pmc/articles/PMC413985/

[28] Xu Y, Yu X, Gu Y, Huang X, Liu G, Liu X. Characterization and genomic study of phage vB_EcoS-B2 infecting multidrug-resistant *Escherichia coli*. Frontiers in Microbiology. 2018;9:793. DOI: 10.3389/fmicb.2018.00793

[29] Thanki AM, Brown N, Millard AD, Clokie MRJ. Genomic characterization of jumbo *Salmonella* phages that effectively target United Kingdom *Salmonella* serotypes. 2019;10:1-14. DOI: 10.3389/fmicb.2019.01491

[30] Abedon ST, Kuhl SJ, Blasdel BG, Kutter EM. Phage treatment of human infections. Bacteriophage. 2011;1:66-85. DOI: 10.4161/bact.1.2.15845

[31] Gigante A, Atterbury RJ. Veterinary use of bacteriophage therapy in intensively-reared livestock. Virology Journal. 2019;16:155. DOI: https://doi.org/10.1186/s12985-019-1260-3

[32] Endersen L, O'Mahony J, Hill C, Ross RP, McAuliffe O, Coffey A. Phage therapy in the food industry. Annual Review of Food Science and Technology. 2014;327-349. DOI: 10.1146/annurev-food-030713-092415

[33] Adriaenssens EM, Rodney Brister J. How to name and classify your phage: An informal guide. Viruses. 2017;9:1-9. DOI: 10.3390/v9040070

[34] Ackermann HW. Phage classification and characterization. In Bacteriophages: Methods and protocols. 2009;1:127-140. DOI: 10.1007/978-1-60327-164-6_13

[35] Clokie MRJ, Kropinski AMB. Bacteriophages: Methods and Protocols. Molecular and applied aspects. Volume 1. Humana Press. 2009. Avaliable at: https://www.springer.com/gp/book/9781588296825

[36] Żbikowska K, Michalczuk M, Dolka B. The use of bacteriophages in the poultry industry. Animals. 2020:10;5. DOI: https://doi.org/10.3390/ani10050872

[37] McLaughlin MR, King RA. Characterization of *Salmonella* bacteriophages isolated from swine lagoon effluent. Current Microbiology. 2008;56;208-213. DOI: 10.1007/s00284-007-9057-9

[38] Goode D, Allen VM, Barrow PA. Reduction of experimental *Salmonella* and *Campylobacter* contamination of chicken skin by application of lytic bacteriophages. Applied and Environmental Microbiology. 2003;69:5032-5036. DOI: 10.1128/AEM.69.8.5032-5036.2003

[39] De Jonge PA, Nobrega FL, Brouns SJJ, Dutilh BE. Molecular and evolutionary determinants of bacteriophage host range. Trends in Microbiology. 2019;27:51-63. DOI: https://doi.org/10.1016/j.tim.2018.08.006

[40] Chan BK, Abedon ST, Loc-carrillo C. Phage cocktails and the future of phage therapy. Future Medicine. 2013;8:769-783. DOI: 10.2217/fmb.13.47

[41] Stern A, Sorek R. The phage-host arms race: shaping the evolution of microbes. Bioessays. 2011;33:43-51. DOI: 10.1002/bies.201000071

[42] Labrie SJ, Samson JE, Moineau S. Bacteriophage resistance mechanisms. Nature Reviews Microbiology. 2010;8:317-27. DOI: https://doi.org/10.1038/nrmicro2315

[43] Berchieri A, Lovell MA, Barrow PA. The activity in the chicken alimentary tract of bacteriophages lytic for *Salmonella* Typhimurium. Research in Microbiology. 1991;142:541-549. DOI: 10.1016/0923-2508(91)90187-f

[44] Gonçalves GAM, Donato TC, Baptista AAS, de Oliveira Corrěa IM, Garcia KCOD, Filho RLA. Bacteriophage-induced reduction in *Salmonella* Enteritidis counts in the crop of broiler chickens undergoing preslaughter feed withdrawal. Poultry Science. 2014;93:216-220. DOI: 10.3382/ps.2013-03360

[45] Toro H, Price SB, Mckee AS, Hoerr FJ, Krehling J, Perdue M, Bauermeister L. Use of bacteriophages in combination with competitive exclusion to reduce *Salmonella* from infected chickens. Avian Diseases. 2005;49:118-124. DOI: 10.1637/7286-100404R

[46] Clavijo V, Baquero D, Hernandez S, Farfan JC, Arias J, Arevalo A, Donado-Godoy P, Vives-Flores M. Phage cocktail SalmoFREE® reduces *Salmonella* on a commercial broiler farm. Poultry Science. 2019;98:5054-5063. DOI: 10.3382/ps/pez251

[47] Clavijo V, Morales LT, Flores MJV, Reyes A. *Salmonella* phage cocktail, its effects and benefits on the gut of chickens in a commercial farm. Research square pre-print. 2020. DOI: 10.21203/rs.3.rs-29114/v1

[48] Sklar IB, Joerger RD. Attempts to utilize bacteriophage to combat *Salmonella enterica* serovar Enteritidis infection in chickens. Journal of Food Safety. 2001;21:15-29. DOI: https://doi.org/10.1111/j.1745-4565.2001.tb00305.x

[49] Atterbury RJ, Gigante AM, Lozano MSR, Medina RDM, Robinson G, Alloush H, Barrow PA, Allen VM. Reduction of *Salmonella* contamination on the surface of chicken skin using bacteriophage. Virology Journal. 2020;17:1-8. DOI: 10.1186/s12985-020-01368-0

[50] Hungaro HM, Mendonça RCS, Gouvêa DM, Vanetti MCD, de O. Pinto CL. Use of bacteriophages to reduce *Salmonella* in chicken skin in comparison with chemical agents. Food Research International. 2013;52:75-81. DOI: https://doi.org/10.1016/j.foodres.2013.02.032

[51] Sukumaran AT, Nannapaneni R, Kiess A, Sharma CS. Reduction of *Salmonella* on chicken meat and chicken skin by combined or sequential application of lytic bacteriophage with chemical antimicrobials. International Journal of Food Microbiology. 2015;207:8-15. DOI: 10.1016/j.ijfoodmicro.2015.04.025

[52] Higgins JP, Higgins SE, Guenther KL, Huff W, Donoghue AM, Donoghue DJ, Hargis BM. Use of a specific bacteriophage treatment to reduce *Salmonella* in poultry products. Poultry Science. 2005;84:1141-1145. DOI: https://doi.org/10.1093/ps/84.7.1141

[53] Duc HM, Son HM, Honjoh KI, Miyamoto T. Isolation and application of bacteriophages to reduce *Salmonella* contamination in raw chicken meat, LWT - Food Science and Technology. 2018;91:353-360. DOI: https://doi.org/10.1016/j.lwt.2018.01.072

[54] Yan T, Liang L, Yin P, Zhou Y, Sharoba AM, Lu Q, Dong X, Liu K, Connerton IF, Li J. Application of a novel phage LPSEYT for biological control of *Salmonella* in foods. Microorganisms. 2020;8:247. DOI: 10.3390/microorganisms8030400

[55] Saez AC, Zhang J, Rostagno MH, Ebner PD. Direct feeding of microencapsulated bacteriophages to reduce *Salmonella* colonization in pigs. Foodborne Pathogens and Disease. 2011;8:1269-1274. DOI: 10.1089/fpd.2011.0905

[56] Gebru E, Lee JS, Son JC, Yang SY, Shin SA, Kim B, Kim MK, Park SC. Effect of probiotic-, bacteriophage-, or organic acid-supplemented feeds or fermented soybean meal on the growth performance, acute-phase response, and bacterial shedding of grower pigs challenged with *Salmonella enterica* serotype Typhimurium. Journal of Animal Science. 2010;88:3880-3886. DOI: 10.2527/jas.2010-2939

[57] Wójcik EA, Stanczyk M, Wojtasik A, Kowalska JD, Nowakowska M, Lukasiak M, Bartnicka M, Kazimierczak J, Dastych J. Comprehensive evaluation of the safety and efficacy of BAFASAL® bacteriophage preparation for the reduction of *Salmonella* in the food chain. Viruses. 2020;12:7. DOI: 10.3390/v12070742

[58] Moye ZD, Woolston J, Sulakvelidze A. Bacteriophage applications for food production and processing. Viruses. 2018;10:1-22. DOI: 10.3390/v10040205

[59] Torres-Barceló C, Hochberg ME. Evolutionary rationale for phages as complements of antibiotics. Trends in Microbiology. 2016;24:249-256. DOI: 10.1016/j.tim.2015.12.011

[60] Tagliaferri TL, Jansen M, Horz HP. Fighting pathogenic bacteria on two fronts: phages and antibiotics as combined strategy. Frontiers in Cellular and Infection Microbiology. 2019;9:22. DOI: 10.3389/fcimb.2019.00022

[61] Jeon G, Ahn J. Assessment of phage-mediated inhibition of *Salmonella* Typhimurium treated with sublethal concentrations of ceftriaxone and ciprofloxacin. FEMS Microbiology Letters. 2020;367:19. DOI: https://doi.org/10.1093/femsle/fnaa159

[62] Jeon G, Ahn J. Evaluation of phage adsorption to *Salmonella* Typhimurium exposed to different levels of pH and antibiotic. Microbial Pathogensis. 2021;150:104726. DOI: https://doi.org/10.1016/j.micpath.2020.104726

[63] Bielke LR, Higgins SE, Donoghue AM, Donoghue DJ, Hargis BM, Tellez G. Use of wide-host-range bacteriophages to reduce *Salmonella* on poultry products. International Journal of Poultry Science. 2007;6:754-757. DOI: 10.3923/ijps.2007.754.757

[64] Santos SB, Fernandes E, Carvalho CM, Sillankorva S, Krylov VN, Pleteneva EA, Shaburova OV, Nicolau A, Ferreira EC, Azeredo J. Selection and characterization of a multivalent *Salmonella* phage and its production in a nonpathogenic *Escherichia coli* strain. Applied and Environmental Microbiology. 2010;76:7338-7342. DOI: 10.1128/AEM.00922-10

[65] Dąbrowska K, Abedon ST. Pharmacologically aware phage therapy: pharmacodynamic and pharmacokinetic obstacles to phage antibacterial action in animal and human bodies," Microbiology and Molecular Biology Reviews. 2019;83:1-25, 2019. DOI: 10.1128/MMBR.00012-19

[66] Danis-Wlodarczyk K, Dąbrowska K, Abedon ST. Phage therapy: the pharmacology of antibacterial viruses. Current Issues in Molecular Biology. 2020;40:81-163. DOI: 10.21775/cimb.040.081

[67] Casey E, Sinderen DV, Mahony J. *In Vitro* characteristics of phages to guide 'real life' phage therapy suitability. Viruses. 2018;30:163. DOI: 10.3390/v10040163

[68] Yongsheng M, Pacan JC, Wang Q, Xu Y, Huang X, Korenevsky A, Sabour PM. Microencapsulation of bacteriophage felix 01 into chitosan-alginate microspheres for oral delivery. Applied and Environmental Microbiology. 2008;74:4799-4805. DOI: 10.1128/AEM.00246-08

[69] Colom J, Cano-Sarabia M, Otero J, Cortés P, Maspoch D, Llagostera M.

Liposome-encapsulated bacteriophages for enhanced oral phage therapy against *Salmonella*. Applied and Environmental Microbiology. 2015;81:4841-4849. DOI: 10.1128/AEM.00812-15

[70] Fister S, Mester P, Witte AK, Sommer J, Schoder D, Rossmanith P. Part of the problem or the solution? Indiscriminate use of bacteriophages in the food industry can reduce their potential and impair growth-based detection methods. Trends in Food Science & Technology. 2019;90:170-174. DOI: https://doi.org/10.1016/j.tifs.2019.02.031

[71] Marco MB, Suarez VB, Quiberoni A, Pujato SA. Inactivation of dairy bacteriophages by thermal and chemical treatments. Viruses. 2019;11:480. DOI: 10.3390/v11050480

[72] Sommer J. Trautner C, Witte AK, Fister S, Schoder D, Rossmanith P, Mester PJ. Don't shut the stable door after the phage has bolted—the importance of bacteriophage inactivation in food environments. Viruses. 2019;11:1-27. DOI: 10.3390/v11050468

[73] Fauconnier A. Phage therapy regulation: From night to dawn. Viruses. 2019:11:352. DOI: 10.3390/v11040352

[74] Brives C, Pourraz J. Phage therapy as a potential solution in the fight against AMR: obstacles and possible futures. Palgrave Communication. 2020;6:1-11. DOI: https://doi.org/10.1057/s41599-020-0478-4

[75] Boeckaerts D, Stock M, Criel B, Gerstmans H, De Baets B, Briers Y. Predicting bacteriophage hosts based on sequences of annotated receptor-binding proteins. Nature Scientific Reports. 2021;11: 1-14. DOI: https://doi.org/10.1038/s41598-021-81063-4

[76] McNair K, Bailey BA, Edwards RA. PHACTS, a computational approach to classifying the lifestyle of phages. Bioinformatics, 2012;28:614-618. DOI: 10.1093/bioinformatics/bts014

[77] Villarroel J, Kleinheinz KA, Jurtz VI, Zschach H, Lund O, Nielsen M, Larsen MV. HostPhinder: A phage host prediction tool. Viruses. 2016;8:116. DOI: 10.3390/v8050116

[78] Ahlgren NA, Ren J, Lu YY, Fuhrman JA, Sun F. Alignment-free d2∗ oligonucleotide frequency dissimilarity measure improves prediction of hosts from metagenomically-derived viral sequences. Nucleic Acids Research. 2017;45:39-53. DOI: 10.1093/nar/gkw1002

[79] Li M, Wang Y, Li F, Zhao Y, Liu M, Zhang S, Bin Y, Smith AI, Webb G, Li J, Song J, Xia J. A Deep Learning-Based Method for Identification of Bacteriophage-Host Interaction," IEEE/ACM Transacations on Computational Biology and Bioinformatics. 2020. DOI: 10.1109/TCBB.2020.3017386

[80] Leite DMC, Brochet X, Resch G, Que YA, Neves A, Peña-Reyes C. Computational prediction of inter-species relationships through omics data analysis and machine learning. BMC Bioinformatics. 2018;19:420. DOI: https://doi.org/10.1186/s12859-018-2388-7

[81] Ren J, Song K, Deng C, Ahlgren NA, Fuhrman JA, Li Y, Xie X, Poplin R, Sun F. Identifying viruses from metagenomic data by deep learning. Quantitative Biology. 2018;8:64-77. DOI: https://doi.org/10.1007/s40484-019-0187-4

[82] Edwards RA, McNair K, Faust K, Raes J, Dutilh BE. Computational approaches to predict bacteriophage-host relationships. FEMS Microbiology Reviews. 2016;40:258-272. DOI: https://doi.org/10.1093/femsre/fuv048

Chapter 11

# Natural Products for Salmonellosis: Last Decade Research

*Nancy Jannet Ruiz-Pérez, Jaime Sánchez-Navarrete and Julia D. Toscano-Garibay*

## Abstract

Salmonellosis is a disease of great relevance in terms of public health given the economic and social impact that causes both in developing and highly industrialized countries. Due to its transmission mechanism, it affects hundreds or thousands of people every year and is considered an acute disease of worldwide distribution. Causative agent of salmonellosis is *salmonella* specie which are small gram-negative bacilli and facultative intracellular pathogen of the Enterobacteriaceae family. Multidrug resistance is reported more frequently in strains of *salmonella*, raising the necessity of new strategies to combat its spread and to treat the disease. Natural products (NPs) derived from traditional medicine knowledge have become an important resource to this end. In this chapter, we present a summary of information published from 2010 to 2020, as a sample of the potentiality of NPs as agents for Salmonellosis. This search was not exhaustive, rather, we aim to obtain a random sample of information using the simplest terms on the matter of natural products for salmonellosis, hopefully, as a reference source for interested researchers.

**Keywords:** *salmonella*, antibacterial activity, natural products, anti-salmonella, Salmonellosis

## 1. Introduction

Salmonellosis is a disease of great relevance in terms of public health given the economic and social impact that causes both in developing and highly industrialized countries. Due to its transmission mechanism, it affects hundreds or thousands of people every year and is considered an acute disease of worldwide distribution [1] with variations in the frequency of serotypes from one country to another [2], being notably more frequent in areas that have not reached adequate sanitation and hygiene conditions or that do not have enough resources and public health infrastructure. There is no distinction in the occurrence of salmonellosis by sex, age, or social and economic status with high incidence at the extremes of life, being the most vulnerable groups, children under 5yo, adults over 60 years of age and immunocompromised individuals [3, 4]. On the other hand, it is also a seasonal disease, so incidence is higher on periods of increased environmental temperature like spring and summer, showing a decrease in autumn and winter [5].



IntechOpen

The raise in salmonellosis at any part of the world is of maximum relevance. For example, an incidence of 0.78–3.8 million cases per year has been estimated in the United States. Natural reservoir is made up of domestic animals (dogs and cats), wild animals (reptiles such as iguanas and turtles) as well as humans (carriers, convalescent). Transmission is through food (with or without manufacture) and water contaminated with human or animal feces and from individual to individual. Salmonellosis presents as sporadic cases or as outbreaks with variable affectations. Incidence rate is dose-dependent in function of the disseminated serotype and is determined by incubation period, symptoms and severity.

Causative agent of salmonellosis are *salmonella* species, numerous disease outbreaks are related to the consumption of eggs, chicken meat and other raw products (mainly dairy). For instance, in an outbreak of enteric salmonellosis serotype Typhimurium (n = 99) induced by consumption of roast porcine meat in an institution for the mentally ill in Konagua it was shown that the incubation period was between 10–12 hours and that the supply of antibiotics prolonged excreta periods. Salmonellosis due to *Salmonella enterica* serotype Enteritidis was detected in an interstate outbreak in the United States in the early 90's, produced by the consumption of ice cream (224,000 cases) and in Canada due to the consumption of commercial packaged cheese (800 cases). Salmonella Javiana (n = 66) has been reported to produce outbreaks as in Boston due to the consumption of chicken sandwiches [6].

## 2. *Salmonella*

*Salmonella* belongs to the Enterobacteriaceae family, which are small gram-negative bacilli varying in sizes ranging in average from 2–3 μm in length and 0.4–0.6 μm in width. These bacilli do not form spores and possess peritrichous flagella hence are mobile microorganisms, although some genera, such as Klebsiella and Shigella, are lacking on these organelles and so on mobility. Traditional grouping classification is carried out using primary biochemical characteristics that allows a further sorting into subgroups based on antigenic structure determinants or using bacteriophage reactions. Currently, with the advances in molecular biology, the differentiation of groups and subgroups can be made using PCR technique for identification, diagnostic and epidemiological purposes.

Regarding its metabolic characteristics, *salmonella* grows in simple synthetic media and can use unique carbon sources, such as glucose in a fermentative way with the subsequent formation of acids and/or gases, reducing nitrates and nitrites, rendering oxidase negative reaction. *Salmonella* also tests positive for methyl red, hydrogen sulfide, indole-ornithine motility (MIO medium), lysine decarboxylase, arginine dihydrolase, ornithine decarboxylase, gas from glucose, and fermentation of numerous carbohydrates such as rhamnose, arabinose, mannitol, etc.

Most enteric microorganisms are resistant to inhibition by the action of certain bacteriostatic dyes, the selective media containing these compounds facilitate considerably isolation from fecal samples, *salmonella* is less sensitive than coliform microorganisms against citrate inhibition action; for instance, SS (Salmonella-Shigella) agar containing both citrate and bile salts is therefore used as a selective medium for the culture of pathogenic species [7, 8].

## 3. Classification

Although controversial and evolving, there is a *salmonella* nomenclature used by the Centers for Disease Control and Prevention (CDC) and recommended by the

Collaborating Center of the World Health Organization (WHO), which according to the differences in their 16S rRNA sequence analysis classifies this genus into two species, *Salmonella enterica* and *Salmonella bongori*. *S. enterica* can also be further classified into six subspecies mainly found in mammals and is responsible for 99% of infections in humans and warm-blooded animals. On the other hand, *S. bongori* is predominantly environmental and on cold-blooded animals [9].

### 3.1 Classification according to Kauffmann-White

Since decades ago, classification of *salmonella* finalizing at the species level are based on its antigenic structure. Although certain strains that have the same antigenic activity could present different metabolic reactions (biotype variants or serotypes), this sorting method is generally accepted and is actively in use.

Surface antigen studies are based on H, O, K and Vi antigens. H is denominated surface or flagellar antigen and participates in host immune response, O (aka somatic antigen) is a lipopolysaccharide located in the cell membrane, K is a capsular antigen and Vi antigen is a subtype of K antigen associated to virulence [9] and the obtention of antisera containing antibodies against all these fractions allows the identification of *salmonella* species. More than 2,500 serotypes have been identified related to the H, O, K and Vi antigens [10] as a result of the numerous absorption tests and cross-reactions studies carried out in Denmark and England by Kauffman and White. Currently, large centers in Copenhagen, London, and Atlanta have the necessary collections of specific antisera for salmonellas typing. In most testing and diagnostic laboratories, *salmonella* strains are identified and classified by their fermentative characteristics and agglutination reactions using group-specific antisera.

In Mexico, for example, a study for the classification and identification of *salmonella* serotypes at public and private health centers and hospitals analyzed 24,394 *salmonella* strains isolated from different sources, 15,843 (64.9%) of human origin and 8,551 (35.1%) non-human demonstrating the usefulness of Kauffmann-White scheme and using antisera produced at the National Institute of Diagnosis and Reference (INDRE) in accordance with the Center for Disease Control and Prevention, Atlanta (GA), showing that most frequent serotypes both in human and non-human samples were *S.* Typhimurium, *S.* Enteritidis, *S.* Derby, *S.* Agona and *S.* Anatum. From the epidemiological point of view, it is interesting to identify which are the circulating and emerging serotypes to implement prevention strategies [8].

## 4. Pathogenicity

*Salmonella* spp. is a highly pathogenic microorganism that presents different pathogenicity mechanisms including adherence, invasiveness, colonization and growth, toxicity and tissue damage [11]. It is a facultative intracellular pathogen causing moderate to severe infections, or even compromising systemic infections risking patients' lives, depending on the serotype, virulence, inoculum and immunological state of involved host, and all of this using only a mixture of toxins and other virulence factors.

Clinical manifestations in humans include enteric fevers, acute gastroenteritis and septicemia in extreme cases. Prototypical enteric fevers are caused by *Salmonella* Typhi, this is also known as typhoid fever, after its incubation period (7–14 days), symptoms such as anorexia, headache, followed by general malaise and fever may occur. The interaction patient-causative agent is essential for the progression of the disease, *salmonella* must find a microhabitat suitable for its establishment, multiplication and virulence factors expression.

*Salmonella* produces at least three toxins: enterotoxin, lipopolysaccharide endotoxin (LPS), and cytotoxin. Enterotoxigenicity, which is a property present in many serotypes of this microorganism, including S. Typhi, is expressed a few hours after contact with the host cell. The pathogenicity mechanisms by which *salmonella* induces diarrhea and septicemia have not yet been clearly elucidated, but it appears to be a complex phenomenon involving numerous virulence factors such as those mentioned above.

The specific virulence factors are encoded by a group of genes for the formation of pathogenicity islands (SPI), with G + C percentages differing from the average of the bacterial genome. Direct repeats are present at the filament ends, carrying genes that encode mobility factors such as integrases, transposases or insertion sequences and are frequently inserted on tRNA. This suggests that they have been obtained from other species by horizontal transfer or by plasmids. There are numerous genes that participate in the invasion and that are present in salmonellas, genes that code for the synthesis of proteins related to the translocation of effector molecules within the cytoplasm of the host cell. Today, it is known that *salmonella* has five islands of pathogenicity: SPI-1, SPI-2, SPI-3, SPI-4 and SPI-5 [10].

## 5. Mechanisms of resistance

Drug resistance and worldwide incidence of *salmonella* infections has been increasingly reported. For example, it has been observed a high incidence among humans, livestock and poultry of *Salmonella enterica* serotype [4, [5],12:i:-], with variants ranging from sensitive- to multi-drug resistant, since the 1990s. Other examples include a strain of *Salmonella enterica* discovered on 2015 that was provided with the gene mcr-1 of plasmid-mediated colistin resistance and clinical isolates from Portugal, China and United Kingdom observed in 2016 with this same gene [12].

Several types of *salmonella* with multi-drug resistance (MDR) are capable of generating diverse types of plasmids, with gene cassettes that provide the property of resistance against antibiotics such as chloramphenicol, tetracycline, ampicillin, and streptomycin [13, 14]. The chromosomal mutation in the regions that determine the resistance to quinolones of the gyrA gene are responsible for the appearance of *salmonella* serotypes with little susceptibility to ciprofloxacin [15]. On the other hand, the mutated genes that code for extended spectrum β-lactamases, are responsible for the serotypes that have begun to develop resistance to cephalosporins [16].

Resistance not only by *salmonella*, but by other microorganisms are currently a public health problem worldwide, which threatens the prevention, control and treatment of innumerable infectious diseases, having as expected consequences in terms of health and economic impact. This problem was recognized by the World Health Organization and in 2001 this organization published the Global Strategy for the Containment of Antimicrobial Resistance, publicizing interventionist actions to delay the appearance and to reduce the spread of resistant microorganisms [17]. For 2012, WHO proposed a series of actions such as strengthening health services and epidemiological surveillance, regulated use of antimicrobials in hospitals and in communities, promoting the development of new drugs and appropriate vaccines, among others [18]. This problem is one of the reasons for the development of new alternatives, being natural products derived from traditional medicine, one of the most used resources.

## 6. Traditional medicine and natural products

The origin of Natural and Traditional Medicine is indisputably linked both to human history and to its fight for survival [19]. Written evidence on plants being used as remedies for disease is as ancient as Mesopotamian tablets, and from there, a nearly endless number of registers in all cultures, supports its essential role on human well-being. Currently, traditional medicine has been delineated as the use of products of natural origin for health preservation, having the so-called Natural Products (NPs) at its focus.

NPs are broadly defined as small molecules produced by a living organism. This definition comprises a wide variety of compounds including the synthesized during basic metabolism (primary metabolites) or as by-products of it (secondary metabolites). Lipids, carbohydrates, proteins and nucleic acids are part of the first kind of NPs, while smaller molecules such as alkaloids, tannins, saponins and flavonoids are examples of secondary metabolites. Many of the latter does not seem to have a metabolic or evolutionary function for the parental organism, but regardless to that, its utility as drugs, preservatives, dyes, food additives and/or antibiotics is undeniable. Its application to counteract the pathogenic microorganisms affecting our specie, alongside side-effects and resistance to antibacterial drugs, is undoubtedly enough motivation for the current formalization and systematization of traditional knowledge, with methodological studies being carried out very frequently nowadays.

There has been an important upturn in the study of compounds of natural origin during the last decade, supported on ethnopharmacological information, folkloric reputation, traditional uses and the existence of previous evidence, and also based on NPs chemical composition and its chemotaxonomic classification. This explosion of information has been enriched primarily through the obtention and separation of crude extracts, essential oils, and/or other types of preparations that are subsequently analyzed for possible biological activities of metabolites or secondary products. Modern experimental strategies have included bioassays (mainly in vitro), development of NP libraries, production of active compounds in cell or tissue cultures, genetic manipulation of organisms, natural combinatorial chemistry, etc. [20]. NPs, being originated in living organisms, are essentially complex mixtures contained within cellular structures, hence the first step into the study of its properties is the separation of such structures. This first step is called extraction, and is generally carried out by liquid solvents at room temperature and atmospheric pressure, along with other well-known and widely used techniques such as steam distillation and the use of supercritical fluids or pressurized gases [21]. The proper choice of an extraction step is necessarily based on the nature, origin and composition of the product to be studied, taking into account the characteristics of the possible solvents (innokenty, reactivity, etc.), toxicity of secondary products, product sufficiency needs and evaluation methods to be followed afterwards, as a whole this step should result suitable to fulfill the objective of a research. Second and third steps are the setting of an adequate model for biological efficacy assessment and the elucidation of individual bioactive components.

In this chapter, we enlisted natural products frequently reported against *salmonella* from bacteria (**Table 1**), fungus (**Table 2**), animal (**Table 3**), plant (**Table 4**) or combined (**Table 5**) origin, organized on a chronologically descending order according to publishing date. To get a glimpse on the universe of information that NPs research has become, we made a fast search on two commonly used and easily accessible databases (PubMed and Google scholar) for the terms: *salmonella, anti-salmonella, salmonellosis, natural product* and *antibacterial activity*, alone or in combinations. Search results without the terms *salmonella* or *salmonellosis* were excluded. From the remaining registers, we selected those corresponding to experimental reports where the extraction step was performed and thoroughly described by authors. Studies on isolated or synthetic NPs were not included and research on infection or tissue damage protection after *salmonella* colonization were also excluded. Review articles or abstracts were not considered, although we accounted

| Parental organism | *Origin* | Extraction method | Segment used | Bioactive compound(s) | Location | *Salmonella* serovar | Year | Ref |
|---|---|---|---|---|---|---|---|---|
| Lactobacillus plantarum ZJ316 | Bacteria | culture supernatant filtration (methanol/ acetonitrile) | bacilli | L-phenyl lactic acid | China | *Salmonella* Paratyphi-A *(CMCC 50093)*, *Salmonella* Paratyphi-B *(CMCC 50094)*, *Salmonella enterica subsp. enterica (ATCC 14028)*, *Salmonella enterica subsp. arizonae (CMCC(B) 47001)*, *Salmonella choleraesuis (ATCC 13312)*, and *Salmonella* Typhimurium *(CMCC 50015)* | 2020 | [22] |
| Lectin (Bifidobacterium adolescentis) from bee honey | Bacteria | crude and purified extracts | honey | lectin | Iraq | *Salmonella* Typhi *(clinical isolates)* | 2019 | [23] |
| Lactobacillus salivarius, L. casei B1, *L. plantarum*, L. delbrueckii and L. delbrueckii | Bacteria | co-culture | co-culture | not specified | Benin | *Salmonella spp., Salmonella* Typhimurium *(ATCC 14028)* | 2019 | [24] |
| Lactococcus lactis subsp.lactis (CNRZ 1427) | Bacteria | not specified | not specified | specific microbial enzymes, perox-ide, weak organic acids anti-bacterial peptides, secretion of bacteriocins protease production | Algeria | *Salmonella spp. (veterinary isolate). Mice tests* | 2014 | [25] |
| Streptomyces spp | Bacteria | crude protein | microbial cells | not specified | India | *Salmonella* Enteritidis | 2014 | [26] |

**Table 1.**
*Summary of frequently reported natural products from bacteria origin against* salmonella.

| Parental organism | *Origin* | Extraction method | Segment used | Bioactive compound(s) | Location | ***Salmonella*** **serovar** | Year | Ref |
|---|---|---|---|---|---|---|---|---|
| Lentinus edodes | Fungus | fermenting/Black rice bran culture | mycelia | bioprocessed polysaccharide | Korea | *Salmonella* Typhimurium *(SL1344)* | 2018 | [27] |
| Coriolus versicolor | Fungus | methanolic extract | not specified (probably full fungi body) | phenolics, polysaccharides, β-glucans, α-glucans, proteins | Serbia | *Salmonella* Enteritidis *(ATCC 13076)* | 2016 | [28] |
| Pleorotus ostreatus (oyster mushroom) | Fungus | ethanolic extract | not specified (probably full fungi body) | not specified | Germany | *Salmonella* Typhi | 2015 | [29] |
| Ganoderma lucidum | Fungus | ethanolic, methanolic, acetone and aqueous extracts | fruiting bodies | not specified | India | *Salmonella* Typhi *(MTCC-531)* | 2010 | [30] |
| Lentinus tuberregium | Fungus | Hexane, Dichloromethane, Chloroform and Ethylacetate extracts | not specified (probably full fungi body) | not specified | India | *Salmonella* Flerineri *(M-1457)* *Salmonella* Typhi *(M-733)* | 2010 | [31] |
| Pichia pastoris X-33 | Yeast | YPD broth supplemented with 1 mg.mL − 1 pancreatin, 0.2% bile salts, and pH adjusted 8 with 0.1 N NaOH | yeast cell | not specified | Brazil | *Salmonella* Typhimurium (strain *29630)* | 2015 | [32] |

**Table 2.**
*Summary of frequently reported natural products from fungi origin against* salmonella.

| **Parental organism** | ***Origin*** | **Extraction method** | **Segment used** | **Bioactive compound(s)** | **Location** | ***Salmonella* serovar** | **Year** | **Ref** |
|---|---|---|---|---|---|---|---|---|
| Apitoxin | Animal | crude apitoxin | apitoxin | Melittin, adolapin, apamin or MCD-peptide, phospholipase A2 or hyaluronidase, histamine, epinephrine | Ecuador | *Salmonella* Anatum, *Salmonella enterica subsp. arizonae*, *Salmonella* Bardo, *Salmonella* Bredeney, *Salmonella* Dabou, *Salmonella* Drac, *Salmonella* Enteritidis, *Salmonella* Infantis, *Salmonella* Isangi, *Salmonella* Montevideo, *Salmonella* Mbandaka, *Salmonella* Ndolo, *Salmonella* Newport, *Salmonella* Rissen, *S. enterica subespecie salamae*, *Salmonella* Seftenberg, *S.* Stanleyville, *S.* Thompson and *Salmonella* Typhimurium | 2020 | [33] |
| Apitoxin | Animal | crude apitoxin | apitoxin | Melittin, adolapin, apamin or MCD-peptide, phospholipase A2 or hyaluronidase, histamine, epinephrine | Ecuador | *Salmonella* Newport, *Salmonella Isangi*, *Salmonella enterica subsp. salame*, *Salmonella* Bardo, *Salmonella* Infantis, *Salmonella* Montevideo, *Salmonella* Stanleyville, *Salmonella* Ndolo, *Salmonella* Dabou, *Salmonella* Typhimurium, *Salmonella* Enteritidis | 2019 | [34] |
| Masske butter | Animal | lactic isolates | microbial cells | lactic acid | Iran | *Salmonella enterica* | 2019 | [35] |
| Propolis | Animal | ethanolic extract | propolis | flavonoids, alkaloids, terpenoids, steroids, saponins, and tannins | Indonesia | *Salmonella spp.* | 2019 | [36] |
| Sarconesiopsis magellanica | Animal | RP-HPLC | larvae | Sarconesin | Colombia | *Salmonella enterica (ATCC 13314)* | 2018 | [37] |
| Dadih dadih | Animal | ice cream | buffalo milk yogurt | not specified | Indonesia | *Salmonella* Typhimurium | 2017 | [38] |
| Donkey's milk | Animal | no extraction | milk | not specified | Serbia | *Salmonella* Enteritidis *(ATCC 13076)* and *Salmonella* Typhimurium *(ATCC 14028)* | 2017 | [39] |
| Colla corii asini | Animal | aqueous and ethanolic extracts | donkey-hide gelatin | glycine, alanine, aspartic acid, glutamic acid, β-amino isobutyric acid | Korea | *Salmonella* Typhimurium *(KCTC 1926)* | 2017 | [40] |
| Bovine natural antibodies | Animal | antibodies | serum | antibodies | The Netherlands | *Salmonella* Typhimurium *(SL3261)* | 2016 | [41] |
| Propolis | Animal | ethanolic extracts | propolis | phenolic acid components. Sinergy with cefixime | India | *Salmonella* Typhimurium *(MTCC 98)* | 2016 | [42] |
| Anguilla spp. | Animal | aqueous dilution | mucus | not specified | Indonesia | *Salmonella* Typhi | 2016 | [43] |

| Parental organism | *Origin* | Extraction method | Segment used | Bioactive compound(s) | Location | *Salmonella* serovar | Year | Ref |
|---|---|---|---|---|---|---|---|---|
| Propolis | Animal | not specified | propolis | phenolic compounds (flavonoids) | Chile | *Salmonella* Enteritidis | 2015 | [44] |
| Honeys | Animal | aqueous dilution | honey | not specified | Pakistan | *Salmonella* Typhi | 2015 | [45] |
| Platelet rich plasma | Animal | whole blood | thrombin PRP/CaCl2 PRP | probably antimicrobial peptides | Iran | *Salmonella enterica* | 2014 | [46] |
| Honey | Animal | aqueous dilution | honey | not specified | Romania | *Salmonella* Enteritidis *(ATCC 13076)* | 2014 | [47] |
| Donkey's milk | Animal | no extraction | milk | not specified | Serbia | *Salmonella* Enteritidis *(ATCC 13076)*, *Salmonella* Typhimurium *(ATCC 14028)*, *Salmonella* Livingstone | 2014 | [48] |
| Propolis | Animal | ethanolic, methanolic and aqueous extracts | propolis | terpenoids, flavonoids, alkaloids, phenols, tannins and saponins | India | *Salmonella* Typhimurium | 2013 | [49] |
| Slovenian Propolis | Animal | 70% and 96% ethanol | propolis | phenolic compounds (probably a synergy) | Slovenia | *Salmonella* Typhimurium *(14028)*, *Salmonella* Enteritidis *(ZM138)* | 2012 | [50] |
| Shrimp Chitosan | Animal | acetic acid 1% | shrimp | not specified | Bangladesh | *Salmonella* Parathypi | 2011 | [51] |
| Honey | animal | saline dilution | honey | not specified | Greece | *Salmonella* Typhimurium *Salmonella enterica subsp. enterica (ATCC 13311)* and *Salmonella* Typhimurium and *Salmonella* | 2011 | [52] |
| Propolis | Animal | ethanolic extract | propolis | quercetin, chrysin, 4′,5-dihydroxy-7-methoxyflavonone and 3,4′,7-trimethoxyflavonone | Turkey | *Salmonella* Enteritidis *(ATCC 13076)* | 2011 | [53] |
| Honey | Animal | aqueous dilution | honey | not specified | India | *Salmonella enterica serovar* Typhi | 2010 | [54] |

**Table 3.**
*Summary of frequently reported natural products from animal origin against* salmonella.

| Parental organism | *Origin* | Extraction method | Segment used | Bioactive compound(s) | Location | *Salmonella* serovar | Year | Ref |
|---|---|---|---|---|---|---|---|---|
| Zanthoxylum Acanthopodium DC | Plant | n-hexane and ethyl acetate extract | Fruit | flavonoids, alkaloids, and saponins | Indonesia | *Salmonella* Typhi | 2020 | [55] |
| *Aleurites moluccana* | Plant | methanol extracts | stem bark | scopoletin | Indonesia | *Salmonella* Typhimurium | 2020 | [56] |
| Combre tummicranthum; Acacianilotica and Phyllanthus pentandrus | Plant | aqueous, ethanol and chloroform extracts | leaves | tannins, flavonoids, saponins, sterols, triterpenes, alkaloids, anthocyanes and free anthraquinones | Niger | *Salmonella* Typh*imurim*, *Salmonella* Typhi, *Salmonella Para*Typhi, *Salmonella* Typh*imurim*, and *Salmonella* Derby | 2020 | [57] |
| Nauclea latifolia | Plant | ethyl acetate and methanol | leaves | tannins, flavonoids and anthraquinones (all are highly polar and polyphenolic) as secondary metabolites but steroids were absent | Indonesia | *Salmonella* Typhi *(clinical isolates, MDR)* | 2020 | [58] |
| *Hippobroma longiflora* | Plant | ethanolic extracts | leaves | alkaloids, flavonoids and saponins | | *Salmonella* Typhi | 2020 | [59] |
| Biarum bovei (cardin) | Plant | ethanol 50% (ultrasound) | leaves | Nerrel, flavonoids and nercernerrel | Iran | *Salmonella* Enteritidis *(CMCC 50041)* | 2020 | [60] |
| *Trema orientalis* L. Blumae (anggrung) | Plant | methanol extracts | leaves | alkaloid, flavonoids, tannins, terpenoids, steroids, saponin, phenolic | Indonesia | *Salmonella spp.* | 2020 | [61] |
| *Agave tequilana* Weber var. azul | Plant | flour | leaves | Fructans | Mexico | *Salmonella* Typhimurium | 2020 | [62] |
| *Clerodendrum fragrans* Vent Willd | Plant | methanol, ethyl acetate and n-hexane (chromatography) | leaves | Tannins and flavonoids | Indonesia | *Salmonella enterica (ATCC 14028)* | 2020 | [63] |
| *Canarium schweinfurthii* | Plant | hydro-ethanolic extract followed by chloroform and ethyl acetate | stem bark | maniladiol, scopoletin, ethyl gallate and Gallic acid | Cameroon | *Salmonella* Typhi, *Salmonella* Enteritidis and *Salmonella* Typhimurium *(clinical isolates)* and *Salmonella* Typhi *(ATCC6539)* | 2020 | [64] |

| Parental organism | *Origin* | Extraction method | Segment used | Bioactive compound(s) | Location | *Salmonella* serovar | Year | Ref |
|---|---|---|---|---|---|---|---|---|
| *Garcinia kola* and Alchornea cordifolia | Plant | hydro-ethanolic and methanolic extracts | leaves, root and stem bark | Anthocyanins, Flavonoids, Glycosides, Phenols, Tannins, Triterpenoids and Steroids | Cameroon | S. Typhi *(collection)*, *S.* Typhimurium and *S.* Enteritidis *(clinical isolates)* | 2020 | [65] |
| *Ziziphus lotus* and Ziziphus mauritianas | Plant | methanolic extracts | leaves, fruits and seeds | Quinic acid, p-coumaric acid, rutin and quercitrin | Tunisia | *Salmonella* Typhimurium *(NRLB4420)* | 2020 | [66] |
| Rhododendron arboreum and *Justicia adhatoda* | Plant | ethanolic and methanolic extracts | leaves | oleanadien-3β-ethan-3-oate | Nepal | *Salmonella enterica subsp. enterica (ATCC 13076)* | 2020 | [67] |
| Uvaria chamae, *Lantana camara* and Phyllantus amarus | Plant | aqueous and ethanolic extracts | leaves and root | not specified | Benin | *Salmonella* Typhimurium *ATCC 14028* and *Salmonella spp. (isolates)* | 2020 | [68] |
| *Vitis vinifera* var. Albariño | Plant | hydro-organic extraction (patented) | fruit | HOL: catechin, epicate-chin and isoquercetin. HOP: phologlucinic acid, miquelianin, rutin, inkaempferol and caftaric acid | Spain | *Salmonella enterica subsp. enterica (CECT 554)* | 2020 | [69] |
| *Citrus hystrix* | Plant | ethanolic extract | peel | not specified | Indonesia | *Salmonella* Typhimurium | 2020 | [70] |
| Olive oil | Plant | ethanolic extract | fruit | polyphenol extracts | China | *Salmonella* Typhimurium *(ATCC 14028)* | 2020 | [71] |
| Agrimonia pilosa Ledeb, *Iris domestica* (L.) Goldblatt and Mabb, Anemone chinensis Bunge, | Plant | aqueous extracts | herb, rhizome, root and tuber | not specified | China | *Salmonella* Enteritidis *(NCTC 0074, 1F6144, LE103* and *QA04/19)* | 2020 | [72] |
| *Litsea cubeba* | Plant | essential oil | fruit | 2,6-octadienal, 3,7-dimethyl-, 2,6-octadien-1-ol, 3,7-dimethyl-, and Z-2,6-octadien-1-ol, 3,7-dimethyl, Z-2,6-Octadienal, 3,7-dimethyl-, Z-citral | China | *Salmonella enterica (CGMCC 1.755)* | 2020 | [73] |

| Parental organism | *Origin* | Extraction method | Segment used | Bioactive compound(s) | Location | *Salmonella* serovar | Year | Ref |
|---|---|---|---|---|---|---|---|---|
| *Quercus infectoria*, *Phyllanthus emblica* | Plant | aqueous, methanolic and ethanolic extracts | gall, fruit | hexadecanoid acid, 9-octadecenoic acid, octadecenoic acid, 2-tert buil-4-isopropil-5 metylphenol | India | *Salmonella* Enteritidis and *Salmonella* Typhi | 2020 | [74] |
| Capparis decidua | Plant | methanolic extract | whole plant | not specified | Pakistan | *Salmonella* Typhi | 2020 | [75] |
| Detarium microcarpum Guill. & Perr. | Plant | ethanolic extract | leaves, twigs, roots, and root bark | flavonoids, sterols, triterpenes, glucosides, coumarins, and saponins | Cameroon | *Salmonella* Typhi *(ATCC 19430)*, *Salmonella* Enteritidis *(ATCC 13076)* | 2020 | [76] |
| Tetrapleura tetraptera | Plant | ethanolic extract | stem | citral, acetic acid, limonene, butanol, 2-hydroxyl-3 butanone, Cis-Verbenol Trans-Verbenol, α-Terpinyl acetate, butanoic acid, 2-methyl butanol | Ghana | *Salmonella* Enteritidis *(CICC 21482)* and *Salmonella* Typhimurium *(CICC 21483)* | 2020 | [77] |
| *Ocimum gratissimum* | Plant | aqueous and ethanolic extracts | leaves | alkaloid, tannins, oxalate, flavonoids and essential oil | Nigeria | *Samonella* Typhi and *Salmonella Para*Typhi *(clinical isolates)* | 2019 | [78] |
| Aeollanthus pubescens | Plant | essential oil (aqueous) | leaves | thymol and carvacrol (anti-radical activity) | Nigeria | *Salmonella spp. (multidrug resistant isolate)* | 2019 | [79] |
| *Annona muricata* L. | Plant | ethanol extracts | flower | secondary metabolites such as alkaloids, phenolic and flavonoid | Indonesia | *Salmonella* Enteritidis | 2019 | [80] |
| *Rhodomyrtus tomentosa* (Ait) Hassk | Plant | N-hexane, ethyl acetate, ethanol | leaves | phenols and flavonoids | Indonesia | *Salmonella* Typhi | 2019 | [81] |
| Morinda lucida | Plant | acetone and aqueous extracts | leaves | not specified | South africa | *Salmonella enterica subsp. enterica including S. enterica serovar* Gallinarum, Dublin, *choleraesuis, Braenderup, Idikan, Kottbus,* Typhimurium and Enteritidis | 2019 | [82] |

| Parental organism | *Origin* | Extraction method | Segment used | Bioactive compound(s) | Location | ***Salmonella*** serovar | Year | Ref |
|---|---|---|---|---|---|---|---|---|
| Zanthoxylum acanthopodium DC (andaliman) | Plant | ethanol extracts | fruit | saponin, tannin, steroid and alkaloid | Indonesia | *Salmonella* Typhi | 2019 | [83] |
| *Physalis peruviana* L | Plant | ethanol extracts | berries and leaves | Phenolic compounds (1-hexanol, eucalyptol and 4-terpenol) | Ecuador | *Salmonella spp. (clinical isolates)* | 2019 | [84] |
| *Carica papaya* L. | Plant | 70% ethanol, followed by n-hexane, ethyl acetate and water | seeds | alkaloids, flavonoids, terpenoids and saponins | Indonesia | *Salmonella* Typhi *(ATCC 1408)* | 2019 | [85] |
| *Psidium guajava* | Plant | methanol and aqueous extracts | leaves and stem bark | alkaloid, saponin, phenol, flavonoids, glycoside, anthraquinones, terpenoid and tannin | Nigeria | *Salmonella* Typhi *(clinical isolates)* | 2019 | [86] |
| *Artocarpus heterophyllus*. Lamk. | Plant | ethanol extracts | leaves | Saponin, flavonoids, terpenoid/steroids and tannin | Indonesia | *Salmonella* Typhi | 2019 | [87] |
| Sesbiana grandiflora L. Press | Plant | 90% ethanol followed by n-hexane, ethyl acetate and aqueous extraction | leaves | Saponin, flavonoids, terpenoid, alkaloids and tannin | Indonesia | *Salmonella* Typhi | 2020 | [88] |
| *Myristica fragrans* | Plant | aqueous extract | seeds | methane, oxybis [dichloro-, 1H-Cyclopenta [c] furan-3-(3aH)-one,6,6a-dihydro-1-(1,3-dioxolan-2-yl)-,(3aR, 1-t, Octadecane, 6-methyl-, Heptadecane, 2,6,10,14-tetramethyl-, BIS (2-Ethylhextl) phthalate, 4H-Pyran-4-one,2,3-dihydro-3,5-dihydroxy- 6-methyl-, 3,4-Dichlorophenethylamine and 1,4-Benzenediol, 2-bromo- | India | *MDR Salmonella* Typhi *isolates (MCASMZU1–13)* | 2020 | [89] |

| Parental organism | *Origin* | Extraction method | Segment used | Bioactive compound(s) | Location | ***Salmonella*** **serovar** | Year | Ref |
|---|---|---|---|---|---|---|---|---|
| *Kalanchoe brasiliensis* Cambess. | Plant | hydroethanolic extract | leaves | flavones and flavonols (3-hydroxyflavones or flavonols with substituted 3-hydroxyl groups (methylated or glycosylated)) | Brazil | *Salmonella* Gastroenteritis | 2019 | [90] |
| White mustard | Plant | essential oil | essential oil | not specified. Synergic with carvacrol and thymol | USA | *Salmonella* Typhimurium | 2019 | [91] |
| *Quercus variabilis* Blume | Plant | 70% ethanol followed by petroleum ether, ethyl acetate, n-butanol and water | valonia and shell | ellagic acid, theophylline, caffeic acid and tannin acid | China | *Salmonella* Paratyphi A, *Salmonella* Typhimurium and *Salmonella* Enteritidis | 2019 | [92] |
| *Melia azedarach* | Plant | ethanol, ethylacetate, hexane, dichloromethane and methanol extracts | leaves | not specified | Syria | *Salmonella* Typhi | 2019 | [93] |
| Ocotea minarum | Plant | 80% ethanol followed by hexane and ethyl acetate | leaves and stem bark | caffeic acid, p-coumaric acid, rosmarinic acid, quercetin and luteolin | Brazil | *Salmonella* Typhimurium *(14028)*, *Salmonella* Enteritidis *(13076)* | 2019 | [94] |
| *Zingiber zerumbet* | Plant | ethanolic extract | rhizome | Alkaloids, terpenoids, and tannins | Indonesia | *S.* Enteritidis *(ATCC 31194)* and *Salmonella* Typhimurium *(ATCC 23564)* | 2019 | [95] |
| Annona muricata | Plant | ethanolic extract | leaves | flavonoids, alkaloids, terpenoids, saponins, coumarins, lactones, | Indonesia | *Salmonella* Typhimurium *(FNCC-0050)* | 2019 | [96] |
| *Ligustrum lucidum* Ait, Lysimachia christinae Hance, Mentha piperita Linn and *Cinnamomum cassia* Presl | Plant | aqueous extracts | fruits, whole plants, leaves, and barks | phenolic acid and flavonoid | China | *S.* Typhimurium *(ST21) (used for prevent contracting infection)* | 2019 | [97] |
| Pectin of *Spondias dulcis* | Plant | aqua, ethanol | Fruit peel | oligosaccharides | Cameroon | *Salmonella* Typhimurium *(ATTC 2680)*, *Salmonella* Typhimurium *(ATTC 2488)* and *Salmonella choleraesuis* | 2019 | [98] |

| Parental organism | *Origin* | Extraction method | Segment used | Bioactive compound(s) | Location | *Salmonella* serovar | Year | Ref |
|---|---|---|---|---|---|---|---|---|
| *Acacia farnesiana* | Plant | hexanic, chloroform, methanolic and aqueous extracts | fruits | methylgallate, gallicacid and (2S)-naringenin-7-O-β-glucopyranoside | Mexico | *Salmonella* Enteritidis *(ATCC857)* | 2019 | [99] |
| *Adansonia digitata* | Plant | aqueous, ethanolic and chloroform extract | leaves and stem bark | alkaloid, flavonoids and tannin | Nigeria | *Salmonella* Typhi *(clinical isolate)* | 2018 | [100] |
| *Cassia occidentalis* | Plant | aqueous extract | leaves | saponin, flavonoids, and tannins, glycoside, cardiac glycosides, steroids, saponin glycoside, anthraquinones and volatile oil (trace) | Nigeria | *Salmonella* Typhimurium | 2018 | [101] |
| *Benincasa hispida* Thunb (Bligo fruit) | Plant | ethanol extracts (96, 70 and 50%) | fruit | not specified (probably a polar molecule) | Indonesia | *Salmonella* Typhi | 2018 | [102] |
| *Citrus sinensis* (L) Osbeck | Plant | aqueous and ethanol (80%) extracts | peel | alkaloid, tannin, saponin, glycoside, flavonoid, terpenoid, and Phenols | Nigeria | *Salmonella* Typhi *(clinical isolate)* | 2018 | [103] |
| Cinammomum cassia | Plant | Sodium bisulfite (1:1), petroleum ether | oil | cinnamaldehyde | Indonesia | *Salmonella* Typhi | 2018 | [104] |
| *Piper aduncum subsp. ossanum* (C. DC.) Saralegui, Piper aduncum *L. subsp*. aduncum, Mentha piperita L., Mentha spicata L., Ocimum basilicum var. genovese L. Ocimum gratissimum L., Rosmarinus officinalis L., Thymus vulgaris L., Melaleuca quinquenervia (Cav) S.T. Blake, *Eugenia axillaris* L., Citrus sinensis (L.) Osbeck, Citrus paradisi Macfad, *Curcuma longa* L., Lippia graveolens (Kunth) | Plant | essential oil (aqueous) | not specified | Probably trans-cinamaldehyde, carvacrol, eugenol and acid 2,4 dihydroxybenzoic | Cuba | *Salmonella* Typhimurium *(ATCC14028), Salmonella enterica subsp. enterica CENLAC (S02, S04, S06, S08, S10), Salmonella enterica (Sc1) isolated from a pig* | 2018 | [105] |

| **Parental organism** | ***Origin*** | **Extraction method** | **Segment used** | **Bioactive compound(s)** | **Location** | ***Salmonella* serovar** | **Year** | **Ref** |
|---|---|---|---|---|---|---|---|---|
| Jacaranda micrantha | Plant | aqueos and 70% ethanolextract | leaves | phenolic compounds, tannins, flavones and saponins | Brazil | *Salmonella choleraesuis (ATCC 10708)* and *Salmonella spp. (food isolated)* | 2018 | [106] |
| *Allium sativum* and Zataria multiflora Boiss | Plant | aqueous oil extract | bulb and whole plant | Allicin and thymol | Egypt | *Salmonella* Typhimurium, *Salmonella* Anatum, *Salmonella* Lagos and *Salmonella* Kentucky | 2018 | [107] |
| *Cinnamomum zeylanicum*, *Eugenia caryophyllata*, *Origanum vulgare*, Thymus vulgaris and Thymus zygis | Plant | essential oil | bark, bud, flowering plant, leaves and flowers | cinnamaldehyde, linalool, eugenol, eugenyl acetate, b-Caryophyllene, carvacrol, thymol, y-Terpinene, geraniol and p-Cymene. | Spain | *Salmonella* Typhimurium *(ATCC14028)*, *Salmonella* Typhimurium and *Salmonella* Enteritidis | 2018 | [108] |
| *Citrus medica*, *Citrus limon* and Citrus microcarpa | Plant | juice (pure extract) | fruit | citric acid, hesperidin, carvacrol and thymol | Korea | *Salmonella* Typhimurium *(ATCC 14028, 19585,* and *DT104 Killercow)* | 2018 | [109] |
| *Equisetum telmateia* | Plant | ethanolic extract followed by petroleum ether, dichloromethane (DCM), ethyl acetate (EtAc) and n-Butanol (n-BuOH). Supercritical extract | stem | Kaempferol 3-O-(6”-O-acetylglucoside), 5-O-Caffeoyl shikimic acid, Catechin | Iran | *Salmonella* Typhi *(PTCC 1609)* | 2018 | [110] |
| Thymus vulgaris L., Rosmarinus officinalis L. | Plant | essential oils | leaves | a-pinene, Thymol, Oxygenated monoterpenes, monoterpene hydrocarbons, borneol, 1,8-cineole | Morocco | *Salmonella* Typhimurium *(ATCC 14028)* | 2018 | [111] |
| *Gracilaria verrucosa* | Plant (algae) | aqueous, methanolic and ethanolic extracts | whole plant | carvacrol, p-cymene and y-terpinene | Indonesia | *Salmonella* Typhimurium | 2018 | [112] |
| Sterculia spp. | Plant | ethanolic extract | bark | flavonoids, alkaloids and saponins | Indonesia | *Salmonella* Typhi | 2018 | [113] |

| Parental organism | *Origin* | Extraction method | Segment used | Bioactive compound(s) | Location | *Salmonella* serovar | Year | Ref |
|---|---|---|---|---|---|---|---|---|
| *Nigella sativa* | Plant | aqueous and methanolic extracts and oil | seed | not specified | Pakistan | *Salmonella enterica* | 2018 | [114] |
| Rice hull smoke extract | Plant | pyrolysis of rice hulls followed by liquefaction | hull | 161 components, bioactive unknown | Korea | *Salmonella* Typhimurium *(CCARM8107)* | 2018 | [115] |
| Basil, ginger, hyssop, caraway, juniper, and sage | Plant | essential oils | several | estragole, cis-pinocamphone, alpha-pinene (in juniper EO), a-thujone (in sage EO), carvone (in caraway EO) and curcumene (in ginger EO) | Serbia | *Salmonella enterica* | 2017 | [116] |
| *Ipomoea aquatica* | Plant | ethanolic and methanolic extracts | leaves | flavonoids | Malaysia | *Salmonella* Typhi | 2017 | [117] |
| *Andrographis paniculata* | Plant | methanolic, ethanolic and acetone extracts | leaves | not specified | India | *Salmonella* Typhi *(clinical isolates)* | 2017 | [118] |
| *Senna occidentalis* | Plant | methanolic extract | root and leaves | flavonoid, tannins, saponins, cardial glycoside | Nigeria | *Salmonella* Typhi | 2017 | [119] |
| *Grewia flava* | Plant | acetone, methanolic, acetylacetate and aqueous extracts | berries, leaves, bark and roots | pelargonidin 3,5-diglucoside, naringenin-7-O-β-D-glucoside, tannins, catechins, and cyanidin-3-glucoside, betulin, lupeol, lupenone and friedelin. | South | *Salmonella* Typhimurium *(ATCC 14028)* | 2017 | [120] |
| *Acacia mearnsii* De Wild., *Aloe arborescens* Mill., *A. striata* Haw., Cyathula uncinulata (Schrad.) Schinz, Eucomis autumnalis (Mill.) Chitt., *E. comosa* (Houtt.) Wehrh., *Hermbstaedtia odorata* (Burch. ex Moq.) T.Cooke, Hydnora africana Thunb, Hypoxis latifolia Wight, Pelargonium | Plant | acetone extract | bark, leaves, bulb, tuber, root and corms | quercetin-3-O-a-l-arabinopyranoside (*P. guajava*) | South Africa | *Salmonella Isangi*, *Salmonella* Typhi, *Salmonella* Typhimurium | 2017 | [121] |

| Parental organism | *Origin* | Extraction method | Segment used | Bioactive compound(s) | Location | *Salmonella* serovar | Year | Ref |
|---|---|---|---|---|---|---|---|---|
| sidoides DC, Psidium guajava L and Schizocarphus nervosus (Burch.) van der Merwe | | | | | | | | |
| *Holarrhena floribunda* | Plant | ethanolic and methanolic extracts | leaves | Alkaloids | Togo | *Salmonella* Typhi *(clinical strains)* | 2017 | [122] |
| *Zanthoxylum caribaeum* Lam. | Plant | ethanolic, methanolic, hexanic, acetone, dichloromethanic, ethylacetate and aqueous extracts | leaves | Germacrene-D, a-Panasinsene and b-Selinene | Brazil | *Salmonella enterica* | 2017 | [123] |
| Rosmarinus officinalis | Plant | essential oil | not specified | not specified | Iran | *Salmonella* Typhimurium *(PTCC 1609)* | 2017 | [124] |
| Myristica fragans | | aqueous extracts | seed | methane, oxybis | India | *Salmonella* Typhi | 2017 | [125] |
| *Cajanus cajan* (Gandul) | Plant | methanolic extract | leaves | flavonoids, phenolics, and steroids (naringenin) | Indonesia | *Salmonella Thypi* | 2017 | [126] |
| Vitex doniana | Plant | aqueous and methanolic extracts | stem-bark and leaves | phytochemicals alkaloid, saponin, tannin, anthraquinone, flavonoid, phenols, terpenoid andglycoside | Nigeria | *Salmonella* Typhi | 2017 | [127] |
| *Hibiscus sabdariffa* | Plant | aqueous water | flower calyx | not specified | Mexico | *Salmonella* Typhimurium and Typhi | 2017 | [128] |
| Ziziphora clinopodioides | Plant | essential oil | leaves | nisin | Iran | *Salmonella* Typhimurium *(ATCC 14028)* | 2017 | [129] |
| Tinospora cordifolia | Plant | aqueous and methanolic extracts | stem | not specified | India | *S.* Typhimurium *(ATCC 23564)* | 2017 | [130] |
| *Sonchus arvensis* L. (tempuyung) | Plant | ethanol extracts | leaves | flavonoids and triterpenoids | Indonesia | *Salmonella* Typhi | 2016 | [131] |

| Parental organism | *Origin* | Extraction method | Segment used | Bioactive compound(s) | Location | *Salmonella* serovar | Year | Ref |
|---|---|---|---|---|---|---|---|---|
| Heliotropium filifolium (Miers) Reiche and of Heliotropium sinuatum (Miers) | Plant | resinous exudate (CH2Cl2) and hexane-ethyl acetate step gradient | fresh parts (cuticular components) | (Filifolinol) (naringenin, 3-O-methylgalanginand pinocembrin) | Chile | *Salmonella* Typhimurium *(ATCC 14028)* | 2016 | [132] |
| *Punica granatum*, oak, Thymus vulgaris and Cinnamomum zeylanicum | Plant | ethanolic and chloroformic extracts | peel, oak trunk, thyme fruit and cortex | gallocatechins, delphinidin, cyanidin, gallicacid,ellagic acid, pelargonidin and sitosterol; hymol, carvacrol and flavonoids; cynamaldheide; (4,5-Di-o-galloyl (+) –protoquercitol) and compound III (3,5-Di-o-galloyl (+)-protoquercito | Iraq | *Salmonella* Typhimurium *(chicken isolate)* | 2016 | [133] |
| Scutellariae radix | Plant | ethanol extracts followed by petroleum ether (PEF), chloroform (CF), ethyl acetate (EAF) and n-butanol (BF) | root | baicalin, wogonoside, baicalein and wogonin | China | *Salmonella* Typhimurium *(CMCC 50041)* | 2016 | [134] |
| *Rhus typhina* and Achillea sintenisii | Plant | not specified | aerial and root parts | not specified | Portugal | *Salmonella* Typhimurium *LT2* | 2016 | [135] |
| Holarrhena antidysentrica (Ha) and Andrographis paniculata (Ap) | Plant | hydroethanolic extract | leaves and stem | alkaloids, flavonoids, saponin, terpenes, phenols, tannins, glycosides carotenoids, anthraquinones, reducing sugars, phlobatannins, sterols | India | *Salmonella* Typhimurium *(MTCC 733)* | 2016 | [136] |
| Black tea (Kombucha) | Plant | Infusion/fermentation | leaves | Catechin and isorhamnetin | India | *Salmonella* Typhimurium *(NCT 572)* | 2016 | [137] |
| Curcuma longa | Plant | 96% ethanol/essential oil | rhizomes | saponin, tannins, alkaloids and flavonoids (probably curcumin and derivatives) | Colombia | *Salmonella spp. (nosocomial isolates)* | 2016 | [138] |

| **Parental organism** | ***Origin*** | **Extraction method** | **Segment used** | **Bioactive compound(s)** | **Location** | ***Salmonella* serovar** | **Year** | **Ref** |
|---|---|---|---|---|---|---|---|---|
| Harungana madagascariensis | Plant | aqueous infusion | leaves | not specified | Cameroon | *Salmonella* Typhimurium | 2016 | [139] |
| *Piper retrofractum*, Phyllanthus emblica, *Terminalia chebula*, *Terminalia bellirica*, Piper sarmentosum, *Plumbago indica*, Piper leptostachyum, *Piper nigrum*, *Zingiber officinale*, *Piper betle*, *Garcinia mangostana* and Caesal piniasappan | Plant | 95% ethanol | fruits, root, stem, rhizome, leaves husk, peduncle and wood | Plumbagin, Piperine, Eugenol, Myristicin, Gingerol, Shogaol and Brazilin | Thailand | *Salmonella spp. (piglet isolates)* | 2016 | [140] |
| Punica granatum | Plant | ethanolic extracts and peel flour | peel, seeds | ellagic acid or ellagic acid derivatives, ellagitannins and HHDP-gallagyl-hexoside | Spain | *Salmonella* Anatum, *Salmonella* Typhimurium | 2016 | [141] |
| *Abrus precatorius* L. | Planta | aqueous extracts | leaves, seed and root | steroids, saponins, phenolics, tannins, flavonoids, terpenoids and alkaloids | Nigeria | *Salmonella* Typhi | 2016 | [142] |
| Piliostigma thonningii | Plant | hexane and aqueous extracts | leaves | Tannins, terpenoids, flavonoids, alkaloids, steroids and phenols | Nigeria | *Salmonella* Typhi | 2015 | [143] |
| *Baillonella toxisperma* | Plant | ethyl acetate, acetone, methanol and hydro-ethanol mixture (2: 8) extracts | leaves and stem bark | terpenoids, tannins, flavonoids, phenols, saponins, steroids and cardiac glycosides. | Cameroon | *Salmonella* Typhi | 2015 | [144] |
| Wood vinegar | Plant | vinegar | natural vinegar | not specified (probably pH 4.15–4.59) | Thailand | *Salmonella* Enteritidis *(DMST15676) Salmonella* Typhimurium *(DMST17242)* | 2015 | [145] |
| Aristolochia indica, Carica papaya, *Eclipta alba* and *Phyllanthus amarus* | Plant | methanol extracts | leaves | n-Hexadecanoic acid | India | *Salmonella* Typhi *(clinical isolate)* | 2015 | [146] |

| Parental organism | *Origin* | Extraction method | Segment used | Bioactive compound(s) | Location | *Salmonella* serovar | Year | Ref |
|---|---|---|---|---|---|---|---|---|
| Curry: *Capsicum annuum*, Citrus hystrix, *Cuminum cyminum* L., Allium ascalonicum L., Allium sativum, Cybopogon citratus, Alpinia galangal, ando coconut milk | Plant | water, UHT coconut milk, and fresh coconut milk were used as extractants. Also ethanolic and aquous extracts (Garlic) | fruit, leaves and peel | not specified | Thailand | *Salmonella* Enteritidis | 2015 | [147] |
| *Portulaca oleracea* | Plant | ethanol extracts | leaves | probably quercetin | Thailand | *Salmonella* Typhi | 2015 | [148] |
| Eucalyptus, mint, cinnamon, garlic, thymus | Plant | oil | bark and leaves | probably cinnamaldehide/ thymol | Egypt | *Salmonella* Enteritidis, *Salmonella* Charity and *Salmonella* Remiremont *(chicken isolates)* | 2015 | [149] |
| Piper crocatum (Red betel vine) | Plant | 70% ethanol, followed by n-hexane, ethyl acetate, chloroform and methanol | leaves | saponin and flavonoids | Indonesia | *Salmonella* Typhi | 2015 | [150] |
| Dionisya revoluta | Plant | methanol extracts | aerial parts | not specified | Iran | *Salmonella* Enteritidis | 2015 | [151] |
| *Achyranthes aspera* | Plant | methanolic extracts followed by chloroform, n-hexane, n-butanol, ethyl acetate and water | leaves | Phenolic compounds, oils, saponins, flavonoids, alkaloids and tannins | Pakistan | *Salmonella* Typhi *(ATCC 19430)* | 2015 | [152] |
| *Alocasia brisbanensis*, *Canavalia rosea*, *Corymbia intermedia*, Hibbertia scandens, *Ipomoea brasiliensis*, Lophostemon suaveolens, *Syncarpia glomulifera*, Smilax australis and Smilax glyciphylla | Plant | hydro-ethanolic (80%) and aqueous extracts | not specified | *L. suaveolens* leaves: α-pinene, β-caryophyllene, aromadendrene, globulol and spathulenol; *S. glomulifera*: α-pinene, aromadendrene and globulol; S. glomulifera leaves wax: eucalyptin and S. glomulifera bark: betulinic acid, oleanolic acid-3-acetate and ursolic acid-3-acetate. | Australia | *Salmonella typhimurium—Group B (clinical isolate)* | 2015 | [153] |

| Parental organism | *Origin* | Extraction method | Segment used | Bioactive compound(s) | Location | *Salmonella* serovar | Year | Ref |
|---|---|---|---|---|---|---|---|---|
| Citrus sinensis | Plant | ethanol, methanol, chloroform, and diethyl ether | peel | saponins, terpenoids, slkaloids, flavonoids, tannins and cardiac glycosides | Pakistan | *Salmonella* Typhimurium *(isolated from spoiled fish)* | 2015 | [154] |
| Nigella sativa | Plant | essential oil | seed | thymoquinone, p-cymene, a-phellandrene, a-pinene, b-pinen, cis-carveol, trans-anethole, thymol, alongipinene and longifolene | Arabia and India | *Salmonella* Paratyphi *A*, *Salmonella* Enteritidis, *Salmonella* Typhimurium, *Salmonella* Heidelberg, *Salmonella* Agona, *Salmonella bongori* | 2015 | [155] |
| Allium sativum L. | Plant | aqueous extracts | bulb | not specified | South Korea | *Salmonella* Typhimurium | 2015 | [156] |
| Spirulina platensis | Plant (algae) | ethanolic and chloroform extracts | cell extracts | not specified | Bangladesh | *Salmonella* Typhi and *Salmonella* Paratyphi | 2015 | [157] |
| Curri = Capsicum annuum, Citrus hystrix, Cuminum cyminum L., Allium ascalonicum L., Allium sativum, Cybopogon citratus, Alpinia galangal, ando coconut milk | Plant | not specified | fruit, peel, seed, bulb, stem, rhizome | especificados por compuesto, reportes previos | Thailand | *Salmonella* Typhimurium *(DT104b)* | 2015 | [158] |
| Vitex doniana | Plant | ethanolic and Acetone extracts | leave, stem bark and root | tannin, saponins, flavonoid, carbohydrate, glycoside, protein and steroid | Nigeria | *Salmonella* Typhi | 2015 | [159] |
| Polygonum odoratum | Plant | essential oil | leaves | Dodecanal 55.49%, Decanal 11.57%, Pentacosane 7.26%, p-Anis aldehyde 6.35% mainly | Thailand | *Salmonella choleraesuis subsp. choleraesuis (ATCC 35640)* | 2015 | [160] |
| Kelussia odoratissima | Plant | aqueous and ethanolic extracts | leaves | not specified | Iran | *Salmonella typhimurium (ATCC 14028)* | 2014 | [161] |

| Parental organism | *Origin* | Extraction method | Segment used | Bioactive compound(s) | Location | ***Salmonella*** **serovar** | Year | Ref |
|---|---|---|---|---|---|---|---|---|
| Coptidis rhizoma (CR), Houttuyniae herba, Taraxaci herba, Glycyrrhizae radix, Puerariae radix, and Rhizoma dioscoreae | Plant | aqueous infusion | herbs | berberine, ginsenoside Rb1, and glycyrrhizin | China | *Salmonella* Typhimurium *(ATCC 6994)* and *ST21 (pig carrier)* | 2014 | [162] |
| Nymphea tetragona | Plant | 50% methanol followed by dichloromethane, ethyl acetate, and butanol | body and root | DFNTE: hydrocarbons (46.46%); EFNTE: methyl gallate (70.44%), 1, 2, 3-benzenetriol or pyrogallol (20.61%), and 6, 8-dimethylbenzocyclooctene (5.90%); BFNTE: 2-hydrazinoquinoline (57.61%), pyro-gallol (20.09%), and methyl gallate (12.77%) | Korea | *Salmonella* Typhimurium *(QC strain KTCC2515* and *clinical isolates ST171, ST482, ST688,* and *ST21)* | 2014 | [163] |
| Virgin coconut oil and palm kernel oil | Plant | essential oil | fruit and seed | not specified | Indonesia | *Salmonella* Typhi *(ATCC 786)* | 2014 | [164] |
| Virgin Coconut Oil | Plant | oil | fruit | not specified | Indonesia | *Salmonella* Typhi *(ATCC 00786)* and *Salmonella* Typhimurium *(ATCC 14028)* | 2014 | [165] |
| Piper nigrum L. | Plant | ethanolic extracts and chloroform extracts | fruit and seed | tannins, alkaloids and Cardiac glycosides, and tannins, alkaloids and flavonoids | India | *Salmonella* Typhi | 2014 | [166] |
| *Morus alba* var. Alba, Morus alba var. Rosa and *Morus rubra* | Plant | hydromethanolic and aqueous extracts | leaves and stem | phenolics and flavonoids | Tunisia | *Salmonella* Typhimurium *(ATCC 14028)* | 2014 | [167] |
| *Khaya senegalensis* | Plant | ethanolic, methanolic and aquous extracts | stem bark | saponins, tannins, reducing sugar, aldehyde, phlobatannins, flavonoids, terpenoids, alkaloids, cardiac glycoside and anthroquinones | Nigeria | *Salmonella* Typhi | 2014 | [168] |

| Parental organism | *Origin* | Extraction method | Segment used | Bioactive compound(s) | Location | *Salmonella* serovar | Year | Ref |
|---|---|---|---|---|---|---|---|---|
| *Mentha longifolia* | Plant | Ethanolic extracts | leaves | not specified | Iran | *Salmonella* Typhimurium | 2014 | [169] |
| Palm oil (Sania), Virgen coconut oil (Palem Mustika) and soybean oil (Mama Suka) | Plant | oils | seeds and fruits | not specified | Indonesia | *Salmonella* Typhi *(ATCC 19943)* | 2014 | [170] |
| Heliotropium | Plant | methanolic extract, 2nd extraction with petroleum ether, ethylacetate and chloroform and aqueous | aerial parts | not specified | Iran | *Salmonella* Enteritidis *(ATCC 13311)* | 2014 | [171] |
| Woad, heartleaf houttuynia herb, baical skullcap, coptidis, andrographitis, | Plant | aqueos extract | bark, leaves, root, rhizome and fruit | not specified | China | *Salmonella* Typhimurium | 2013 | [172] |
| *K. senegalensis* bark and leaves, *S. alexandrina* leaves, S. argel leaves, *T. indica* L. fruits and T. foenum, graecum seeds | Plant | methanolic extract | bark, leaves and seed | not specified | Sudan | *Salmonella* Typhi *(ATCC19430)* and *Salmonella Paratyphyphi-A (ATCC 9150 / SARb42)* | 2013 | [173] |
| Phylanthus amarus | Plant | aqueous and ethanolic extracts | leaves | Phyllanthin, Nirtetralin, Linalool, phytol | India | *Salmonella* Typhi | 2013 | [174] |
| Carissa opaca | Plant | 95% methanol followed by n-hexane, ethyl acetate, chloroform, butanol and water | fruits | orientin, isoquercetin, myricetin and apigenin (and probably other secondary metabolites) | Pakistan | *Salmonella typhy (ATCC 0650)* | 2013 | [175] |
| *Mangifera indica* | Plant | acetone extract | leaves | mangiferin | Pakistan | *Salmonella* Typhi *(clinical isolates)* and *Salmonella (ATCC 14028)* | 2013 | [176] |
| *Sinapis alba* L. | Plant | essential oil | seeds | 4-hydroxybenzyl isothiocyanate | USA | *Salmonella spp. (isolates) Salmonella* Typhimurium *(ATCC 14028), Salmonella* Abaetuba and *Salmonella* Dessau | 2013 | [177] |

| Parental organism | *Origin* | Extraction method | Segment used | Bioactive compound(s) | Location | ***Salmonella*** **serovar** | Year | Ref |
|---|---|---|---|---|---|---|---|---|
| Origanum vulgare | Plant | essential oil | seed | not specified | USA | *Salmonella* Newport *(LAJ160311)* | 2013 | [178] |
| Annona comosus and Citrus senensis | Plant | ethanolic extract | peel | alkaloids, flavonoids, saponins, tannins | Nigeria | *Salmonella paratyphi-B*, and *Salmonella* Typhi | 2013 | [179] |
| Carthamus nctoricus L., Poncirus trifollata Raf., Scutellaria balcalensis Georgi, Prunus sargentii, *Cucurbita moschata*, *Allium cepa* L., Portulaca oleracea L., Xanthium strumarium L., Duchesnea chrysantha, *Cudrania tricuspidata* and *Juniperus chinensis* | Plant | ethanolic extract | leaves, peel | not specified | Korea | *Salmonella* Gallinarum | 2013 | [180] |
| Herba pogostemonis | Plant | aqueous extract | leaves | acetol,D-sphignosin, 5-aminoimidazole-1-carboxyamie, caffeic acid, chlorogenic acid, neohesperedin,O-acetylsalicylic acid, quinic acid,3,4-dihydroxybenzoic acid, andDL-hydroxyphenylglycol | Korea | *Salmonella* Typhimurium *(ATCC140)* | 2012 | [181] |

| Parental organism | *Origin* | Extraction method | Segment used | Bioactive compound(s) | Location | *Salmonella* serovar | Year | Ref |
|---|---|---|---|---|---|---|---|---|
| Enicostemma littorale | Plant | chloroform, methanol and acetone by soxhlet apparatus | leaves, stem and root | not specified | India | *Salmonella* Typhi | 2012 | [182] |
| *Hibiscus rosa-sinensis* | Plant | aqueous and ethanolic extracts | flower extract | cyanidin, quercetin, hentriacontane, calcium oxalate, thiamine, riboflavin, niacin and ascorbic acids | India | *Salmonella spp.* | 2012 | [183] |
| Capsicum annuum and *Capsicum frutescens* | Plant | aqueous and methanolic extracts | fruit | alkaloids, flavonoids, polyphenols, and sterols | Ivory Coast | *Salmonella* Typhimurium *(ATCC 13311)* | 2012 | [184] |
| Coriandrum sativam (L.) | Plant | essential oil | fruit dry | Bicycle(4.1.0),heptanes,3,7,7-trimethyl-(1a,6a,3a), propanoic acid,2-methyl-3,7-dimethyl octadiennyl ester, (E)-, 2- undecenal, 2-Napthalenemethanol, decahydro-a,a,4a-trimethyl-8-methylene- [2R-(2a,4aa,8aa)] | India | *Salmonella* Typhi | 2012 | [185] |
| Berberis baluchistanica, Seriphidium quettense, Iphionaaucheri, Ferula costata | Plant | crude methanol extracts | roots, aerial parts | not specified | Pakistan | *Salmonella* Typhimurium | 2012 | [186] |
| *Oenothera rosea* | Plant | aqueous and ethanolic extracts | aerial parts | not specified | Mexico | *Salmonella* Enteritidis *(clinical isolate)* | 2012 | [187] |
| Ocimum gratissimum and Gongronema latifolium | Plant | aqueous and ethanolic extracts | leaves and stem | not specified | Nigeria | *Salmonella* Typhi | 2012 | [188] |
| Curry: Capsicum annuum, Citrus hystrix, Cuminum cyminum L., Allium ascalonicum L., Allium sativum, Cybopogon citratus, | Plant | Kaeng Kathi (UHT coconut milk | fruit, peel, seed, bulb, stem, rhizome | not specified | Indonesia | *Salmonella* Typhimurium *U302 (DT104b)* | 2012 | [189] |

| Parental organism | *Origin* | Extraction method | Segment used | Bioactive compound(s) | Location | *Salmonella* serovar | Year | Ref |
|---|---|---|---|---|---|---|---|---|
| Alpinia galangal, ando coconut milk | | | | | | | | |
| *Averrhoa bilimbi* Linn | Plant | ethanolic extract | leaves | not specified | Indonesia | *Salmonella* Typhi | 2012 | [190] |
| Ocimum gratissimum | Plant | ethanol extracts | leaves | alkaloids, cardiac glycosides, saponins, tannins and steroids | Nigeria | *Salmonella* Typhi *(clinical isolates)* | 2011 | [191] |
| *Ardisia elliptica* Thumb | Plant | 95% ethanol | fruit | anythocyanins and syringic acid | Thailand | *Salmonella sp.* | 2011 | [192] |
| Quercus | Plant | ethanol by soxhlet apparatus | acorn | not specified | Iran | *Salmonella* Typhi *(MDR)* | 2011 | [193] |
| Sonchus spp. (6 sp) *S. arvensis*, *S. oleraceus*, S. Lingianus, S. Brachyotus, *S. asper*, *S. uliginosus* | Plant | methanolic extract | aerial parts | phenols and flavonoids | China | *Salmonella enterica* | 2011 | [194] |
| York cabbage, Brussels sprouts, broccoli and white cabbage | Plant | methanolic extract | whole plant | Hydroxybenzoic acid, hydroxycinnamic acid, flavone, polymethoxylated flavone, glycosylated flavonoid and anthocyanin | Ireland | *Salmonella Abony (NCTC 6017)* | 2011 | [195] |
| Achyrocline satureioides | Plant | ethanolic extract | aerial parts | 23-methyl-6-Odesmethylauricepyrone | Argentina | *Salmonella* Enteritidis | 2011 | [196] |
| *Trapa bispinosa* Roxb | Plant | methanolic extract | fruit | not specified | Bangladesh | *Salmonella* Typhi | 2011 | [197] |
| *Acalypha indica* | Plant | methanolic extract | leaves and roots | not specified | India | *Salmonella* Typhi | 2011 | [198] |
| Punica granatum | Plant | ethanolic extract | peel | not specified | Korea | *Salmonella* Typhi *(ATCC 19943)*, *S.* Dublin *(ATCC 39184)*, *S.* Derby *(ATCC 6960)*, *S. choleraesuis (ATCC 7001) y S.* Gallinarum *(ATCC 9184)*, *S.* Enteritidis, *S.* | 2011 | [199] |

| Parental organism | *Origin* | Extraction method | Segment used | Bioactive compound(s) | Location | *Salmonella* serovar | Year | Ref |
|---|---|---|---|---|---|---|---|---|
| | | | | | | Typhimurium, *S.* Gallinarum *y S. Paratyphi A* | | |
| Punica granatum L.. Eugenia jambolana Lam., *Eugenia uniflora* L., Caryophyllus aromaticus L., Psidium araca Raddi, Achyrocline satureioides (Lam.), Rosmarinus officinalis L., Cynara scolymus L., Salvia officinalis L., Laurus nobilis L., Bidens pilosa L., Baccharis trimera (Less.) DC, *Plectranthus barbatus* Andrews, *Sonchus oleraceus* L., Mikania glomerata Spreng., *Taraxacum officinale* F.H. Wigg, Emiia sonchifolia (L) DC, *Plantago australis* Lam., Maytenus ilicifolia (Schrad) Planch, Aloe arborescens Mill., *Malva sylvestris* L. | Plant | hydromethanolic extracts | leaves, fruit, package content, aerial and flowered aerial portions. | not specified | Brazil | *Salmonella* Agona, *Salmonella* Anatum, *Salmonella* Cerro, Cerro*Cubana, Salmonella* Derby, *Salmonella* Enteritidis, *Salmonell Give, Salmonella* Heidelberg, *Salmonella* Infantis, *Salmonella* London, *Salmonella* Manhattan, *Salmonella* Meleagridis, *Salmonella* Montevideo, *Salmonella* Newport, *Salmonella* Oranienburg, *Salmonella* Panama, *Salmonella* Pullorum, *Salmonella* Typhimurium | 2011 | [200] |
| *Cucurbita pepo* | Plant | methanolic and ethanolic extracts | seed | saponins, flavonoids, Tannins, alkaloids, and steroids | Nigeria | *Salmonella* Typhi | 2011 | [201] |
| *Aloe vera* | Plant | methanolic and ethanolic extracts | leaves | Anthraquinone, Alkaloids, Saponins, Balsams, Flavonoids and Tannins | Nigeria | *Salmonella* Typhi | 2011 | [202] |
| Gynostemma pentaphyllum | Plant | ethanolic extract | leave, stem | not specified | Thailand | *Salmonella* Typhi *Salmonella* Typhimurium | 2011 | [203] |

| Parental organism | *Origin* | Extraction method | Segment used | Bioactive compound(s) | Location | ***Salmonella*** **serovar** | Year | Ref |
|---|---|---|---|---|---|---|---|---|
| Terminalia stenostachya y Terminaliaspinosa | Plant | dichloromethanic, methanolic, acetone and ethanolic extracts | stem barks and roots | not specified | Tanzania | *Salmonella* Typhi | 2011 | [204] |
| Hofmeisteria schaffneri | Plant | infusion and essential oil | aerial parts | hofmeisterin III, thymyl isovalerate and 8,9-epoxy-10-acetoxythymyl angelate | Mexico | *Salmonella* Typhi *(ATCC9992)* | 2011 | [205] |
| Ficus polita Vahl. | Plant | methanolic extract | roots | euphol-3- O -cinamato C 39 H 56 O 2, lupeol C 30 H 50 O, taraxar-14-eno C30 H 50 O 1 | Cameroon | *Salmonella* Typhi *(ATCC6539)* | 2011 | [206] |
| *Aegle marmelos* (L.) Corr. Serr. (Rutaceae), *Cassia fistula* L., Moringa oleifera Lam., Melia azedarach L., Bombax ceiba L. and *Brassica rapa ssp. campestris* L. | Plant | aqueous and methanolic extracts | vegetables, seeds | not specified | Pakistan | *Salmonella* Typhi | 2011 | [207] |
| 75 plants (Healianthus annum Linn.) | Plant | ethanolic extracts | leaves | not specified | India | *Salmonella* Typhosa | 2010 | [208] |
| *Syzygium cumini* | Plant | aqueous and ethanolic extracts | leaves | flavonoids, alkaloids, | India | *Salmonella* Enteritidis, *Salmonella* Typhi, *Salmonella* Typhi *A,Salmonella para*Typhi *A, Salmonella para*Typhi*B* | 2010 | [209] |
| Black pepper (Piper nigrum Linn.) | Plant | acetone extract; dichloromethanic extract; | fruit | piperine | India | *Salmonella* Typhi | 2010 | [210] |
| Abrus precatorius L. | Plant | methanolic and petroleum ether extract | leaves, seeds and roots | Methanolic and petroleum ether extract | India | *Salmonella* Typhi, *Salmonella Paratyphi A, Salmonella Paratyphi B* | 2010 | [211] |
| Psidium guajava | Plant | methanolic extract | leaves | flavonoids: morin-3-Olyxoside, morin-3-O-arabinoside, quercetin-3-Oarabinoside and quercetin. | Thailand | *Salmonella enterica (ATTC 8326)* | 2010 | [212] |

| Parental organism | *Origin* | Extraction method | Segment used | Bioactive compound(s) | Location | *Salmonella* serovar | Year | Ref |
|---|---|---|---|---|---|---|---|---|
| | | | | Anthocyans, alkaloids, tannins, and terpenoids. | | | | |
| Origanum vulgare L. | Plant | essential oils | aerial parts | not specified | Turkey | *Salmonella* Enteritidis *RSKK 96046,* | 2010 | [213] |
| *Tectona grandis* | Plant | methanolic extract | leaves | not specified | India | *Salmonella* Typhimurium *(MTCC 98* | 2010 | [214] |
| *Ocimum canum*, Acalypha indica, Eclipta alba and *Lawsonia inermis* | Plant | chloroform and methanol | whole plant | not specified | India | *Salmonella para*Typhi | 2010 | [215] |
| *Adiantum capillus-veneris* L. (Adiantaceae), Adiantum incisum forsk. (Adiantaceae), Adiantum lunulatum Burm. F. (Adiantaceae), Actiniopteris radiata (Swartz.), Enlace (Actiniopteridaceae), Araiostegia pseudocystopteris Copel. (Davalliaceae), Athyrium pectinatum (Wall ex Mett.) T. Moore (Athyriaceae), Chelienthes albomarginata Clarke (Sinopteridaceae), *Cyclosorus dentatus* (Forsk.) Ching (Thelypteridaceae), Dryopteris cochleata (Don.) C. Chr. (Dryopteridaceae), hipodematio crenatum (Forsk.) Kuhn (Hypodematiaceae), *Marsilea minuta* L. (Marsileaceae) y Tectaria coadunata (J. Smith) | Plant | aqueous and methanolic extracts | leaves | not specified | India | *Salmonella arizonae (MTCC No. 660)*, *Salmonella* Typhi *(MTCC No. 734)* | 2010 | [216] |

| Parental organism | *Origin* | Extraction method | Segment used | Bioactive compound(s) | Location | *Salmonella* serovar | Year | Ref |
|---|---|---|---|---|---|---|---|---|
| Ecklonia cava | Plant (algae) | EtOH followed by n-hexane, CH2Cl2, EtOAc, n-BuOH (10 g) and aqueous | not specified (probably full algae) | Eckol | Korea | *Salmonella* Typhi *(ATCC 19943), Salmonella* Dublin *(ATCC 39184), Salmonella* Derby *(ATCC 6960), Salmonella choleraesuis (ATCC 7001), Salmonella* Gallinarum *(ATCC 9184), Salmonella* Enteritidis, *Salmonella* Typhimurium, *S.* Gallinarum, and *Salmonella* Paratyphi *A* | 2010 | [217] |
| Thymus vulgaris L., Ocimum basilicum L., Coriandrum sativum L., Rosmarinus officinalis L., Salvia officinalis L., Foeniculum vulgare L., Mentha spicata L., Carum carvi L. | Plant | essential oil | not specified | not specified | Romania | *Salmonella enterica serovar* Enteritidis Cantacuzino *CICC10878, Salmonellaenterica serovar* Enteritidis | 2010 | [218] |
| Pikutbenjakul = *Piper longum*, Piper sarmentosum, Piper interruptum, Plumbago indica y Zingiber | Plant | ethanolic extract | not specified | not specified | Thailand | *Salmonella sp. Salmnella typhy amd salmonella* Typhimurium | 2010 | [219] |
| Quercus infectoria, *Kaempferia galanga*, Coptis chinensis and Glycyrrhiza uralensis | Plant | DMSO | galls, roots, rhizomes, | not specified | Thailand | *Salmonella* Typhi *(DMST 5784)* | 2010 | [220] |
| Eugenol | Plant | essential oil of clove | flower extract | 177 peaks and HHDP-gallagyl-hexoside | Indonesia | *Salmonella* Typhi | 2010 | [221] |
| *Sida rhombifolia* Linn. | Plant | methanolic extract | not especified | polyphenols, alkaloids and steroids | Cameroon | *Salmonella* Typhi, *Salmonella* Enteritidis | 2010 | [222] |

**Table 4.**
*Summary of frequently reported natural products from plant origin against* salmonella.

| Parental organism | *Origin* | Extraction method | Segment used | Bioactive compound(s) | Location | *Salmonella* serovar | Year | Ref |
|---|---|---|---|---|---|---|---|---|
| Goat milk kefir (Lactococcus cremoris, Streptococcus cremoris | Bacteria and Yeast | no extraction | microbial cells | lactic acid, ethanol and $CO_2$, diacetyl acetaldehyde, ethyl and | Indonesia | *Salmonella* Typhimurium *(ATCC 14028)* | 2019 | [223] |
| Epicoccum nigrum, *Entada abyssinica* | Fungus and plant | ethyl acetate extract | leaves | not specified | Cameroon | *Salmonella* Typhimurium | 2017 | [224] |
| Origanum vulgare, Lactococcus lactis (Nisin), EDTA | Plant and Bacteria | essential oil | seeds | carvacrol, p-cymene and y-terpinene | Brazil | *Salmonella* Enteritidis | 2016 | [225] |
| Allium sativum, Nigella sativa, *Azadirachta indica*, *Ficus carica*, *Trigonella foenum-graecum* and honey | Plant and Animal | aqueous extracts | bulb, seed, leaves and fruit | not specified | Pakistan | *Salmonella spp.* | 2014 | [226] |
| Apis mellipodae honey and Allium sativum | Animal and plant | macerated and aqueous dilution | honey/bulb | Honey: high, osmolarity, hydrogen peroxidase, acidity and Allium sativum: allicin | Ethiopia | *Salmonella* Typhi *(clinical isolate)* and *Salmonella spp.* *(NCTC 8385)* | 2013 | [227] |

**Table 5.**
*Summary of reported natural products of combined origins against* salmonella.

congress and meeting proceedings where useful data were present. NPs and bioactive principles were registered according to the molecules isolated by the authors and/or in contrast to the literature. This search was not exhaustive, rather, we aim to obtain a random sample of information using the simplest terms on the matter of natural products for salmonellosis.

All these works were developed on all continents, being Asia the most active, followed by Africa, America, Europe and Oceania (**Figure 1A** and **B**). It is noteworthy that much of the research was developed in equatorial locations where biodiversity is abundant. Country-wise, there is a remarkable number of publications from India and Indonesia, where incidence of *salmonella* is high. The map constructed for the distribution of publishing frequencies, in fact, resulted fairly similar to a previously reported *salmonella* incidence map (**Figure 1A** versus [9]). The number of articles per year showed an upward trend though it stabilizes in the last five years (**Figure 1C**).

The spectrum of biological activities evaluated are as diverse as the application to which they are oriented, from the study of antimutagenic, antioxidant,

| Animal species | Vegetable species | Microorganisms | Solvents | Bioactive compound |
|---|---|---|---|---|
| Identification of genus and, if possible, species | Geographical site | Identification of genus, and species | Explanation for its selection | Isolation technique |
| | Harvesting data | Reference strain identification | No-reactivity assessment | Structure determination method (MS–GC, for instance) |
| | Ethnobiological identification | Identification of origin:<br>• Clinical isolate<br>• Food<br>• Soil ... | No-interference assessment | |
| | | Identification by PCR | No-toxicity assessment | |

**Table 6.**
*Checklist proposed for NPs research.*

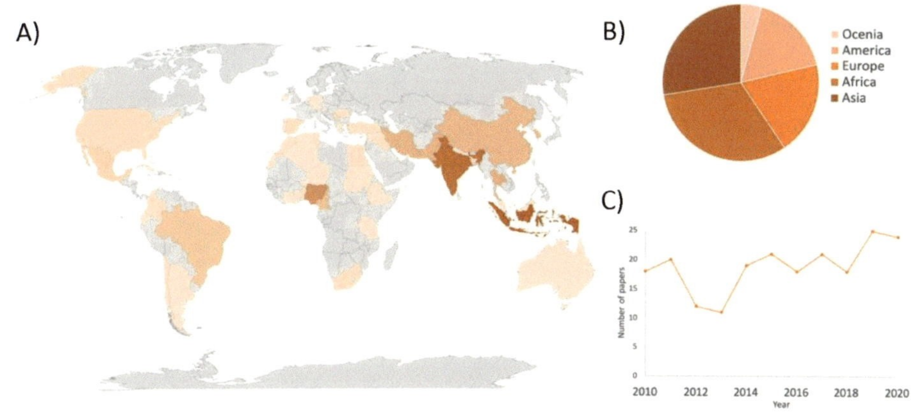


**Figure 1.**
*(A) Distribution map of publishing frequencies. (B) Continental frequency. (C) Publications per year.*

anticancer, anthelmintic, antiviral, antifungal activities to its antibacterial potential, being its activity against *salmonella* spp. one of the most studied activities.

The analysis of the last decade research render studies exploring the antibacterial activity against *salmonella* serovars of crude extracts and essential oils, from compounds of natural origin, as well as their components. A wide variety of these NPs have been evaluated from commercial formulations, products of animal origin such as honey, propolis, milk and chitosan, through complete plants and/or their components (roots, stem, leaves and flowers), up until products of microbial metabolism as crude protein extracts, membrane and cell wall glycosides, natural antibiotic peptides (nisin). Several chemical compounds such as water, ethanol, methanol, acetone, formaldehyde, hexane, ethyl acetate and chloroform were used as solvents by direct maceration extraction rather than vapor distillation or more complex methods.

Nonetheless, we believe the description of methodological conditions could further standardized with the inclusion of a fixed set of data. According to our observation, the list of items enlisted in **Table 6** could be a minimal checklist when performing NP research.

## 7. Conclusions

Salmonellosis, caused by *salmonella* serovars, is still an uneradicated disease both in industrialized and developing countries. Multidrug resistance is a phenomenon increasingly widespread and alternative tools for disease control are urgently necessary. Natural products research based on traditional medicine is nowadays a consolidated study field full of vitality, *salmonella* research in particular has an upward trend with work being develop worldwide. Authors cited within this chapter explored biological activities of local organisms for the solution of salmonellosis for their communities, although a minority showed interested in foreign resources or commercial formulations. We observed a higher number of active researches on countries with diverse and abundant natural resources coincidentally also with high salmonellosis incidence. Even though our search is a minimal sample from the whole work being published on NPs and salmonellosis, it reveals certain features of the field.

Most of the works displayed in here are initial screening in vitro studies, maybe due to the scarce number of sources for funding in vivo applications. In perspective, NPs studies for clinical applications is a potential goal in order to control this disease.

## Acknowledgements

Authors acknowledge Hospital Juarez of Mexico for giving the facilities for writing this chapter. This work was developed with no funding.

## Conflict of interest

Authors declare no conflict of interests.

## Author details

Nancy Jannet Ruiz-Pérez, Jaime Sánchez-Navarrete and Julia D. Toscano-Garibay*
Research Direction, Hospital Juarez of Mexico, Mexico City, Mexico

*Address all correspondence to: julia.toscano@gmail.com

## References


[1] Gutiérrez-Cogco L, González-Bonilla C, Giono-Cerezo S, Beltrán LG. Principal serotypes of Salmonella identified in 10703 strains in Mexico from 1982 to 1993. Rev Latinoam Microbiol. 1994;36(4):221–6.

[2] De Jong B, Ekdahl K. The comparative burden of salmonellosis in the European Union member states, associated and candidate countries. BMC Public Health. 2006;6(1):4.

[3] SSA. Boletin de Epidemiología [Internet]. Secretaria de Salud. 1999. Available from: https://www.gob.mx/salud/acciones-y-programas/boletin-epidemiologia-1999

[4] Direccion General Epidemiologia. Direccion General de epidemiología [Internet]. Sistema Nacional de Vigilancia epidemiologica. 1994. Available from: https://www.sinave.gob.mx/

[5] Goldberg MB, Rubin RH. The spectrum of Salmonella infection. Infect Dis Clin North Am. 1988;2(3):571–598.

[6] Chin J. El control de las enfermedades transmisibles. Pan American Health Org; 2001.

[7] Davis BD, Dulbecco R, Eisen HN, Ginsberg HS, Wood Jr WB. Principles of microbiology and immunology. Princ Microbiol Immunol. 1968;41.

[8] Gutiérrez-Cogco L, Montiel-Vázquez E, Aguilera-Pérez P, González-Andrade M del C. Serotipos de Salmonella identificados en los servicios de salud de México. Salud Publica Mex. 2000;42:490–495.

[9] Eng S-K, Pusparajah P, Ab Mutalib N-S, Ser H-L, Chan K-G, Lee L-H. Salmonella: a review on pathogenesis, epidemiology and antibiotic resistance. Front Life Sci. 2015;8(3):284–293.

[10] CAE. Evaluación del riesgo asociado a la presencia de los serovares zoonóticos de Salmonella en huevo fresco producido en la CAE Asociado a la infección alimentaria en la población por consumo de huevo fresco [Internet]. 2008. Available from: http://www.elika.net/datos/riesgos/Archivo20/2008-05-Salmonella WEB.pdf

[11] Polotsky Y, Dragunsky E, Khavkin TH. Morphologic Evaluation of the Pathogenesis of Bacterial Enteric Infections-Part II. Crit Rev Microbiol. 1994;20(3):185–208.

[12] Arnott A, Wang Q, Bachmann N, Sadsad R, Biswas C, Sotomayor C, et al. Multidrug-Resistant Salmonella enterica 4,[5], 12: i:-Sequence Type 34, New South Wales, Australia, 2016–2017. Emerg Infect Dis. 2018;24(4):751.

[13] Guerra B, Soto SM, Argüelles JM, Mendoza MC. Multidrug resistance is mediated by large plasmids carrying a class 1 integron in the emergent Salmonella enterica serotype [4, 5, 12: i: –]. Antimicrob Agents Chemother. 2001;45(4):1305–1308.

[14] Guerra B, Soto S, Helmuth R, Mendoza MC. Characterization of a self-transferable plasmid from Salmonella enterica serotype Typhimurium clinical isolates carrying two integron-borne gene cassettes together with virulence and drug resistance genes. Antimicrob Agents Chemother. 2002;46(9):2977–2981.

[15] Chiu C-H, Wu T-L, Su L-H, Chu C, Chia J-H, Kuo A-J, et al. The emergence in Taiwan of fluoroquinolone resistance in Salmonella enterica serotype Choleraesuis. N Engl J Med. 2002;346(6):413–419.

[16] Carattoli A, Tosini F, Giles WP, Rupp ME, Hinrichs SH, Angulo FJ, et al. Characterization of plasmids carrying CMY-2 from expanded-spectrum cephalosporin-resistant Salmonella strains isolated in the United States between 1996 and 1998. Antimicrob Agents Chemother. 2002;46(5):1269–72.

[17] (WHO) WHO. WHO Global Strategy for Containment of Antimicrobial Resistance (WHO/CDS/CSR/DRS/2001.2 a). 2001. Geneva World Heal Organ. 2013;

[18] Organization WH. The evolving threat of antimicrobial resistance. Options action Geneva WHO Libr Cat Data. 2012;

[19] Plain Pazos C, Pérez de Alejo Plain A, Rivero Viera Y. La Medicina Natural y Tradicional como tratamiento alternativo de múltiples enfermedades. Rev Cuba Med Gen Integr. 2019;35(2).

[20] Sarker SD, Nahar L. An introduction to natural products isolation. Methods Mol Biol. 2012; 864:1–25.

[21] Houghton P, Raman A. Laboratory handbook for the fractionation of natural extracts. Springer Science & Business Media; 2012.

[22] Zhou Q, Gu R, Li P, Lu Y, Chen L, Gu Q. Anti-Salmonella mode of action of natural l-phenyl lactic acid purified from Lactobacillus plantarum ZJ316. Appl Microbiol Biotechnol. :1–10.

[23] Kadhem BM, Mater HN, Alsaadi LG, Mahdi LH, Auda IG, Zwain AH, et al. Antibacterial Activity of a novel Lectin produced by bee honey Bifidobacterium adolescentis against Multidrug Resistant Salmonella typhi. J Pharm Sci Res. 2019;11(3):1102–1106.

[24] Agbankpe AJ, Dougnon TV, Balarabe R, Deguenon E, Baba-Moussa L. In vitro assessment of antibacterial activity from Lactobacillus spp. strains against virulent Salmonella species isolated from slaughter animals in Benin. Vet World. 2019;12(12):1951.

[25] Najett M, Fadela C, Snoussi M, Abderrahim C. The antibacterial activity expressed in vitro and in vivo by Lactococcus lactis subsp. lactis CNRZ 1427 against Salmonella sp. South Asian J Exp Biol [Internet]. 2014;4(5):240–6. Available from: http://sajeb.org/index.php/sajeb/article/view/19513/9681

[26] Francis PM, Manickam R, Venkataesan BP. Crude protein extract of Actinobacteria exhibits antibacterial activity against Salmonella typhi. Int J Curr Microbiol App Sci. 2014;3(7):319–326.

[27] Kim SP, Lee SJ, Nam SH, Friedman M. The composition of a bioprocessed shiitake (Lentinus edodes) mushroom mycelia and rice bran formulation and its antimicrobial effects against Salmonella enterica subsp. enterica serovar Typhimurium strain SL1344 in macrophage cells and in mice. BMC Complement Altern Med. 2018 Dec;18(1):322.

[28] Matijašević D, Pantić M, Rašković B, Pavlović V, Duvnjak D, Sknepnek A, et al. The antibacterial activity of Coriolus versicolor methanol extract and its effect on ultrastructural changes of Staphylococcus aureus and Salmonella Enteritidis. Front Microbiol. 2016;7:1226.

[29] Meilina L, Hanizar E, Sari DNR. Antibacterial Activity of Pleurotus ostreatus grey oyster Variety Against Pathogen Bacterial of Salmonella typhi. Explor Conserv Biodivers. 2015;48.

[30] Quereshi S, Pandey AK, Sandhu SS. Evaluation of antibacterial activity of different Ganoderma lucidum extracts. J Sci Res. 2010;3:9–13.

[31] Manjunathan J, Kaviyarasan V. Solvent based effectiveness of antibacterial activity of edible mushroom Lentinus tuberregium (Fr.). Int J PharmTech Res. 2010;2(3):1910–1912.

[32] França RC, Conceição FR, Mendonça M, Haubert L, Sabadin G, de Oliveira PD, et al. Pichia pastoris X-33 has probiotic properties with remarkable antibacterial activity against Salmonella Typhimurium. Appl Microbiol Biotechnol. 2015;99(19):7953–7961.

[33] Lamas A, Arteaga V, Regal P, Vázquez B, Miranda JM, Cepeda A, et al. Antimicrobial activity of five apitoxins from apis mellifera on two common foodborne pathogens. Antibiotics. 2020; 9(7):1–9.

[34] Arteaga V, Lamas A, Regal P, Vázquez B, Miranda JM, Cepeda A, et al. Antimicrobial activity of apitoxin from Apis mellifera in Salmonella enterica strains isolated from poultry and its effects on motility, biofilm formation and gene expression. Microb Pathog. 2019;137:103771.

[35] Nasrollahzadeh A, Khomeiri M, Mahmoudi M. Genetic identification and evaluation of antibacterial activity of lactic isolates derived from Masske butter against pathogenic bacteria Staphylococcus aureus and Salmonella enterica. Food Sci Technol. 2019;16(93): 23–33.

[36] Hasan Z. Antibacterial activity of propolis Trigona spp. from bukittinggi west Sumatera against Salmonella sp. Chem Prog. 2019;4(2).

[37] Díaz-Roa A, Patarroyo MA, Bello FJ, Da Silva PIJ. Sarconesin: Sarconesiopsis magellanica Blowfly Larval Excretions and Secretions With Antibacterial Properties. Front Microbiol. 2018;9: 2249.

[38] Rezvi S, Putra AE, Faadhila T. In Vivo Antibacterial Activity of Dadih and Dadih Ice Cream Toward Salmonella typhimurium Development. Ann Glob Heal. 2017;83(1).

[39] Šarić L, Pezo L, Šarić B, Plavšić D, Jovanov P, Karabasil N, et al. Calcium-dependent antibacterial activity of donkey's milk against Salmonella. Ann Microbiol. 2017;67(2):185–194.

[40] Park K, Lee M, Oh T, Kim K-Y, Ma J. Antibacterial activity and effects of Colla corii asini on Salmonella typhimurium invasion in vitro and in vivo. BMC Complement Altern Med. 2017;17(1):520.

[41] van Altena SEC, Peen MA, van der Linden FH, Parmentier HK, Savelkoul HFJ, Tijhaar EJ. Bovine natural antibodies in antibody-dependent bactericidal activity against *Escherichia coli* and Salmonella Typhimurium and risk of mastitis. Vet Immunol Immunopathol [Internet]. 2016;171:21–7. Available from: http://dx.doi.org/10.1016/j.vetimm.2016.01.009

[42] Kalia P, Kumar NR, Harjai K. Studies on the therapeutic effect of propolis along with standard antibacterial drug in Salmonella enterica serovar Typhimurium infected BALB/c mice. BMC Complement Altern Med. 2016 Nov;16(1):485.

[43] Nurtamin T, Nurman RY, Hafizah I. Antibacterial Activity of Eel (Anguilla spp.) Mucus against Salmonella typhi. Indones Biomed J. 2016;8(3):179–182.

[44] Nina N, Quispe C, Jiménez-Aspee F, Theoduloz C, Feresín GE, Lima B, et al. Antibacterial activity, antioxidant effect and chemical composition of propolis from the Región del Maule, Central Chile. Molecules. 2015;20(10):18144–18167.

[45] Hussain MB, Hannan A, Akhtar N, Fayyaz GQ, Imran M, Saleem S, et al.

Evaluation of the antibacterial activity of selected Pakistani honeys against multi-drug resistant Salmonella typhi. BMC Complement Altern Med. 2015;15 (1):32.

[46] Mahdizade E, Mohabatkar H, Behbahani M. Antibacterial activity of platelet rich plasma *against Salmonella enterica*. Iran J Public Health. 2014;43 (2):225.

[47] Vica ML, Glevitzky M, Dumitrel G-A, Junie LM, Popa M. Antibacterial activity of different natural honeys from Transylvania, Romania. J Environ Sci Heal Part B. 2014;49(3):176–181.

[48] Šarić LĆ, Šarić BM, Mandic AI, Tomic J, Torbica A, Nedeljkovic N, et al. Antibacterial activity of donkey milk against Salmonella. Agro FOOD Ind Hi Tech. 2014;25:5.

[49] Kalia P, Kumar NR, Harjai K. Phytochemical screening and antibacterial activity of different extracts of propolis. Int J Pharm Biol Res. 2013;3(6):219–222.

[50] Mavri A, Abramovič H, Polak T, Bertoncelj J, Jamnik P, Smole Možina S, et al. Chemical properties and antioxidant and antimicrobial activities of Slovenian propolis. Chem Biodivers. 2012 Aug;9(8):1545–1558.

[51] Islam MDM, Masum SM, Mahbub KR. In vitro antibacterial activity of shrimp chitosan against salmonela paratyphi and staphylococcus aureus. J Bangladesh Chem Soc. 2011;24 (2):185–190.

[52] Voidarou C, Alexopoulos A, Plessas S, Karapanou A, Mantzourani I, Stavropoulou E, et al. Antibacterial activity of different honeys against pathogenic bacteria. Anaerobe. 2011;17 (6):375–379.

[53] Temiz A, Şener A, Tüylü Aö, Sorkun K, saliH B. Antibacterial activity of bee propolis samples from different geographical regions of Turkey against two foodborne pathogens, Salmonella enteritidis and Listeria monocytogenes. Turkish J Biol. 2011;35(4):503–511.

[54] Mandal S, DebMandal M, Pal NK, Saha K. Antibacterial activity of honey against clinical isolates of Escherichia coli, Pseudomonas aeruginosa and Salmonella enterica serovar Typhi. Asian Pac J Trop Med. 2010;3(12):961–964.

[55] Susanti N, Situmorang E, Fitri W. Effectiveness of the Antibacterial Activity of n-Hexane Andaliman (Zanthoxylum Acanthopodium DC) Extract Against Bacillus subtilis, Salmonella typhi, and *Staphylococcus aureus*. J Phys Conf Ser. 2020;1462(1).

[56] Prabowo WC, Agustina R. Antibacterial Activity of Scopoletin from Stem Bark of Aleurites moluccana Against Salmonella typhi Wisnu. J Trop Pharm Chem. 2020;5(1):29–32.

[57] Sanda A, Kader A, Maârouhi IM, Oumarou SS. Evaluation of antibacterial activity from extract of three plant species on strains of *salmonella* in niger. WORLD J Pharm Pharm Sci. 2020;9(3): 462–481.

[58] Okunye O, Idowu P, Kolade T. Antibacterial activity of crude extracts of the leaves of Nauclea latifolia Smith (Rubiaceae) and some selected conventional antibiotics on clinical isolates of Salmonella typhi. Ann Heal Res. 2020;6(3):258–264.

[59] Anjelina SH. Antibacterial Activity of Ethanolic Extract of Kitolod (Hippobromalongiflora) Leaf Against Staphylococcus aureus and Salmonella typhi. Asian J Pharm Res Dev. 2020;8 (1):52–54.

[60] Kordjazi A, Farahmandfar R. Antibacterial activity of hydroalcoholic extract of Cardin leaf on Listeria

monocytogenes, Salmonella enteritidis and *Pseudomonas aeruginosa*. Food Sci Technol. 2020;16(97):29–35.

[61] Wijaya EB. Antibacterial Activity Test of Trema orientalis L. Blumae (Anggrung) Granule Extract Against Bacteria Salmonella sp with The Wet Granulation Method. Indones Nat Res ... [Internet]. 2020;5(1):45–53. Available from: http://journal.uta45jakarta.ac.id/index.php/INRPJ/article/view/1751

[62] Martínez-Ortega EA, López-Briones JS, Rodríguez-Hernández G, Ramírez-Orozco RE, Franco-Robles E. Antibacterial activity of agave fructans against Salmonella typhimurium. Nat Prod Res [Internet]. 2020;34(18):2639–41. Available from: https://doi.org/10.1080/14786419.2018.1548446

[63] Simorangkir M, Maha AP. Antibacterial Activity And Phytochemical Screening From Chromatography Fraction Of Ethanol Extract Of Sarang Banua (Clerodendrum fragrans Vent Willd) Against Salmonella enterica. Indones J Chem Sci Technol. 2020;3(2):42–48.

[64] Sokoudjou JB, Atolani O, Njateng GSS, Khan A, Tagousop CN, Bitombo AN, et al. Isolation, characterization and in vitro anti-salmonellal activity of compounds from stem bark extract of Canarium schweinfurthii. BMC Complement Med Ther. 2020 Oct;20(1):316.

[65] Djague F, Lunga PK, Toghueo KRM, Melogmo DYK, Fekam BF. Garcinia kola (Heckel) and Alchornea cordifolia (Schumach. & Thonn.) Müll. Arg. from Cameroon possess potential antisalmonellal and antioxidant properties. PLoS One. 2020;15(8):e0237076.

[66] Yahia Y, Benabderrahim MA, Tlili N, Bagues M, Nagaz K. Bioactive compounds, antioxidant and antimicrobial activities of extracts from different plant parts of two Ziziphus Mill. species. PLoS One [Internet]. 2020;15(5):1–16. Available from: http://dx.doi.org/10.1371/journal.pone.0232599

[67] Joshi B, Panda SK, Jouneghani RS, Liu M, Parajuli N, Leyssen P, et al. Antibacterial, Antifungal, Antiviral, and Anthelmintic Activities of Medicinal Plants of Nepal Selected Based on Ethnobotanical Evidence. Evid Based Complement Alternat Med. 2020;2020:1043471.

[68] Legba B, Dougnon V, Chabi Y, Gbaguidi C, Aniambossou A, Deguenon E, et al. Evaluation of in-vivo anti-Salmonella activity of Uvaria chamae, Lantana camara and Phyllantus amarus used in Benin, West Africa. BMC Vet Res. 2020 Feb;16(1):49.

[69] Rama J-LR, Mallo N, Biddau M, Fernandes F, de Miguel T, Sheiner L, et al. Exploring the powerful phytoarsenal of white grape marc against bacteria and parasites causing significant diseases. Environ Sci Pollut Res Int. 2020 Jan;

[70] Ulhaq ZS, Hendyatama TH, Hameed F, Santosaningsih D. Antibacterial activity of Citrus hystrix toward Salmonella spp. infection. Enferm Infecc Microbiol Clin. 2020;

[71] Guo L, Gong S, Wang Y, Sun Q, Duo K, Fei P. Antibacterial activity of olive oil polyphenol extract against Salmonella Typhimurium and Staphylococcus aureus: Possible Mechanisms. Foodborne Pathog Dis. 2020;17(6):396–403.

[72] McMurray RL, Ball MEE, Tunney MM, Corcionivoschi N, Situ C. Antibacterial Activity of Four Plant Extracts Extracted from Traditional Chinese Medicinal Plants against Listeria monocytogenes, Escherichia coli, and Salmonella enterica subsp. enterica serovar Enteritidis. Microorganisms. 2020;8(6):962.

[73] Mei C, Wang X, Chen Y, Wang Y, Yao F, Li Z, et al. Antibacterial activity and mechanism of Litsea cubeba essential oil against food contamination by Escherichia coli and Salmonella enterica. J Food Saf. 2020;e12809.

[74] Nair A, Balasaravanan T, Jadhav S, Mohan V, Kumar C. Harnessing the antibacterial activity of Quercus infectoria and Phyllanthus emblica against antibiotic-resistant Salmonella Typhi and Salmonella Enteritidis of poultry origin. Vet World. 2020;13(7): 1388.

[75] Mirza AS, Baig MT, Huma A, Ibrahim S, Shahid U, Jabeen A, et al. Antibacterial activity of methanol extract of Capparis decidua edgew (forssk.) against *Staphylococcus aureus, Bacillus cereus,Salmonella typhi,* and *Escherichia coli*. Pharmacophore. 2020; 11(4).

[76] Mbock MA, Fouatio WF, Kamkumo RG, Fokou PVT, Tsofack FN, Lunga P-K, et al. In vitro and in vivo anti-salmonella properties of hydroethanolic extract of Detarium microcarpum Guill. & Perr. (Leguminosae) root bark and LC-MS-based phytochemical analysis. J Ethnopharmacol. 2020;113049.

[77] Agyemang K, Ofori Donkor P, Ayim I, Adzitey F, Lin L, Cui H. Antibacterial activity and mechanism of Tetrapleura tetraptera stem extract against Salmonella strains and its application in raw chicken meat. J Food Process Preserv. 2020;e14489.

[78] Jodi SM, Farouq AA, Magashi AM, Muomora GD, Nata'ala MK, Gambo A, et al. Phytochemical Properties and Antibacterial Activity of Leaf Extract of Ocimum gratissimum on Salmonella Species. J Adv Biol Biotechnol. 2019;22 (2):1–8.

[79] Nestor AO, Cyrille BK, Philippe S, Mahudro Y, Gwladys KS, Yannick A, et al. Antibacterial Activity of Essential Oil of Aeollanthus pubescens on Multidrug Resistant Strains of Salmonella and Escherichia coli Isolated from Laying Hens Farming in Benin. Adv Microbiol. 2019;9(09):804.

[80] Zuraidawati Z, Dewi M, Darmawi D, Sugito S, Hamzah A, Winaruddin W. 18. Phytochemistry and Antibacterial Activity Test of Ethanol Extracts of Soursop flower (Annona muricata L.) against Salmonella enteritidis. J Med Vet. 2019;13(1):132–137.

[81] Salni S, Marisa H. Evaluation on Antibacterial Activity of Karamunting Leaf Extract (Rhodomyrtus tomentosa (Ait) Hassk) with Various Solvents to Shigella Dysenteriae and Salmonella typhi. Malaysian J Fundam Appl Sci. 2019;15(5):671–674.

[82] Olawuwo OS, Aro AO, Erhabor JO, Eloff JN, McGaw LJ. The in vitro antibacterial activity and safety of Morinda lucida leaf extracts against Salmonella serovars relevant in livestock infections. Planta Med. 2019;85(18): PV-13.

[83] Sihombing NS, Sihombing F, Juwitaningsih T, Roza D, Pasaribu DM. Antibacterial Activity Analysis of Zanthoxylum acanthopodium DC Extract on Bacteria of Bacillus Subtilis, and Salmonella Typhi. In: AISTSSE 2018: Proceedings of The 5th Annual International Seminar on Trends in Science and Science Education, AISTSSE 2018, 18-19 October 2018, Medan, Indonesia. European Alliance for Innovation; 2019. p. 221.

[84] Favian Bayas-Morejón, Angélica Tigre, Ma. Bernarda Ruilova, Oderay Merino, Miriam Merino, Moisés Arreguín, Sonia Salazar, Riveliño Ramón, José Sánchez-Morales, Luis Verdezoto, Iván Moreno And Cecilia Gómez. Antibacterial Activity of Uvilla (Physalis peruviana L) Extracts Against

Salmonella Sp. Strains. Asian J Microbiol Biotechnol Environ Sci [Internet]. 2019; 21((November Suppl.) : 2019):S32–7. Available from: www.infostat.ar

[85] Hidayati DN, Hidayati N, Evinda E, Fitriana NR, Kusumadewi AP. Antibacterial activity of fractions from papaya seeds (Carica papaya L.) extract against Escherichia coli and Salmonella typhi and the contributing compounds. Pharmaciana. 2019;9(1):183–190.

[86] Abdullah MS, Nas FS, Ali M. Antibacterial Activity of Psidium guajava Leaf and Stem Bark Extracts against Clinical Isolates of Staphylococcus aureus and Salmonella typhi.

[87] Siahaan D, Gurning K, Iksen. Uji Aktivitas Antibakteri Ekstrak Etanol Daun Nangka (Artocarpus heterophyllus Lamk) Terhadap Bakteri Staphylococcus aureus, Escherichia coli, Staphylococcus epidermis dan Salmonella typhi. J Pharm Sci. 2019;2 (2):49–54.

[88] Arimaswati A, Safitri Woaw, Hartati H. Aktivitas Antibakteri Ekstrak Daun Turi (Sesbania grandiflora (L.) Press) terhadap Pertumbuhan Bakteri Salmonella thypi dan Streptococcus mutans (Antibacterial Activity of Turi Leaf Extract (Sesbania grandiflora (L.) Press) against Salmonella thypi and Str. Medula. 2020;7(1).

[89] Balakrishnan S, Ibrahim KS, Duraisamy S, Sivaji I, Kandasamy S, Kumarasamy A, et al. Antiquorum sensing and antibiofilm potential of biosynthesized silver nanoparticles of Myristica fragrans seed extract against MDR Salmonella enterica serovar Typhi isolates from asymptomatic typhoid carriers and typhoid patients. Environ Sci Pollut Res Int. 2020 Jan;27(3):2844–2856.

[90] Mayorga OAS, da Costa YFG, da Silva JB, Scio E, Ferreira ALP, de Sousa OV, et al. Kalanchoe brasiliensis Cambess., a Promising Natural Source of Antioxidant and Antibiotic Agents against Multidrug-Resistant Pathogens for the Treatment of Salmonella Gastroenteritis. Oxid Med Cell Longev. 2019;2019:9245951.

[91] Porter JA, Monu EA. Evaluating the Antimicrobial Efficacy of White Mustard Essential Oil Alone and in Combination with Thymol and Carvacrol against Salmonella. J Food Prot. 2019 Dec;82(12):2038–2043.

[92] Zhou D, Liu Z-H, Wang D-M, Li D-W, Yang L-N, Wang W. Chemical composition, antibacterial activity and related mechanism of valonia and shell from Quercus variabilis Blume (Fagaceae) against Salmonella paratyphi a and Staphylococcus aureus. BMC Complement Altern Med. 2019 Oct;19 (1):271.

[93] Mwale C, Makunike KN, Mangoyi R. Antibacterial Activity of Melia azedarach Leaves against Salmonella typhi and Streptococcus pneumoniae. Int Ann Sci. 2019;8(1): 47–53.

[94] Rodrigues AB, de Almeida-Apolonio AA, Alfredo TM, Dantas FG da S, Campos JF, Cardoso CAL, et al. Chemical Composition, Antimicrobial Activity, and Antioxidant Activity of Ocotea minarum (Nees & Mart.) Mez. Oxid Med Cell Longev. 2019;2019: 5736919.

[95] Rahayu ID, Widodo W, Prihartini I, Winaya A. Antibacterial activity of ethanolic extracts from Zingiber zerumbet rhizome against Salmonella spp. Biodiversitas J Biol Divers. 2019;20 (11).

[96] Maesaroh U, Martien R, Dono ND. Antibacterial activity and characterization of Annona muricata Linn leaf extract-nanoparticles against Escherichia coli FNCC-0091 and

Salmonella typhimurium FNCC-0050. In: IOP Conference Series: Earth and Environmental Science. IOP Publishing; 2019. p. 12055.

[97] Chang C-H, Fu J-H, Su C-H, Yin M-C, Hsu Y-M. Four spices prevent mice from contracting Salmonella enterica serovar Typhimurium. Exp Ther Med. 2019;18(4):2956.

[98] Zofou D, Shu GL, Foba-Tendo J, Tabouguia MO, Assob J-CN. In vitro and in vivo anti-Salmonella evaluation of pectin extracts and hydrolysates from “Cas Mango”(Spondias dulcis). Evidence-Based Complement Altern Med. 2019;2019.

[99] Hernández-García E, García A, Garza-González E, Avalos-Alanís FG, Rivas-Galindo VM, Rodríguez-Rodríguez J, et al. Chemical composition of Acacia farnesiana (L) wild fruits and its activity against Mycobacterium tuberculosis and dysentery bacteria. J Ethnopharmacol. 2019 Feb;230:74–80.

[100] Abdallah M, Ali M. Antibacterial Activity of Leaf and Stem Bark Extracts of Adansonia Digitata Against Escherichia Coli and Salmonella Typhi Grown in Potiskum, Yobe State Nigeria. Ann Microbiol Infect Dis. 2018;1(4):1–7.

[101] Bagega AI, Usman AA, Dankaka SM, Amina M, Mani AU. Biorem Sci Tech. Antibacterial Activity and Phytochemical Analysis of Cassia occidentalis Leaf Extract on Salmonella Typhimurium. 2018;6(2):5–8.

[102] Mubarak F, Sartini S, Purnawanti D. Effect of Ethanol Concentration on Antibacterial Activity of Bligo Fruit Extract (Benincasa hispida Thunb) to Salmonella typhi. Indones J Pharm Sci Technol. 2018;5(3):76.

[103] Ali M, Aureus S, Typhi S. Phytochemical Screening and Antibacterial Activity of Citrus Sinensis Peel Extracts on Clinical Isolates of Staphylococcus Aureus and Salmonella Typhi. J Allied Pharm Sci. 2018;50–54.

[104] Ngurah BIG., Tokan MK, Saputra A. Characterization of Cinnamadehyde Compound Isolated from Cinnamon Oil and Its Salmonella Typhy. J Appl Chem Sci. 2018;5(2):469–472.

[105] Rubio A, Travieso M del C, Riverón Y, Martínez A, Peña J, Espinosa I, et al. Actividad antibacteriana de aceites esenciales de plantas cultivadas en Cuba sobre cepas de Salmonella enterica. Rev Salud Anim. 2018;40(3):1–10.

[106] Drebes T, Ethur EM, Avancini CAM. Triagem Fitoquímica e ativdiade antibacteriana dos extratos aquosos e hidroalcoólicos brutos de *Jacaranda micrantha Staphylococcus* coagulase positiva e *Salmonella* spp. padrões e isoladas em produtos de. Arch Vet Sci [Internet]. 2018;23(2):27–39. Available from: https://revistas.ufpr.br/veterinary/article/view/51197

[107] El-Azzouny MM, El-Demerdash AS, Seadawy HG, Abou-Khadra SH. Antimicrobial Effect of Garlic (Allium sativum) and Thyme (Zataria multiflora Boiss) Extracts on Some Food Borne Pathogens and Their Effect on Virulence Gene Expression. Cell Mol Biol (Noisy-le-grand). 2018 Jul;64(10):79–86.

[108] Solarte AL, Astorga RJ, de Aguiar FC, De Frutos C, Barrero-Domínguez B, Huerta B. Susceptibility Distribution to Essential Oils of Salmonella enterica Strains Involved in Animal and Public Health and Comparison of the Typhimurium and Enteritidis Serotypes. J Med Food. 2018 Sep;21(9):946–950.

[109] Chung D, Cho TJ, Rhee MS. Citrus fruit extracts with carvacrol and thymol eliminated 7-log acid-adapted Escherichia coli O157:H7, Salmonella typhimurium, and Listeria

monocytogenes: A potential of effective natural antibacterial agents. Food Res Int. 2018 May;107:578–588.

[110] Yeganegi M, Tabatabaei Yazdi F, Mortazavi SA, Asili J, Alizadeh Behbahani B, Beigbabaei A. Equisetum telmateia extracts: Chemical compositions, antioxidant activity and antimicrobial effect on the growth of some pathogenic strain causing poisoning and infection. Microb Pathog. 2018 Mar;116:62–67.

[111] Fadil M, Fikri-Benbrahim K, Rachiq S, Ihssane B, Lebrazi S, Chraibi M, et al. Combined treatment of Thymus vulgaris L., Rosmarinus officinalis L. and Myrtus communis L. essential oils against Salmonella typhimurium: Optimization of antibacterial activity by mixture design methodology. Eur J Pharm Biopharm. 2018;126:211–220.

[112] Dayuti S. Antibacterial activity of red algae (Gracilaria verrucosa) extract against Escherichia coli and Salmonella typhimurium. In: IOP Conference Series: Earth and Environmental Science. IOP Publishing; 2018. p. 12074.

[113] Kapitan LAV. Antibacterial activity test of ethyl ascetic fraction ethanol extract faloak tree bark (Sterculia sp.) on Salmonella thyposa bacteria. In: Proceeding 1st International Conference Health Polytechnic of Kupang. 2018. p. 510–518.

[114] Ashraf S, Anjum AA, Ahmad A, Firyal S, Sana S, Latif AA. In vitro activity of Nigella sativa against antibiotic resistant Salmonella enterica. Environ Toxicol Pharmacol. 2018;58:54–58.

[115] Kim SP, Lee SJ, Nam SH, Friedman M. Mechanism of Antibacterial Activities of a Rice Hull Smoke Extract (RHSE) Against Multidrug-Resistant Salmonella Typhimurium In Vitro and in Mice. J Food Sci. 2018 Feb;83(2):440–445.

[116] Kocić-Tanackov S, Blagojev N, Suturović I, Dimić G, Pejin J, Tomović V, et al. Antibacterial activity of essential oils against Escherichia coil, Salmonella enterica and Listeria monocytogenes. J Food Saf Food Qual Fur Leb. 2017;68(4):88–95.

[117] Padmavathy A, Rasny MRM, Reyadh R, Khan J. Evaluation of Antibacterial Activity of Different Extracts of Ipomoea Aquatica Leaves against Escherichia Coli and Salmonella Typhi. J Manag Sci. 2017;15(2).

[118] Farooq U, Sharma Y. Antibacterial Activity of Andrographis paniculata against Salmonella typhi. UGC Approv List J Abstr Index by NISCAIR Indian Sci Abstr Website www ujpah. 2017;17.

[119] Ummasalma AS, Yakubu M. Antibacterial activity of root and leaves extracts of Senna occidentalis against Salmonella typhi. Katsina J Nat Appl Sci. 2017;6(1):36–41.

[120] Lamola SM, Dzoyem JP, Botha F, van Wyk C. Anti-bacterial, free radical scavenging activity and cytotoxicity of acetone extracts of Grewia flava. Afr Health Sci. 2017 Sep;17(3):790–796.

[121] Bisi-Johnson MA, Obi CL, Samuel BB, Eloff JN, Okoh AI. Antibacterial activity of crude extracts of some South African medicinal plants against multidrug resistant etiological agents of diarrhoea. BMC Complement Altern Med. 2017 Jun;17(1):321.

[122] Hoekou YP, Tchacondo T, Karou SD, Yerbanga RS, Achoribo E, Da O, et al. Therapeutic potentials of ethanolic extract of leaves of Holarrhena floribunda (g. Don) dur. And schinz (Apocynaceae). African J Tradit Complement Altern Med AJTCAM. 2017;14(2):227–233.

[123] Souza JG de L de, Toledo AG, Santana CB, Santos CV dos, Mallmann AP, Silva JPB da, et al.

Chemical composition and antibacterial activity of essential oil and leaf extracts of Zanthoxylum caribaeum Lam. against serotypes of Salmonella. Rev Bras Saúde e Produção Anim. 2017;18(3):446–453.

[124] Hadian M, Rajaei A, Mohsenifar A, Tabatabaei M. Encapsulation of Rosmarinus officinalis essential oils in chitosan-benzoic acid nanogel with enhanced antibacterial activity in beef cutlet against Salmonella typhimurium during refrigerated storage. LWT. 2017; 84:394–401.

[125] Balakrishnan S, Sivaji I, Kandasamy S, Duraisamy S, Kumar NS, Gurusubramanian G. Biosynthesis of silver nanoparticles using Myristica fragrans seed (nutmeg) extract and its antibacterial activity against multidrug-resistant (MDR) Salmonella enterica serovar Typhi isolates. Environ Sci Pollut Res. 2017;24(17):14758–14769.

[126] Agus S, Achmadi SS, Mubarik NR. Antibacterial activity of naringenin-rich fraction of pigeon pea leaves toward Salmonella thypi. Asian Pac J Trop Biomed. 2017;7(8):725–728.

[127] Ali M, Aminu F, Ibrahim IS. In-vitro assessment of antibacterial activity and phytochemical screening of Vitex doniana on clinical isolate of Salmonella typhi. IN-VITRO. 2017;3(1).

[128] Rangel-Vargas E, Gutiérrez-Alcántara EJ, Gómez-Aldapa CA, Falfán-Cortés RN, Segovia-Cruz JA, Salas-Rangel LP, et al. Antibacterial activity of roselle calyx extracts, sodium hypochlorite, colloidal silver and acetic acid against multidrug-resistant salmonella serotypes isolated from coriander. J Food Saf. 2017;37(2):e12320.

[129] Shahbazi Y, Shavisi N, Mohebi E. Antibacterial activity of Ziziphora clinopodioides essential oil and nisin against Bacillus subtilis and Salmonella Typhimurium in commercial barley soup. Bulg J Vet Med. 2017;20(1).

[130] Alsuhaibani S, Khan MA. Immune-stimulatory and therapeutic activity of Tinospora cordifolia: Double-edged sword against salmonellosis. J Immunol Res. 2017;2017.

[131] Yanuarisa R, Agustina D, Santosa A. Antibacterial Activity of Ethanolic Extract from Tempuyung Leaf (Sonchus arvensis L.) againts Salmonella typhi by In Vitro Study. J Agromedicine Med Sci. 2016;2(2):1.

[132] Parra M, Valenzuela B, Soto S, Modak B. Antibacterial activity of cuticular components from heliotropium species, against staphylococcus aureus and salmonella tiphymurium. Bol Latinoam y del Caribe Plantas Med y Aromat. 2016;15(6):422–428.

[133] Abdal-hadei Ali Al-Nasrawi H. Antibacterial Activity of Some Medicinal Plants Extracts Against Escherichia Coli and Salmonella Typhimurium Isolated From Domestic Chicken in Al-Qadissiya Province. Basrah J Vet Res. 2016;15(2):192–210.

[134] Xu J-F, Hou Y-Y, Chen L. Antibacterial Activity of Scutellariae Radix Extracts and the Bioactive Compounds against E.coli and Salmonella. 2016;1056–1065.

[135] Barreiros RM. Tests of antibacterial activity by naphthoquinones derivatives and extracts from Rhus typhina and Achillea sintenisii in Salmonella typhimurium LT2. 2016.

[136] Tanwar A, Chawla R, Chakotiya AS, Thakur P, Goel R, Basu M, et al. Effect of Holarrhena antidysentrica (Ha) and Andrographis paniculata (Ap) on the biofilm formation and cell membrane integrity of opportunistic pathogen Salmonella typhimurium. Microb Pathog. 2016 Dec; 101:76–82.

[137] Bhattacharya D, Bhattacharya S, Patra MM, Chakravorty S, Sarkar S,

Chakraborty W, et al. Antibacterial Activity of Polyphenolic Fraction of Kombucha Against Enteric Bacterial Pathogens. Curr Microbiol. 2016 Dec;73 (6):885–896.

[138] Méndez Álvarez N, Angulo Ortíz A, Contreras Martínez O. [In vitro antibacterial activity of Curcuma longa (Zingiberaceae) against nosocomial bacteria in Montería, Colombia]. Rev Biol Trop. 2016 Sep;64(3):1201–1208.

[139] Kengni F, Fodouop SPC, Tala DS, Djimeli MN, Fokunang C, Gatsing D. Antityphoid properties and toxicity evaluation of Harungana madagascariensis Lam (Hypericaceae) aqueous leaf extract. J Ethnopharmacol. 2016 Feb;179:137–145.

[140] Ketpanyapong W, Itharat A. Antibacterial Activity of Thai Medicinal Plant Extracts Against Microorganism Isolated from Post-Weaning Diarrhea in Piglets. J Med Assoc Thai. 2016 Jul;99 Suppl 4:S203–S210.

[141] Gullon B, Pintado ME, Pérez-Álvarez JA, Viuda-Martos M. Assessment of polyphenolic profile and antibacterial activity of pomegranate peel (Punica granatum) flour obtained from co-product of juice extraction. Food Control. 2016;59:94–98.

[142] Sunday OJ, Babatunde SK, Ajiboye AE, Adedayo RM, Ajao MA, Ajuwon BI. Evaluation of phytochemical properties and in-vitro antibacterial activity of the aqueous extracts of leaf, seed and root of Abrus precatorius Linn. against Salmonella and Shigella. Asian Pac J Trop Biomed. 2016;6(9):755–759.

[143] Ewansiha JU, Okafor AC, Doughari J, Busari MB. Antibacterial activity of the leaf extract of Piliostigma thonningii against Salmonella typhi and Shigella dysenteriae. Adv Med Plant Res. 2015;3(October):151–154.

[144] Fodouop M, Tamo SPB, Pegnyemb DE, Etoa FX, de Paula RA, Barbosa AAT, et al. In vitro antibacterial activity of Baillonella toxisperma (Pierre) extracts against Staphylococcus aureus, Salmonella typhi, Proteus mirabilis and Bacillus cereus F3748. African J Microbiol Res. 2015;9(39): 2088–2094.

[145] Sirimongkolvorakul S, Paditporn K, Lertpatarakomol R. Antibacterial activity of wood vinegar against Salmonella Enteritidis and Salmonella Typhimurium Sunisa. In: Climate Smart Sustainable Animal Agriculture for Food Security. 2015. p. 3.

[146] Rajathangam C, Gobu P, Jeyadevan J. Screening of medicinal plants for their potential antibacterial activity against Salmonella typhi. 2015;2 (1):75–82.

[147] Yasurin P, Saenghiruna T. In-vitro individual antibacterial activity of Thai red curry paste ingredients against Salmonella enterica Enteritidis (human) and Listeria monocytogenes 10403S. Biomed Pharmacol J. 2015;8(1):1–8.

[148] Sakinah EN, Mufida DC. Antibacterial activity of ethanol leaf extract of purslane ( Portulaca oleracea) against Salmonella typhi and Shigella dysentriae. 2015;

[149] Abd El Tawab A, El -Hofy F, Belih S, El Shemy M. Antibacterial activity of some medicinal plant oils against Escherichia coli and Salmonella species in -vitro. Benha Vet Med J. 2015; 28(2):163–168.

[150] Rachmawaty FJ, Tamhid H., Hidayah R. Antibacterial activity comparison between the fraction of ethyl acetate, chloroform and methanol of red betel vine (Piper crocatum) leaves ethanol extract toward Salmon. 2015.

[151] Ahani M, Rahimifsard N, Shojaii A. Antibacterial activity of Ddfferent extracts Of Aerial Parts Of Dionysia Revoluta Boiss. against Escherichia coli,

Staphylococcus aureus, Pseudomonas aeruginosa, Salmonella Enteritidis, Enterococcus Maryam. In 2015.

[152] Khuda F, Iqbal Z, Khan A, Zakiullah, Shah WA, Shah Y, et al. Report: Antibacterial and antifungal activities of leaf extract of Achyranthes aspera (Amaranthaceae) from Pakistan. Pak J Pharm Sci. 2015 Sep;28(5):1797–1800.

[153] Packer J, Naz T, Harrington D, Jamie JF, Vemulpad SR. Antimicrobial activity of customary medicinal plants of the Yaegl Aboriginal community of northern New South Wales, Australia: a preliminary study. BMC Res Notes. 2015 Jun;8:276.

[154] Mehmood B, Dar KK, Ali S, Awan UA, Nayyer AQ, Ghous T, et al. Short communication: in vitro assessment of antioxidant, antibacterial and phytochemical analysis of peel of Citrus sinensis. Pak J Pharm Sci. 2015 Jan;28(1):231–239.

[155] Sarwar A, Latif Z. GC–MS characterisation and antibacterial activity evaluation of Nigella sativa oil against diverse strains of Salmonella. Nat Prod Res. 2015;29(5):447–451.

[156] Lee S-Y, Nam S-H, Lee H-J, Son S-E, Lee H-J. Antibacterial activity of aqueous Garlic extract against Escherichia coli O157: H7, Salmonella typhi murium and Staphylococcus aureus. J Food Hyg Saf. 2015;30(2):210–216.

[157] Ahsan S, Arefin MS, Munshi JL, Begum MN, Maliha M, Rahman S, et al. In vitro antibacterial activity of Spirulina platensis extracts against clinical isolates of Salmonella enterica serovars Typhi and Paratyphi (SUBP03). Stamford J Microbiol. 2015;5 (1):22–25.

[158] Sapabguy C, Yasurin P. Natural antibacterial activity of Thai red curry paste in coconut milk based curry; Kang-Kati, model on Salmonella sp. and *Listeria monocytogenes*. Walailak J Sci Technol. 2015;12(5):473–480.

[159] Kuta FA, Onochei I, Garba S, Damisa D. an in-vitro and in-vivo Antibacterial activity of Vitexdoniana crude extracts on Salmonella typhi. IJSRSET; 2015.

[160] Fujita K, Chavasiri W, Kubo I. Anti-Salmonella Activity of Volatile Compounds of Vietnam Coriander. Phyther Res. 2015;29(7):1081–1087.

[161] Tabatabaei Yazdi F, Heidari Sureshjani M, Alizadeh Behbahani B, Mortazavi SA. In-vitro antibacterial activity of aqueous and ethanolic extracts of Kelussia odoratissima on Salmonella typhimurium. In: 4th national congress on agriculture & aquatic & food. 2014.

[162] Chang C-H, Yu B, Su C-H, Chen DS, Hou Y-C, Chen Y-S, et al. Coptidis rhizome and Si Jun Zi Tang can prevent Salmonella enterica serovar Typhimurium infection in mice. PLoS One. 2014;9(8):e105362.

[163] Hossain MA, Park J-Y, Kim J-Y, Suh J-W, Park S-C. Synergistic effect and antiquorum sensing activity of Nymphaea tetragona (water lily) extract. Biomed Res Int. 2014;2014: 562173.

[164] Loung FS, Silalahi J, Suryanto D. Antibacterial activity of enzymatic hydrolyzed of virgin coconut oil and palm kernel oil against Staphylococcus aureus, Salmonella thypi and *Escherichia coli*. Int J PharmTech Res. 2014;6(2): 628–633.

[165] Harahap U, Silalahi J. Antibacterial activity of enzymatic hydrolysis of Virgin Coconut Oil against Salmonella. Int J PharmTech Res. 2014;6(2):589–599.

[166] Ganesh P, Kumar RS, Saranraj P. Phytochemical analysis and antibacterial

activity of Pepper (Piper nigrum L.) against some human pathogens. Cent Eur J Exp Biol. 2014;3(2):36–41.

[167] Thabti I, Elfalleh W, Tlili N, Ziadi M, Campos MG, Ferchichi A. Phenols, flavonoids, and antioxidant and antibacterial activity of leaves and stem bark of Morus species. Int J food Prop. 2014;17(4):842–854.

[168] Ugoh SC, Agarry OO, Garba SA. Studies on the antibacterial activity of Khaya senegalensis [(Desr.) A. Juss)] stem bark extract on Salmonella enterica subsp. enterica serovar Typhi [(ex Kauffmann and Edwards) Le Minor and Popoff]. Asian Pac J Trop Biomed. 2014; 4:S279–S283.

[169] Bakht J, Shaheen S, Shafi M. Antimicrobial potentials of Mentha longifolia by disc diffusion method. Pak J Pharm Sci. 2014;27(4).

[170] Silalahi J, Manurung R, Sitompul E. Antibacterial Activity of Hydrolyzed Oils of Different Fatty Acid Composition against Salmonella Thypi and Lactobacillus plantarum. Int J PharmTech Res. 2014;7(2):233–237.

[171] Rahimifard N, Bagheri E, Asgarpanah G, Kabiri Balajadeh B, Yazdi HR. Antibacterial activity of total extract, petroleum ether, chloroform, ethyl acetate and aqueous fractions of aerial parts of Heliotropium bacciferum. J Med Plants. 2014;4(52):122–135.

[172] ZHANG D, CHEN Y, NAN H, DU T, FU X, LIU M. Measurement of antibacterial activity of 10 Chinese herbal medicines against Salmonella isolated from pullorum disease. Chinese J Vet Med. 2013;3.

[173] Hassouna RA, Khalid AS, Khalid HS. In vitro Antibacterial Activity of Five Sudanese Medicinal Plants Against Salmonella species. Univ Africa J Sci. 2013;36–56.

[174] Dabanka CP. Antibacterial Activity of Phyllanthusamarus (Schumand Thonn) Extract against Salmonella Typhicausative Agent Of Typhoid Fever. KWAME NKRUMAH UNIVERSITY OF SCIENCE AND TECHNOLOGY; 2013.

[175] Sahreen S, Khan MR, Khan RA, Shah NA. Estimation of flavoniods, antimicrobial, antitumor and anticancer activity of Carissa opaca fruits. BMC Complement Altern Med. 2013 Dec;13: 372.

[176] Hannan A, Asghar S, Naeem T, Ikram Ullah M, Ahmed I, Aneela S, et al. Antibacterial effect of mango (Mangifera indica Linn.) leaf extract against antibiotic sensitive and multi-drug resistant Salmonella typhi. Pak J Pharm Sci. 2013 Jul;26(4):715–719.

[177] David JRD, Ekanayake A, Singh I, Farina B, Meyer M. Effect of white mustard essential oil on inoculated Salmonella sp. in a sauce with particulates. J Food Prot. 2013 Apr;76 (4):580–587.

[178] Moore-Neibel K, Gerber C, Patel J, Friedman M, Jaroni D, Ravishankar S. Antimicrobial activity of oregano oil against antibiotic-resistant Salmonella enterica on organic leafy greens at varying exposure times and storage temperatures. Food Microbiol. 2013;34 (1):123–129.

[179] Lawal D, Yunusa I, Bala I. A Study of the Phytochemical Properties and Synergistic Antibacterial Activity of Annona comosus (LINN) Merr. Peel and Citrus sinensis Peel Extracts on Aeromonas hydrophila and Salmonella species. Bayero J Pure Appl Sci. 2013;6 (1):40–45.

[180] Ham Y-J, Yang J-H, Na C-S. Screening of antibacterial activity against to Staphylococcus aureus, Listeria monocytogens, Mannhemia haemolytica and Salmonella gallinarum

using different plant extracts. Korean J Org Agric. 2013;21(1):105–113.

[181] Kim SP, Moon E, Nam SH, Friedman M. Composition of Herba Pogostemonis water extract and protection of infected mice against Salmonella Typhimurium-induced liver damage and mortality by stimulation of innate immune cells. J Agric Food Chem. 2012 Dec;60(49):12122–12130.

[182] Abirami P, Gomathinayagam M, Panneerselvam R. Preliminary study on the antimicrobial activity of Enicostemma littorale using different solvents. Asian Pac J Trop Med. 2012 Jul; 5(7):552–555.

[183] Ruban P, Gajalakshmi K. In vitro antibacterial activity of Hibiscus rosa–sinensis flower extract against human pathogens. Asian Pac J Trop Biomed. 2012;2(5):399–403.

[184] Koffi-Nevry R, Kouassi KC, Nanga ZY, Koussémon M, Loukou GY. Antibacterial activity of two bell pepper extracts: Capsicum annuum L. and Capsicum frutescens. Int J food Prop. 2012;15(5):961–971.

[185] Suganya S, Bharathidasan R, Senthilkumar G, Madhanraj P, Panneerselvam A. Antibacterial activity of essential oil extracted from Coriandrum sativam (L.) and GC-MS analysis. J Chem Pharm Res. 2012;4(3): 1846–1850.

[186] Kakar SA, Tareen RB, Kakar MA, Jabeen H, Kakar SUR, Al-Kahraman Y, et al. Screening of antibacterial activity of four medicinal plants of Balochistan-Pakistan. Pak J Bot. 2012; 44:245–250.

[187] Gomez-Flores R, Reyna-Martínez R, Tamez-Guerra P, Quintanilla-Licea R. Antibacterial activity of Oenothera rosea (L'Hér) leaf extracts. J Adv Med Med Res. 2012; 396–404.

[188] Bankole HA, Anjorin AA, Kazeem MI, Ogbeche ME, Agbafor U. Antibacterial activity of Ocimum gratissimum AND Gongronema latifolium ON Staphylococcus aureus AND Salmonella typhi. East African J Sci Technol. 2012;2(1):114–128.

[189] Lazuardi I, Saenghiruna T, Yasurin P. Natural antimicrobial activity of Thai red curry's herbs on Salmonella typhimurium DT104b. AU J Technol. 2012;16(1):1–6.

[190] Hasdiana F, Kuswarini S, Koendhari EB. Antibacterial activity of Belimbing wuluh (Averrhoa Bilimbi Linn) extract on Salmonella Typhi growth. Folia Medica Indones. 2012;48 (4):144.

[191] Adewumi AAJ, Aina VO, Saratu AS, Nkechi LU. Antibacterial activity of crude extract of Ocinum gratissimum (African basil) on clinical strains of Escherichia coli, Salmonella typhi and Klebsiella pneumoniae. J Pharm Allied Sci. 2011;8(3):1348–1354.

[192] Dejsungkranont M. Development of anti-Salmonella activity natural colorant powder from ripe fruits of Ardisia elliptica Thumb. 2011;

[193] Mohebi R, Ghafourian S, Sekawi Z, Khosravi A, Galehdari EA, Hushmandfar R, et al. In vitro and in vivo antibacterial activity of acorn herbal extract against some Gram-negative and Gram-positive bacteria. Roum Arch Microbiol Immunol. 2011;70 (4):149–152.

[194] Xia D-Z, Yu X-F, Zhu Z-Y, Zou Z-D. Antioxidant and antibacterial activity of six edible wild plants (Sonchus spp.) in China. Nat Prod Res. 2011;25(20): 1893–1901.

[195] Jaiswal AK, Rajauria G, Abu-Ghannam N, Gupta S. Phenolic composition, antioxidant capacity and antibacterial activity of selected Irish

Brassica vegetables. Nat Prod Commun. 2011;6(9):1934578X1100600923.

[196] Joray MB, del Rollán MR, Ruiz GM, Palacios SM, Carpinella MC. Antibacterial activity of extracts from plants of central Argentina—isolation of an active principle from Achyrocline satureioides. Planta Med. 2011;77(01):95–100.

[197] Razvy MA, Faruk MO, Hoque MA. Environment friendly antibacterial activity of water chestnut fruits. J Biodivers Environ Sci. 2011;1(1):26–34.

[198] Shanmugapriya R, Ramanathan T, Thirunavukkarasu P. Evaluation of antioxidant potential and antibacterial activity of Acalypha indica Linn. using in vitro model. Asian J Biomed Pharm Sci. 2011;1(1).

[199] Choi J-G, Kang O-H, Lee Y-S, Chae H-S, Oh Y-C, Brice O-O, et al. In vitro and in vivo antibacterial activity of Punica granatum peel ethanol extract against Salmonella. Evidence-Based Complement Altern Med. 2011;2011.

[200] Voss-Rech D, Klein CS, Techio VH, Scheuermann GN, Rech G, Fiorentin L. Antibacterial activity of vegetal extracts against serovars of Salmonella. Ciência Rural. 2011;41(2): 314–320.

[201] Chonoko UG, Rufai AB. Phytochemical screening and antibacterial activity of Cucurbita pepo (Pumpkin) against Staphylococcus aureus and Salmonella typhi. Bayero J Pure Appl Sci. 2011;4(1):145–147.

[202] Kuta FA, Damisa D, Adabara NU, Bello IM, Samuel S. Antibacterial activity of Aloe vera plant extract against Salmonella typhi. Extraction. 2011;6(1):83–86.

[203] Srichana D, Taengtip R, Kondo S. Antimicrobial activity of Gynostemma pentaphyllum extracts against fungi producing aflatoxin and fumonisin and bacteria causing diarrheal disease. Southeast Asian J Trop Med Public Heal. 2011;42(3):704.

[204] Mbwambo ZH, Erasto P, Nondo ROS, Innocent E, Kidukuli AW. Antibacterial and cytotoxic activities of Terminalia stenostachya and Terminalia spinosa. Tanzan J Health Res. 2011;13(2): 103–105.

[205] Pérez-Vásquez A, Capella S, Linares E, Bye R, Angeles-López G, Mata R. Antimicrobial activity and chemical composition of the essential oil of Hofmeisteria schaffneri. J Pharm Pharmacol. 2011;63(4):579–586.

[206] Kuete V, Kamga J, Sandjo LP, Ngameni B, Poumale HMP, Ambassa P, et al. Antimicrobial activities of the methanol extract, fractions and compounds from Ficus polita Vahl. (Moraceae). BMC Complement Altern Med. 2011;11(1):1–6.

[207] Ali NH, Faizi S, Kazmi SU. Antibacterial activity in spices and local medicinal plants against clinical isolates of Karachi, Pakistan. Pharm Biol. 2011; 49(8):833–839.

[208] Yadav HL, Yadav RN, Gupta A, Akram J, Saxena AR. Investigations on antibacterial activity of higher plants against the human pathogenic bacterium Salmonella typhosa. Indian J Sci Res. 2010;1(2):111–115.

[209] Gowri SS, Vasantha K. Phytochemical screening and antibacterial activity of Syzygium cumini (L.)(Myrtaceae) leaves extracts. Int J Pharm Tech Res. 2010;2(2):1569–1573.

[210] Karsha PV, Lakshmi OB. Antibacterial activity of black pepper (Piper nigrum Linn.) with special reference to its mode of action on bacteria. 2010;

[211] Mistry K, Mehta M, Mendpara N, Gamit S, Shah G. Determination of

antibacterial activity and MIC of crude extract of Abrus precatorius L. Adv Biotech. 2010;10(2):25–27.

[212] Rattanachaikunsopon P, Phumkhachorn P. Contents and antibacterial activity of flavonoids extracted from leaves of Psidium guajava. J Med Plants Res. 2010;4(5): 393–396.

[213] Özkalp B, Sevgi F, Özcan M, Özcan MM. The antibacterial activity of essential oil of oregano (Origanum vulgare L.). J Food Agric Env. 2010;8 (2):6–8.

[214] Purushotham KG, Arun P, Jayarani JJ, Vasnthakumari R, Sankar L, Reddy BR. Synergistic in vitro antibacterial activity of Tectona grandis leaves with tetracycline. Int J pharmtech Res. 2010;2(1):519–523.

[215] Devi K, Karthikai GD, Thirumaran G, Arumugam R, Anantharaman P. Antibacterial activity of selected medicinal plants from Parangipettai coastal regions; Southeast coast of India. Acad J Plant Sci. 2010;3: 122–125.

[216] Parihar P, Parihar L, Bohra A. In vitro antibacterial activity of fronds (leaves) of some important pteridophytes. J Microbiol Antimicrob. 2010;2(2):19–22.

[217] Choi J-G, Kang O-H, Brice O-O, Lee Y-S, Chae H-S, Oh Y-C, et al. Antibacterial activity of Ecklonia cava against methicillin-resistant Staphylococcus aureus and Salmonella spp. Foodborne Pathog Dis. 2010;7(4): 435–441.

[218] Lixandru B-E, Drăcea NO, Dragomirescu CC, Drăgulescu EC, Coldea IL, Anton L, et al. Antimicrobial activity of plant essential oils against bacterial and fungal species involved in food poisoning and/or food decay. Roum Arch Microbiol Immunol. 2010; 69(4):224–230.

[219] Kondo S, Sattaponpan C, Phongpaichit S, Srijan A, Itharat A. Antibacterial activity of Thai medicinal plants Pikutbenjakul. J Med Assoc Thailand= Chotmaihet thangphaet. 2010;93:S131–S135.

[220] Mekseepralard C, Kamkaen N, Wilkinson JM. Antimicrobial and Antioxidant Activities of Traditional Thai Herbal Remedies for Aphthous Ulcers. Phytother Res. 2010;24:1514–1519.

[221] Devi KP, Nisha SA, Sakthivel R, Pandian SK. Eugenol (an essential oil of clove) acts as an antibacterial agent against Salmonella typhi by disrupting the cellular membrane. J Ethnopharmacol. 2010;130(1):107–115.

[222] JP AA, Dzoyem JP, Pieme CA, Penlap VB. In vitro antibacterial activity and acute toxicity studies of aqueous-methanol extract of Sida rhombifolia Linn.(Malvaceae). BMC Complement Altern Med. 2010;10(1):40.

[223] Said N, Fahrodi DU, Malaka R, Maruddin F. Assessment of the antibacterial activity of goat milk kefir on Escherichia coli ATCC 8739 and Salmonella enteric subsp. enterica serovar typhimurium ATCC 14028 using a well diffusion method. In: IOP Conference Series: Earth and Environmental Science. IOP Publishing; 2019. p. 12051.

[224] Dzoyem JP, Melong R, Tsamo AT, Maffo T, Kapche DGWF, Ngadjui BT, et al. Cytotoxicity, antioxidant and antibacterial activity of four compounds produced by an endophytic fungus Epicoccum nigrum associated with Entada abyssinica. Rev Bras Farmacogn. 2017;27(2):251–253.

[225] da Silva JPL, de Souza EF, Della Modesta RC, Gomes IA, Freitas-Silva O,

de Melo Franco BDG. Antibacterial activity of nisin, oregano essential oil, EDTA, and their combination against Salmonella Enteritidis for application in mayonnaise. Vigilância Sanitária em Debate Soc Ciência Tecnol. 2016;4(1): 83–91.

[226] Khan AY, Latif Z. Screening of medicinal natural extracts for their antibacterial activity against Salmonella species. Pak J Bot. 2014;46(6):2269–2275.

[227] Andualem B. Combined antibacterial activity of stingless bee (Apis mellipodae) honey and garlic (Allium sativum) extracts against standard and clinical pathogenic bacteria. Asian Pac J Trop Biomed. 2013 Sep;3(9):725–731.

Chapter 12

# An Overview of *Salmonella* Biofilms and the Use of Bacteriocins and Bacteriophages as New Control Alternatives

*Alexandre Lamas, Patricia Regal, Laura Sanjulián, Aroa López-Santamarina, Carlos Manuel Franco and Alberto Cepeda*

## Abstract

*Salmonella* is a major food-borne pathogen around the world. In the European Union (EU), this pathogen is responsible of more than 90,000 human cases of salmonellosis every year. Salmonellosis in normally linked to the consumption of contaminated food, especially poultry products as meat, eggs and the products elaborated with them. Several control measures have been implemented in the EU to reduce the prevalence of *Salmonella* in the food chain. However, the ability of *Salmonella* to form biofilm along the food chain difficult its eradication. Also, ineffective cleaning and disinfection measures favors biofilm formation. The widespread use of biocides along the food chain has led to the emergence of resistant *Salmonella* strains. Therefore, it is necessary to look for alternatives to biocides to eradicate *Salmonella* biofilms. In this chapter we evaluate the use of bacteriocins and bacteriophages and their derivatives as a new alternative to eliminate *Salmonella* biofilms along the food chain.

**Keywords:** *Salmonella*, biofilms, control, bacteriocins, bacteriophages

## 1. Introduction

*Salmonella* genus is composed only by two species, *S. enterica* and *S. bongori* and more than 2600 different serotypes. *S. bongori* is composed of about 20 different serotypes and strains of this species are rarely isolated. Most of the serotypes belong to *S. enterica*. This species is subdivided in six different subspecies: *S. enterica* subsp. *enterica* (I), *S. enterica* subsp. *salamae* (II), *S. enterica* subsp. *arizonae* (IIIa), *S. enterica* subsp. *diarizonae* (IIIb), *S. enterica* subsp. *houtenae* (IV) and *S. enterica* subsp. *indica* (VI) (**Table 1**) [1]. The subspecies *enterica* attracts most of the attention of researchers as it is responsible for more than 99% of *Salmonella* infections in humans. Although the other *S. enterica* subspecies can also cause infections in humans, these infections tend to occur mainly in people with a very weakened immune system. The non-*enterica* subspecies of *Salmonella enterica* are usually isolated mainly from cold-blooded animals such as reptiles [2].

| Species | Subspecies | Number of serotypes |
|---|---|---|
| *S. enterica* | | 2637 |
| | *S. enterica* subspecies *enterica* | 1586 |
| | *S. enterica* subspecies *salamae* | 522 |
| | *S. enterica* subspecies *arizonae* | 102 |
| | *S. enterica* subspecies *diarizonae* | 338 |
| | *S. enterica* subspecies *houtenae* | 76 |
| | *S. enterica* subspecies *indica* | 13 |
| *S. bongori* | | 22 |

**Table 1.**
*Number of serotypes present in each* Salmonella *species and subspecies.*

*Salmonella* is important because it is one of the world's leading food-borne pathogens. In the European Union (EU), *Salmonella* is the second food-borne pathogen in number of human infections only behind the genus *Campylobacter*. In the year 2018, *Salmonella* was responsible of 91,857 human cases of salmonellosis and 119 deaths in the EU. Most infections are due to the consumption of food contaminated with *Salmonella* [3]. Thus, this pathogen can be isolated from different type of animals and their food derived products as bovine, porcine, ovine, fish or seafood [4–6]. But the largest number of human infections are related to the consumption of poultry products, especially meat and eggs as well as derived products [3]. As a consequence, the EU has developed legislation for member states to implement national control plans for *salmonella* in poultry production [7, 8]. The objective of this legislation is to reduce annually the prevalence of *Salmonella* in different types of farms including breeder farms, layer farms and broiler farms. Furthermore, this legislation also establishes that those serotypes that are of major epidemiological importance will be subject to special surveillance. For example, in broiler flocks *S.* Typhimurium and *S.* Enteritidis are subjected to this control. The ultimate goal of the European Union is for the combined prevalence of *S.* Typhimurium and *S.* Enteritidis to be less than 1% [9]. This due to these two serotypes are responsible of more than the 70% of human infections in the EU [3]. Cleaning and disinfection processes are of great importance to reduce the prevalence of *Salmonella* in the food chain. The implementation of inadequate control measures may result in *Salmonella* being able to resist in the food chain environment and contaminate different batches of food [10]. One of these bacterial resistance mechanisms is the formation of biofilms. For decades, biocidal substances such as quaternary ammoniums have been used to eliminate the presence of biofilms in the food industry [11]. However, the presence of multidrug-resistant strains is increasing [12]. This is a major concern as it may hinder the removal of biofilms from the food chain. Therefore, the development of alternative substances to combat food pathogen biofilms is necessary [13]. A brief description of *Salmonella* biofilms and the use of natural alternatives such as bacteriocins and bacteriophages to combat biofilms will be given throughout this chapter.

## 2. *Salmonella* biofilms

### 2.1 Basic concepts on biofilms

Costerton et al. [14] were the first researchers in stablish the term biofilm in paper published in *Scientific American* in 1978. They propose that most bacteria in aquatic ecosystems growth attached to surfaces in a closed self-produced matrix. Researchers also postulates that sessile cells (biofilm) differ from the planktonic cells (floating). It is important to note that the authors include the reference to aquatic environment because it was the first place where bacterial biofilms were observed. But, at present it is known that biofilms are the predominant style of life of bacterial in environment and its related with 80% of bacterial infections. Actually, biofilm is defined as a community of bacterial cells enclosed in a self-produced polymeric matrix and adhered to biotic (plant surfaces, epithelial cells, gallstones) or abiotic surfaces (plastic, rubber, glass, stainless steel). Biofilms have a great importance in the food production chain and human health because cells enclosed in this matrix are extremely difficult to eradicate because are more resistant to environmental stressors as antibiotics, disinfectants, host immune system [15–18].

There are four different steps of biofilm formation: 1) bacterial attachment, 2) microcolony formation, 3) bacterial maturation and 4) dispersion (**Figure 1**). The initial adhesion of bacterial cells is highly influenced by surface properties (roughness, hydrophobic interactions), environmental changes and bacterial regulation. Biofilm maturation and architecture is regulated by the signals of bacteria cells that compose biofilm and its stability depends on the accumulation of specific proteins, eDNA and polysaccharides. The presence of disruptive factors as proteases and nucleases and other enzymes activates biofilm dispersion. Factors as quorum sensing play an important role in this last step which function is the colonization of new niches [19].

### 2.2 Biofilm formation steps

#### *2.2.1 Adhesion*

*Salmonella* cells adhesion can be active or passive according the motility of bacteria or gravitational transport of planktonic cells. Both surfaces of bacterial cells and substrate surface highly influence the initial cell attachment. At this point bacterial cells have small quantities of extracellular polymeric substance (EPS) and maintain independent movement from other bacterial cells. Adhesion is reversible during this phase and cells do not present the morphological changes associated with biofilm cells and they can return to its planktonic state [16].

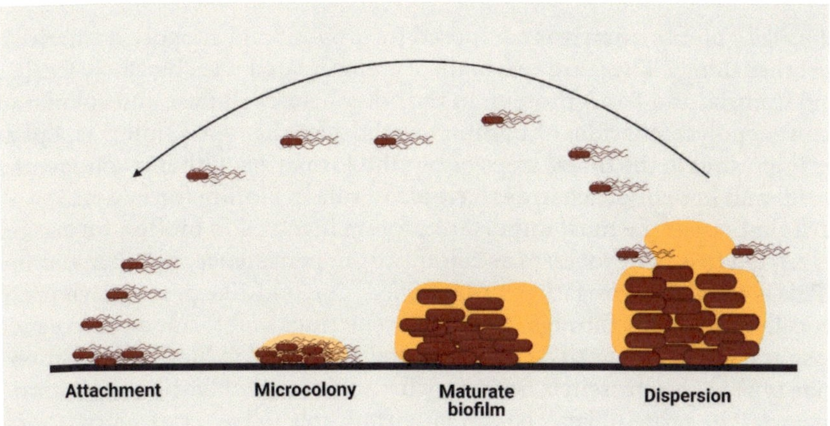


**Figure 1.**
*Steps involved in* Salmonella *biofilm formation. Created with biorender.com.*

*2.2.2 Irreversible adhesion*

The change from a weak interaction to a strong interaction between surface and bacterial cells is responsible to the switch from a reversible adhesion to an irreversible adhesion step. This change can happen in minutes and the production of EPS is key. The secretion of this polymeric substance by bacterial cells enhances the cell-surface interaction being necessary shear forces or chemical substances to break the adhesion [16, 20].

*2.2.3 Microcolony formation*

The formation of biofilm microcolony results from the accumulation of bacteria growth and the production and association with EPS. As a result, the bond between bacteria and substrate increases and protect bacteria from different environmental stressors. The cell-to-cell communication mechanism play an important role in this step of biofilm formation by regulating the expression of biofilm related genes. This results in an increased EPS production and caption of planktonic cells [21].

*2.2.4 Maturation*

The small microcolonies formed join to form the mature biofilm and its characteristic three-dimensional structure. The production of EPS and union between cells permits that mechanical pressure do not detach the biofilm from the surface. There are three different parts in mature biofilm. The bottom layer is a biofilm that forms a network structure that did not completely covers the surface that supports the biofilm. The intermediate layer is composed by a compact basement membrane. Finally, in the outer layer are located the planktonic cells [16].

*2.2.5 Dispersion*

The last step of biofilm formation is dispersion. In this phase the biofilm cells revert to their planktonic form. There are different factors that influences biofilm dispersion including external disturbance, starvation, endogenous enzymes, the release of EPS or surface binding proteins. This is an important step for the colonization of new niches by bacterial cells [22].

### 2.3 Structural components of *Salmonella* biofilms

*Salmonella* biofilm matrix is composed by proteins and exopolysaccharides among other things. There are two main proteins related with biofilms. Curli, an amyloid fimbria, and BapA protein. In the other hand, cellulose and colonic acid are the main exopolysaccharides of biofilm matrix. Also the type I fimbriae, Lpf and Pef are important in the initial steps of biofilm formation. Other components as fatty acids and lipopolysaccharides have also a role in biofilm formation.

Curli fimbriae is the most important protein involved in biofilm formation. Also is related to other processes as colonization, persistence, motility and invasion. This is a highly aggregative, unbranched, amyloid-like protein that promote cell-to-cell interactions through surfaces interactions and forms a complex with cellulose and O-capsule antigen. Other protein involved in biofilm formation is fimbriae type I. This protein is necessary for adhesion and biofilm formation in enterocytes. The protein BapA has an important role in bacterial aggregation and biofilm formation in air-liquid interface through homophilic interaction between bacterial cells [23–26].

Cellulose is the main polysaccharide involved in *Salmonella* biofilm formation. It is necessary for biofilm maturation phase in different surfaces, and it is inversely correlated with virulence as its production is suppressed in *Salmonella* enterocyte colonization phase. Another exopolysaccharide is the lipid bound O-antigenic capsule, with importance in resistance to desiccation and environmental persistence. This exopolysaccharide has demonstrated a role in biofilm formation in gallstones and plants but lower importance in adhesion to abiotic surfaces as glass or plastic. In other hand, cholinic acid is important for three-dimensional structure formation in enterocytes but not in abiotic surfaces, gallstones or alfalfa seeds. Therefore, some polysaccharides are only important for some types of biofilm formation [27–33].

Flagella, which are basic for cell movement and swarming in *Salmonella* also play a role in biofilm formation. In the initial step of reversible and irreversible adhesion, motility is important. Also, motility is necessary for 3D biofilm structure and the dispersion phase. But in other steps of biofilm formation the expression of flagella is inhibited. There is switch mechanism system that causes a reduction of flagella function and increased the expression of cellulose, resulting in the inhibition of flagellar rotation. This demonstrates the ambivalent role of flagella in biofilm formation. Fatty acids have also a role in *Salmonella* biofilm formation, especially in hydrophilic surface such as glass but not in hydrophobic surfaces as gallstones [34–36].

### 2.4 Genetic control of *Salmonella* biofilms

The change from a planktonic to a biofilm cell lifestyle needs some physiological changes. This switch is controlled by a complex genetic machinery that regulates the production of substances that conform the biofilm extracellular matrix, bacterial metabolism and the response to environmental signals. The transition between planktonic to biofilm cells and the expression of specific biofilm matrix-associated components is the master regulator of biofilm formation CsgD. It forms part of the operon that control the synthesis of curli fimbriae and acts as a transcriptional activator of the quorum sensing LuxR family. CsgD expression respond to different environmental signals as nutrient concentration, temperature, growth phase, oxygen tension, osmolarity, membrane integrity, tryptophan, and indole. CsgD positively regulates cellulose biosynthesis in *Salmonella* through direct stimulation of adrA transcription. AdrA synthetize c-di-GMP, a signaling molecule, that also activates the cellulose synthase BcsA, resulting in increased production of cellulose. Although it is the most important, there are other enzymes involved in cellulose synthesis [37–40].

RpoS and Crl are other important regulators of *Salmonella* biofilm formation regulating the expression of several components. Gene *rpoS* encodes a sigma factor called σS that regulates genes involved in stress response and stationary phase. It has been observed that almost the 25% of genes regulated with this sigma factor are overexpressed in biofilm cells of *S.* Typhimurium. For example, RpoS increases the expression of *csgD* and biofilm formation in environments with limited iron availability and regulate the expression of *adrA* in some steps of biofilm formation and is involved in the expression of genes related with motility. In other hand, the transcriptional regulator Crl protein regulates the activity of σS. RpoS and Crl have an effect in each other and their concentration are negatively correlated. Crl is necessary for maximal expression of *csgB, csgD* or *bcsA* and increased the expression of other genes related to RpoS. It is also remarkable that its effect are higher at 28°C than at 37°C. This indicates that this transcriptional regulator acts as a temperature sensor of *Salmonella* biofilm formation [41–44].

The bacterial messenger molecule c-di-GMP regulates several biological functions as virulence, motility, cell cycle regulation, differentiation, and biofilm formation. This molecule promotes *Salmonella* biofilm formation by regulating the production of some important components of biofilm matrix as cellulose and curli fimbriae. The c-di-GMP has a positive feedback on *csgD* expression. Thus, high levels of c-di-GMP increased the levels of CsgD, this increased the levels of AdrA and therefore c-di-GMP and cellulose synthesis [45, 46].

Other regulatory system implicated in motility and biofilm formation is the two-component system BarA/SirA. This system is modulated by factors as external pH, metabolic end products (formate, acetate), short chain fatty acids or bile salts. SirA modulates the *Salmonella* Csr system, an important regulator of motility, virulence, carbon storage, secondary metabolism and biofilm formation. CsrA control the change between sessile cells and motility, mainly activating motility. SirA activate the transcription of small RNAs CsrB and CsrC that inhibits CsrA activity and motility related genes. This increases type I fimbriae production and therefore biofilm formation [47, 48].

## 2.5 Quorum sensing

Another mechanism implicated in biofilm formation is Quorum sensing (QS). This is a cell-to-cell communication mechanism used by bacteria to adapt to environmental changes and implant a common bacterial strategy to respond to environmental stressors. QS is implicated in responsive defense against eukaryotic host cells, nutrient access, growth restriction environments, survive in hostile environments as well as cell differentiation to other form of life as biofilm cells. This communication is based in small molecules called autoinducers and that diffuse through bacterial membranes. Autoinducers are secreted at a basal level during bacterial growth. The concentration of this molecules increases with the growth of bacterial population until reach a threshold level and modulate the expression of QS target genes (**Figure 2**) [49, 50].

Gram-negative bacteria QS is divide into three categories: (i) N-acyl homoserine lactones (AHLs) called AI-1; (ii) furanosyl borate diester derived from the recycling of S-adenosyl-homocysteine to homocysteine called Autoinducer II (AI-2) for interspecies and intraspecies communication; and (iii) Autoinducer (AI-3) related to the recognition of host catecholamines epinephrine and norepinephrine. In the

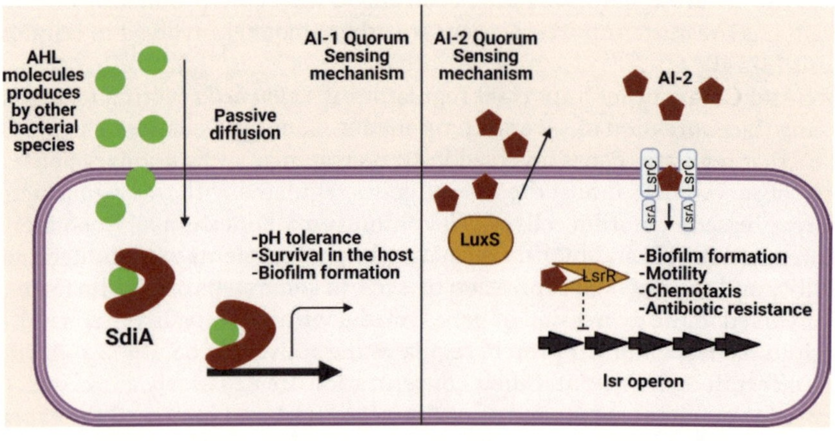


**Figure 2.**
*Schematic representation of quorum sensing mechanisms AI-1 and AI-2 in* Salmonella. *Created with biorender.com.*

case of *Salmonella*, only encode the receptor for AHLs but do not produce AI-1 molecules. But *Salmonella* can recognize AHLs produced by other bacterial genera as *Pseudomonas aeruginosa* or *Yersinia enterocolitica*. QS is basic in the formation of healthy biofilms and have a role in every stage. Genes regulated by the AI-1 receptor SdiA promote *Salmonella* cell adhesion and the production of extracellular proteins that compose biofilm matrix. Thus, *Salmonella* can response to the presence of AHLs molecules produced by other bacteria and increase biofilm formation. In the same way, AI-2 LuxS also can increase the expression of motility and biofilm related genes. Therefore QS is key a component of biofilm formation regulation [51–53].

## 3. Biofilms in the food industry

Nowadays, it is totally accepted that most bacteria grow in biofilm in the environment. Biofilms can have beneficial effects. For example, biofilm formation by *Lactobacillus* and *Lactococcus* results in more efficient fermentation processes and in the case of human health protect against the adhesion of pathogenic bacteria in the gut. But biofilm formation by undesirable bacteria, as food-borne pathogens, has a negative impact on food industry. Also, bacteria growing in biofilm can cause deterioration in the machinery as corrosion, efficiency reduction in heat transfer or clogging filters [54, 55].

Biofilms are a persistent source of contamination in the food industry. This cause hygiene and economic issues due to the spoilage of different food product batches with bacteria that persist in biofilms [56]. This is especially important in today's globalized world where food is globally distributed. Also, in the last years consumers demand fresh and minimally processed food products. Hygiene measures must therefore be strict to avoid contamination of food products. The presence of food-borne pathogen biofilms in the food processing environment can result in large number of food batches contaminated and outbreaks worldwide [57]. A good example was the salmonellosis outbreak caused by contamination of different batches of infant formula manufactured in a single facture and causing an outbreak that affected different countries around the world. Poor cleaning and disinfection procedures of food industry surfaces results in the presence of food residues that in the presence of humidity favors the development of bacterial biofilms as *Salmonella.* Cross-contamination occurs when food contact with surfaces with bacterial biofilms or also through aerosols from contaminated equipment. Until now, there is limited information of the real presence of *Salmonella* biofilms in the food processing environment. But *in vitro* studies have demonstrated that *Salmonella* can attach to different material commonly present in the food industry as plastic, glass, or stainless steel [57, 58].

Biofilm formation is influences but different factors as bacterial genus, species and even strains. But surface have a high influence on the ability of bacteria to adhere and form biofilm [59, 60]. Different type of material as stainless steel, glass, rubber, polystyrene and polyurethane, Teflon, nitrile and rarely wood are present in the food industry [61–63]. Physical properties have influence on biofilm formation, especially surface tension. Bacterial adhesion is favor by moist, energy free surfaces. Bacterial cells have better adherence to hydrophilic surfaces in comparison to hydrophobic surfaces. Surface roughness also influence cell adherence [57, 64]. In this sense, polished stainless steel showed less bacterial adherence than unpolished stainless steel [65]. Also, a study that compared stainless steel, glass and wood found that this latter surface favor biofilm formation because its porosity and ability to hold organic matter [66]. But also, surface influences biofilm formation in food industry. In this sense, welds, joints, corners or equipment design could enhance initial bacterial cell adherence [67]. But the presence of organic molecules on food industry surfaces is one of the major factors that influences biofilm formation. The presence of a layer of molecules as milk or meat proteins, EPS produced by other bacteria, favor the initial adhesion of bacterial cells. Diverse studies have observed that the presence of chicken juice macromolecules in stainless steel surfaces favor the initial adhesion of *C. jejuni* or *S.* Typhimurium. However, in some occasion macromolecules have the opposite effect. In this sense, an study observed that milk proteins reduced the initial adhesion of *L. monocytogenes* [68–70].

In the food production chain, there are different environmental conditions that can modulate *Salmonella* biofilm formation ability through modulation of initial adherence. Nutrient availability is one of these environmental conditions to which bacteria have to adapt. Under specific conditions, *Salmonella* has to persist under limited nutrient availability [71]. Biofilm formation is one strategy used for *Salmonella* cells to survive under this environmental stress conditions [72]. *In vitro* studies have demonstrated that *Salmonella* enhance a biofilm under limited nutrient conditions. These studies used common laboratory media as Tryptic Soy Broth or peptone water. These studies are a first approximation of the possible behavior of *Salmonella* under nutrient-limited conditions [71]. Temperature is another factor that changes through the food production chain. Several studies have demonstrated that *Salmonella* strains showed different biofilm formation amount under different temperatures tested. Interestingly, temperatures below 37°C and specially temperatures of 20°C favored *Salmonella* biofilm formation. The pH also influences *Salmonella* biofilm formation. A study that evaluated a total of 60 *S. enterica* strains under different pH, NaCl concentrations and temperature concluded that pH was the environmental factor that most influenced biofilm formation in *S. enterica* strains tested. This is probably due to the different ability of strains to adapt to acidic pH through an acid tolerance response mechanism [60, 73]. In the same way, another study found that weak acidic pHs (6) increased initial adhesion to stainless steel surfaces in comparison to neutral pHs. But curiously, acidic pHs reduced the number of cells present in mature biofilms due among other things to a lower presence of biofilm matrix components as polysaccharides and proteins [74]. Gene expression showed that acidic pH caused changes in the expression of virulence and biofilm related genes [75]. The environmental conditions under biofilms are formatted also influences its resistance to disinfectants. In this sense, biofilms formed under refrigeration temperatures showed higher sensitivity to disinfectants than those produced at 25°C under nutrient restriction as well as biofilm formed under acidic pH. In the other hand, mature biofilm are more resistant to substances such as quaternary ammonium compounds, peroxyacetic acid or organic acids. This is probably due a higher presence of matrix compounds as cellulose and curli fimbriae [76].

Although monospecies biofilm studies are interesting to understand the mechanism involved in biofilm formation under different environmental conditions of a specific bacteria, in nature biofilms are commonly composed by bacteria of different species and genera. These different bacteria communicate with each other through diverse mechanism as quorum sensing stablishing synergistic interactions that increase the resistance of biofilm to stressful environments. Also, genetic exchanges between different bacteria can occur in the biofilm environment [77]. This is specially interesting when resistance genes are transmitted. Dual biofilm studies are the first step to study multi-species biofilms. In this kind of studies, the biofilm formation ability of each bacterial group is studied individually, and then conjunct studies are carried out to determine the synergic mechanism stablished between the different groups [78]. In this sense, a study observed that *Salmonella* and *E. coli* mixed biofilms are more sensitive to disinfectants that biofilm of only one species [79]. In other hand, *S.* Enteritidis and *P. aeruginosa* mixed biofilms are more resistant to chlorine treatments [80]. In the same way, it was observed that mixed biofilms of *S.* Typhimurium and cultivable lettuce microorganism increased resistance to cold oxygen plasma treatments [81]. These studies provide a first clue of mixed biofilms. These studies are a first approach to multi-species studies. But undoubtedly the study of biofilms composed of hundreds of different bacterial genera will provide valuable information to fully understand how biofilms behave in nature. Such studies supported by genomics, metabolomics and high-resolution imaging will be the trend of the coming years in this field of microbiology.

## 4. Bacteriocins

### 4.1 Briefly definition and characteristics

Bacteriocins are defined as a group of ribosomally produced antimicrobial peptides synthesized by both Gram-positive and Gram-negative bacteria. These molecules are characterized by its ability to act against closely related bacteria (narrow spectrum) or a diverse group of bacteria (broad spectrum) [82]. Bacteriocins can be dived in two general groups: Class I composed by peptides with post translational modifications and Class II composed by unmodified peptides. The production of bacteriocins is considered as a competition mechanism that allows bacteria to kill other bacteria that can compete with it for a certain niche or for nutrients. This suggests that many bacterial groups produce at least one bacteriocin, which means that there are still many bacteriocins to be discovered [83, 84]. Bacteriocins have a great antimicrobial capacity against their targets at nanomolar concentrations and exerts its activity by membrane permeabilization [85].

In recent years these molecules have received much interest in general and in particular their application in the food chain. The main reason is the search for alternatives to antibiotics due to the emergence of antimicrobial resistance [86]. While the use of antibiotics to treat enteric pathogens can cause harm to commensal bacteria in the intestinal microbiota, narrow-spectrum bacteriocins can be used in such a way that only the target bacteria are affected by the treatment [86]. On the other hand, the bacteriocins can be used as modular of the intestinal microbiota. For example, they can be used to establish a microbiota that favors the fattening of the chickens and therefore as natural substitutes for antibiotics as growth promoters [87]. In addition, today consumers are demanding food products where the use of chemicals is reduced to a minimum, and natural alternatives such as bacteriocins would be welcomed. Finally, another advantage is that the bacteriocins can be used directly or bacteriocin-producing probiotic cultures can be used resulting in the production of these molecules in situ. This would eliminate the process of production and purification of bacteriocins making their application more economical. But bacteriocins can be also useful to inhibit and eradicate biofilm biofilms in the food production chain (**Figure 3**).

### 4.2 Applied studies on *Salmonella* biofilms

One of the first studies in this field, two concentrations of enterocin AS-48 (25 and 50 mg/L) produced by *Enterococcus* were tested in combation with antibiotics and biocides against four *Salmonella* strains [88]. Concentrations of 25 mg/L of bacteriocin in combination with antimicrobials highly inhibited the growth of *Salmonella*. This bacteriocin also have effects on sessile biofilm cells. Preformed biofilms were treated with different combinations of bacteriocin and antimicrobials. Enterocin AS-48 at 50 mg/L had a synergic effect in combination with some

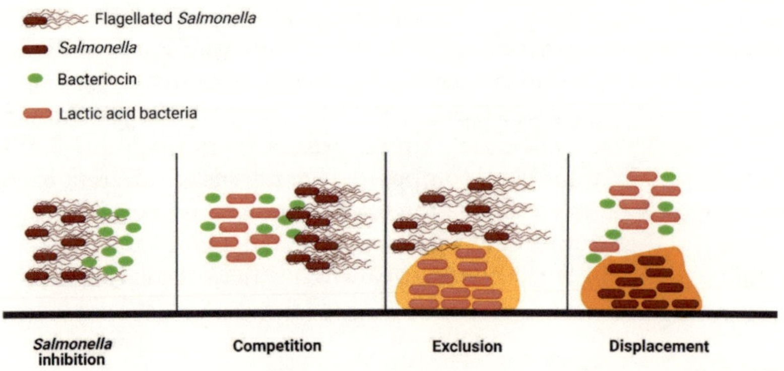


**Figure 3.**
*Mode of action of bacteriocin and bacteriocing-producing bacteria to inhibit and/or eradicate* Salmonella *biofilms.*

biocides. But the results differ between strain tested. In another study, Bag and Chattopadhyay [89] tested the antibiofilm activity of nisin alone or in combination with essential oil components. *S.* Typhimurium preformed biofilms were treated with MIC doses of nisin alone or in combination with ρ-coumaric acid. MIC doses of nisin only reduced in 20% biofilm formation. However, in combination with ρ-coumaric acid were reduced in almost 80%. This study demonstrated that nisin by itself have a low antibiofilm activity. Kim et al. [90] tested the crude bacteriocin DF01 derived from *Lactobacillus brevis* DF01 against *S.* Typhimurium biofilms. The incubation of this pathogen with bacteriocin DF01 reduced *S.* Typhimurium biofilm formation in almost 47%. However, the treatment of preformed biofilms with bacteriocin did not reduce biofilm mass. Therefore, the main action of DF01 bacteriocin is interfere in the biofilm formation process. In a similar study, Seo and Kang [91] evaluated the antibiofilm effect of bacteriocins purified from *Pediococcus acidilactici* K10 and HW01 in *S.* Typhimurium biofilm formed in stainless steel and chicken meat. Crystal violet staining method and fluorescence microscopy showed that those two bacteriocins reduces *S.* Typhimurium biofilm formation. In contrast to previous studies, this work demonstrates the ability of bacteriocins to also reduce the formation of biofilms in the food matrix itself.

In addition, instead of bacteriocins, the bacteriocin-producing bacteria themselves can also be used as alternative way to reduce *Salmonella* biofilm formation through competition, exclusion and displacement [92]. Das et al. observed that *L. plantarum* KSBT 56 isolated from Indian traditional food reduces in 2 log CFU/mL the cells present in *S.* Enteritidis biofilms [93]. Gómez et al. [94] used potential probiotic lactic acid bacteria (LAB) to inhibit the formation of food-borne pathogens biofilms. In this study they evaluated both bacteriocinogenic (sakacine and nisin producer strains) and non-bacteriocinogenic *Lactobacillus* and *Lactocococcus* strains against *S.* Typhimurium. The researchers preformed biofilms of LAB and after formation added a culture of *S.* Typhimurium. Preformed biofilms of LAB significantly reduced the attachment and biofilm formation of *Salmonella* in comparison to control. However, it is important to note that this reduction was not influenced by the production of bacteriocins. In another interesting study, the adhesion of food-borne pathogens as *S.* Typhimurium to wood commonly used in traditional cheese production in Sicilia was evaluated. The results showed that indigenous milk LAB highly adhere to wood surfaces while in samples artificially contaminated with *S.* Typhimurium, no adherence of this food-borne pathogen was observed. The researchers propose that biofilms formed by LAB in wood surfaces have a protective effect in biofilm formation by food-borne pathogens [95].

## 5. Bacteriophages and derived protein endolysin

### 5.1 Briefly definition and characteristics

Bacteriophages are viruses that infects bacterial cells with a high specificity. The life cycle of bacteriophages can be classified in two general categories: the lytic cycle (virulent) and the lysogenic cycle (temperate phage). In the lytic cycle the infection process starts with the irreversible attachment of the phage tail proteins to a receptor of the bacterial cell surface (protein or lipopolysaccharides). The ability of the bacteriophages to recognize and attach to molecules of the bacterial cell surface defines its host range. Once the phage DNA is in the host cell, specific enzymes are synthetized to drive host cell to the production of proteins necessary for the generation of new phage particles and cell lysis enzymes. At the end of the phage cycle, cell lysis, release of progeny phage and infection of neighboring susceptible cells occurs. Temperate phages combine its capacity to carried out the lytic cycle with the ability to persist as a prophage in the genome of the host cell and replicate with them. Diverse environmental signal can result in the prophage entering in the lytic cycle [96]. The use of temperate phages in medical and food applications is avoided because can cause transduction of genetic material between bacteria including virulence genes. In addition, due to its cycle, they do not kill all the bacteria that infect [97].

Lytic phages are those chosen for being used in phage therapy because they can replicate exponentially on bacterial culture and can eliminate multidrug resistant bacteria [86]. Based on their activity spectrum can be defined as monovalent phages when they are specific to one type of bacterial species and polyvalent phages when they are able to attack two or more bacterial species. But normally phages have a narrow host range, strains specific in most cases, and therefore cocktails composed by two or more phages are normally used to broaden the antimicrobial spectrum and reduce phage resistance [98].

Although bacteriophages have been known for over a century, the development of antibiotics resulted in their use not being explored in the Western world. However, the global problem of antimicrobial resistance and the need to seek alternatives has resulted in bacteriophages being brought back into the spotlight. Its applications in the food chain are very wide. They can be used for the treatment of bacterial diseases of production animals, for the disinfection of facilities and the elimination of biofilms or they can be added to food or packaging to inhibit the growth of food pathogens [86, 97]. In fact, there are different commercially available bacteriophage solutions to be applied to food or food processing facilities. Some examples are ListShield™, SalmoFresh™ and EcoShield PX™ commercialized by Intralytix or PhageGuard Listex and PhageGuard S commercialized by PhageGuard.

The bacteriophages synthetized at the end of the phage multiplication cycle peptidoglycan hydrolases commonly called endolysin. Its function is to lyse the host bacterial cell by directly target bonds in the bacterial cell wall peptidoglycan structure. This result in the degradation of the rigid murein layer and the release of newly assembled bacteriophage virions [99]. While endolysins can act as exolysins in the Gram-positive bacterial peptidoglycan layer, they cannot degrade the bacterial outer membrane of Gran-negative bacterial cells. Therefore, the outer membrane can prevent the access and the effect of endolysins [100]. For that reason, it is necessary to combine endolysins with other treatments for the lysis of Gram-negative bacteria. The combination of endolysin with outer membrane disruptors in one the main options for the application of enzymes in Gram-negative bacteria. Gram-positive phages endolysins have a modular structured formed by a cell-wall-binding domain that specifically recognizes the cell wall-associated ligand molecules and an enzymatically active domain that cleaves the peptidoglycan structure. Although gram-negative bacteriophage endolysins may also have this structure, they usually have a globular structure that only possesses a an enzymatically active domain [101, 102].One of the main advantages of the use of endolysins is that a very small amount of purified enzyme is enough to lyse in minutes or even seconds a dense suspension of bacterial cells. This in combination with their substrate specificity makes them have great potential for application in food science [103]. Endolysins are considered to be safe and also have some advantages compared to the use of bacteriophages because do not create gene transduction issues and therefore not contribute to the emerging problem of antimicrobial resistant bacteria [104]. Its applications in the food industry are very wide. They can be added directly to food, can be part of bioactive packaging or can even be used to remove biofilms in the food industry environment Furthermore, due to their specificity, they can be applied directly to treat intestinal infections in farm animals without causing alterations in the intestinal microbiota [103].

### 5.2 Applied studies on *Salmonella* biofilms

Tiwari et al. [105] tested a specific *S.* Enteritidis virulent phage called SE2 against planktonic and biofilm cells of an antimicrobial resistant *S.* Enteritidis strain. The phage showed a high bacteriolytic effect. This phage reduced in 4.2 log UFC/mL the count of *S.* Enteritidis after incubation of 4 h at 37°C and 2.5 log UFC/mL after incubation at 4°C. These results demonstrate that this phage can also be used effectively at refrigeration temperature. Also, biofilm studies showed that treatments with phage SE2 concentrations of $10^{11}$ PFU/mL reduced in 97% viable cells present in biofilms formed in glass. Also this phage showed that could maintain its activity at different ranges of pH and temperature. It has been also proposed that phage predation could increase biofilm formation by bacteria in some specific conditions. Hosseinidoust et al. [106] carried out and study to evaluate this theory in different pathogens including *S.* Typhimurium and to determinate if the increase of biofilm formation is due to the development of phage resistance or to non-evolutionary mechanism as spatial refuge. The results indicate that phage resistance was the mechanism implicated in increased biofilm formation in *P. aeruginosa.* However, in the case of *S.* Typhimurium it was due to non-evolutionary mechanisms [106]. Karaca et al. [107] evaluated the effect of phage P22 in *S.* Typhimurium biofilm formation in polystyrene and stainless-steel surfaces. The authors evaluated both the incubation of phage particles with *Salmonella* in biofilm studies and the treatment of preformed biofilms. *S.* Typhimurium biofilm formation was significantly reduced at high phage titer ($\leq 10^6$ PFU/mL). Also, all phage titers were effective against biofilm formation in 24 h incubation period but only higher phage titers were effective in 48–72 h incubation time. In addition, the ability to reduce biofilm formation was lower in polystyrene than in stainless steel. In the other hand, phage treatment was not effective in eradicating pre-formed *Salmonella* biofilms. This is probably due to the presence of extracellular matrix components that prevent bacteriophages from binding to specific receptors on the bacterial surface. In this sense, Yüksel et al. [108] combined phage P22 with EDTA and nisin to improve the antibiofilm activity of phage. The combination of the three inhibit in 93% *S.* Typhimurium biofilm formation at low phage titer concentrations but only reduced 70% mature biofilms. Therefore, the combination of phages with other antimicrobial substances could enhance antibiofilm activity. But it is still difficult to reduce biofilm in mature stages, when high quantities of extracellular matrix substances are present.

Garcia et al. [109] tested a cocktail of lytic bacteriophages biofilm to eradicate biofilms formed by different *Salmonella* serotypes in different surfaces (stainless steel, glass, and polyvinyl chloride) at short and long incubation times. Preformed biofilms were treated with $10^8$ PFU/mL during 3, 6 and 9 h. The results were not very promising and had a lot of variation between different surfaces and *salmonella* serotypes. In the same way, Gong et al. [110] tested different phage concentration ($10^4$–$10^8$ PFU/mL) against hard *Salmonella* biofilms formed in microtiter plates. Phages were selected based in its range activity against the different *Salmonella* serotypes included in the study. The reduction of biofilm formation was of 90% when *Salmonella* was incubated in combination with phages and 66% in pre-formed biofilms. Milho et al. [111] tested the phage PVP-SE2 against *S.* Enteritidis biofilms formed in food contact surfaces polystyrene and stainless steel. This phage caused reductions of 2 to 5 log CFU $cm^2$ at room temperature of 24 h and 48 h old *Salmonella* biofilms, showing its efficacy to control *S.* Enteritidis biofilms. Also, it was observed that this phage inhibited the growth of *S.* Enteritidis in poultry skin, even in freezing phage-pretreated poultry skin. The same research group evaluated the antibiofilm effect of phages in *E. coli* and *S. Enteritidis* dual-species biofilms [112]. The results of this study showed that phages were more effective to eradicate mono-species biofilms than dual-species biofilms. It is important to consider this when designing products that include phages to eradicate biofilms as biofilms in the food industry are often composed of various bacterial species. Kosznik-Kwasnicka et al. [113] evaluated three phages vB_SenM-1, vB_SenM-2, and vB_SenS-3 with lytic activity against different *Salmonella* serotypes. The phages were able to reduce biofilm cells and biomass in different strains tested and under different temperatures. This is important as there are different temperatures in the food chain and this study would indicate that phage treatment could be used over a wide temperature range. In the same way, Esmael et al. [114] tested to *S.* Typhimurium lytic phages against 72 h-old biofilms formed in microtiter plates. Concentrations of 8 log10 PFU/mL reduced more than three times biofilm formation. However, most of the studies conducted so far focus on specific *Salmonella* serotypes. One of the main characteristics of phages is their specificity. Thus, phages usually show activity against specific species, serotypes or even strains. This leads to a number of studies evaluating phage cocktails. Even so, it is difficult to find a phage cocktail effective against all *Salmonella* serotypes. This is one of the main problems to be solved with the use of phages in the food industry.

Using a food model, Sadekuzzaman et al. [115] evaluated the efficacy of 2 h bacteriophage treatment against *Salmonella* biofilms formed in lettuce surface. Although effective, phage treatment only reduced 1.0 log CFU/cm the count of *Salmonella*. Another alternative is the use of the active parts of the phages, for example the phage-encoded proteins. Altought some of the functions performed by proteins can be also performed by the phage itself, the use of proteins can have advantages in consumer acceptance and in terms of regulation. In this sense. Zhang et al. [116] tested endolysin LysSTG2 against *S.* Typhimurium biofilms. One hour treatment with 100 μg/mL of this endolysin, reduces 72 h biofilm in 13%. However, the combination of this endolysin with slightly acidic hypochlorous water containing 40 mg/L available chlorine reduces *S.* Typhimurium biofilm cells in 99%. Therefore, the combination of endolysin with other antimicrobial substances is a potential alternative against *Salmonella* biofilms.

## 6. Conclusion

*Salmonella* biofilm formation in the food production chain is a major public health problem. Mechanisms regulating biofilm formation in *Salmonella* are complex and is regulated by a wide range of environmental factors. The ability of

*Salmonella* to form biofilm in a wide temperature or pH range as well as in other stressful situations poses a major problem for its eradication. Also of concern is the increase of *Salmonella* strains with resistance to multiple biocides. Both bacteriocins and bacteriophages are a potential alternative to eliminate *Salmonella* biofilms. In addition, they can be combined synergistically with traditional antimicrobials, thus reducing the amount of antimicrobials used. One of the main limiting factors in its application is its range of activity. Normally bacteriocins and bacteriophages present a narrow spectrum of activity. They are therefore very useful for use against a specific pathogen. But in order to have a broad spectrum of activity to prevent different bacterial groups in the food chain, formulations combining a cocktail of bacteriocins and phages are needed. Studies evaluating such products as an alternative to traditional biocides are still limited, but future research and the use of recombinant technologies will make it possible to obtain products with high efficacy against *Salmonella* biofilms.

## Acknowledgements

Alexandre Lamas has a postdoctoral fellowship from Xunta de Galicia (Axudas de apoio á etapa de formación posdoutoral IN606B (Modalidade A)).

## Conflict of interest

The authors declare no conflict of interest.

## Author details

Alexandre Lamas*, Patricia Regal, Laura Sanjulián, Aroa López-Santamarina, Carlos Manuel Franco and Alberto Cepeda
LHICA, Department of Analytical Chemistry, Nutrition and Bromatology, School of Veterinary, Universidade de Santiago de Compostela, Lugo, Spain

*Address all correspondence to: alexandre.lamas@usc.es

## References

[1] Grimont PAD, Weill FX. Antigenic Formulae of Salmonella Serovars. Paris, France: 2007.

[2] Lamas A, Miranda JM, Regal P, Vázquez B, Franco CM, Cepeda A. A comprehensive review of non-enterica subspecies of Salmonella enterica. Microbiol Res. 2018;**206**:60-73.

[3] European Food Safety Authority and European Centre for Disease Prevention and Control (EFSA and ECDC). The European Union One Health 2018 Zoonoses Report. EFSA Journal. 2019;**17**:e05926.

[4] Xu Z, Wang M, Zhou C, Gu G, Liang J, Hou X, et al. Prevalence and antimicrobial resistance of retail-meat-borne Salmonella in southern China during the years 2009-2016: The diversity of contamination and the resistance evolution of multidrug-resistant isolates. Int J Food Microbiol. 2020;**333**.

[5] Pires AFA, Stover J, Kukielka E, Haghani V, Aminabadi P, de Melo Ramos T, et al. Salmonella and Escherichia coli Prevalence in Meat and Produce Sold at Farmers' Markets in Northern California. J Food Protection. 2020;**83**:1934-1940.

[6] Prabhakar P, Lekshmi M, Ammini P, Nayak BB, Kumar S. Salmonella contamination of seafood in landing centers and retail markets of Mumbai, India. J AOAC Int. 2020;**103**:1361-1365.

[7] European Commissiom. Regulation (EC) No 2160/2003 of the European Parliament and of the council of 17 November 2003 on the control of salmonella and other specified food-borne zoonotic agents. Official Journal of the European Union. 2003 12.12.;**L325**:1-15.

[8] European Commission. Commission regulation (EC) No 1177/2006of 1 August 2006implementing Regulation (EC) No 2160/2003 of the European Parliament and of the Council asregards requirements for the use of specific control methods in the framework of the nationalprogrammes for the control of salmonella in poultry. Official Journal of the European Union. 2006 2.8.;**L 212**:3-5.

[9] European Commission. Commission regulation (EU) No 200/2012 of 8 March 2012 concerning a Union target for the reduction of Salmonella enteritidis and Salmonella typhimurium in flocks of broilers, as provided for in Regulation (EC) No 2160/2003 of the European Parliament and of the Council. Official Journal of the European Union. 2012 9.3.;**L71**:31-6.

[10] Lamas A, Regal P, Vázquez B, Miranda JM, Cepeda A, Franco CM. Salmonella and Campylobacter biofilm formation: a comparative assessment from farm to fork. J Sci Food Agric. 2018;**98**:4014-4032.

[11] Galie S, García-Gutiérrez C, Miguélez EM, Villar CJ, Lombó F. Biofilms in the food industry: health aspects and control methods. Frontiers in microbiology. 2018;**9**:898.

[12] Cheng G, Ning J, Ahmed S, Huang J, Ullah R, An B, et al. Selection and dissemination of antimicrobial resistance in Agri-food production. Antimicrobial Resistance & Infection Control. 2019;**8**:1-13.

[13] Kumar M, Sarma DK, Shubham S, Kumawat M, Verma V, Nina PB, et al. Futuristic Non-antibiotic Therapies to Combat Antibiotic Resistance: A Review. Frontiers in Microbiology. 2021;**12**:16.

[14] Costerton JW, Geesey GG, Cheng K. How bacteria stick. Sci Am. 1978;**238**:86-95.

[15] Hall-Stoodley L, Costerton JW, Stoodley P. Bacterial biofilms: From the natural environment to infectious diseases. Nat Rev Microbiol. 2004;**2**:95-108.

[16] Stoodley P, Sauer K, Davies DG, Costerton JW. Biofilms as complex differentiated communities [Internet]; 2002 [cited 14 April 2021]. Available from: https://www.scopus.com/inward/record.uri?eid=2-s2.0-0036407233&doi=10.1146%2fannurev.micro.56.012302.160705&partnerID=40&md5=56270cb23f1b0956eb2ef33d086fc0e5.

[17] Stewart PS, Costerton JW. Antibiotic resistance of bacteria in biofilms. Lancet. 2001;**358**:135-138.

[18] Costerton JW, Stewart PS, Greenberg EP. Bacterial biofilms: A common cause of persistent infections. Science. 1999;**284**:1318-1322.

[19] Zhao X, Zhao F, Wang J, Zhong N. Biofilm formation and control strategies of foodborne pathogens: Food safety perspectives. RSC Adv. 2017;7: 36670-36683.

[20] Palmer J, Flint S, Brooks J. Bacterial cell attachment, the beginning of a biofilm. J Ind Microbiol Biotechnol. 2007;**34**:577-588.

[21] Donlan RM, Costerton JW. Biofilms: Survival mechanisms of clinically relevant microorganisms. Clin Microbiol Rev. 2002;**15**:167-193.

[22] Srey S, Jahid IK, Ha S. Biofilm formation in food industries: a food safety concern. Food control. 2013;**31**:572-585.

[23] Barnhart MM, Chapman MR. Curli biogenesis and function. Annu.Rev. Microbiol. 2006;**60**:131-147.

[24] White AP, Gibson DL, Collinson SK, Banser PA, Kay WW. Extracellular polysaccharides associated with thin aggregative fimbriae of Salmonella enterica serovar Enteritidis. J Bacteriol. 2003;**185**:5398-5407.

[25] Boddicker JD, Ledeboer NA, Jagnow J, Jones BD, Clegg S. Differential binding to and biofilm formation on, HEp-2 cells by Salmonella enterica serovar Typhimurium is dependent upon allelic variation in the fimH gene of the fim gene cluster. Mol Microbiol. 2002;**45**:1255-1265.

[26] Latasa C, Roux A, Toledo-Arana A, Ghigo J, Gamazo C, Penadés JR, et al. BapA, a large secreted protein required for biofilm formation and host colonization of Salmonella enterica serovar Enteritidis. Mol Microbiol. 2005;**58**:1322-1339.

[27] Römling U, Rohde M, Olsén A, Normark S, Reinköster J. AgfD, the checkpoint of multicellular and aggregative behaviour in Salmonella typhimurium regulates at least two independent pathways. Mol Microbiol. 2000;**36**:10-23.

[28] Prouty AM, Gunn JS. Comparative analysis of Salmonella enterica serovar Typhimurium biofilm formation on gallstones and on glass. Infect Immun. 2003;**71**:7154-7158.

[29] Barak JD, Jahn CE, Gibson DL, Charkowski AO. The role of cellulose and O-antigen capsule in the colonization of plants by Salmonella enterica. Mol Plant-Microbe Interact. 2007;**20**:1083-1091.

[30] Pontes MH, Lee E, Choi J, Groisman EA. Salmonella promotes virulence by repressing cellulose production. Proceedings of the National Academy of Sciences. 2015;**112**: 5183-5188.

[31] Snyder DS, Gibson D, Heiss C, Kay W, Azadi P. Structure of a capsular polysaccharide isolated from Salmonella

enteritidis. Carbohydr Res. 2006;**341**:2388-2397.

[32] Gibson DL, White AP, Snyder SD, Martin S, Heiss C, Azadi P, et al. Salmonella produces an O-antigen capsule regulated by AgfD and important for environmental persistence. J Bacteriol. 2006;**188**:7722-7730.

[33] Crawford RW, Gibson DL, Kay WW, Gunn JS. Identification of a bile-induced exopolysaccharide required for Salmonella biofilm formation on gallstone surfaces. Infect Immun. 2008;**76**:5341-5349.

[34] Verstraeten N, Braeken K, Debkumari B, Fauvart M, Fransaer J, Vermant J, et al. Living on a surface: swarming and biofilm formation. Trends Microbiol. 2008;**16**:496-506.

[35] Simm R, Ahmad I, Rhen M, Le Guyon S, Römling U. Regulation of biofilm formation in Salmonella enterica serovar Typhimurium. Future microbiology. 2014;**9**:1261-1282.

[36] Zorraquino V, García B, Latasa C, Echeverz M, Toledo-Arana A, Valle J, et al. Coordinated cyclic-di-GMP repression of Salmonella motility through YcgR and cellulose. J Bacteriol. 2013;**195**:417-428.

[37] Gerstel U, Römling U. The csgD promoter, a control unit for biofilm formation in Salmonella typhimurium. Res Microbiol. 2003;**154**:659-667.

[38] Römling U, Sierralta WD, Eriksson K, Normark S. Multicellular and aggregative behaviour of Salmonella typhimurium strains is controlled by mutations in the agfD promoter. Mol Microbiol. 1998;**28**:249-264.

[39] Hamilton S, Bongaerts RJ, Mulholland F, Cochrane B, Porter J, Lucchini S, et al. The transcriptional programme of Salmonella enterica serovar Typhimurium reveals a key role for tryptophan metabolism in biofilms. BMC Genomics. 2009;**10**:1-21.

[40] Gerstel U, Park C, Römling U. Complex regulation of csgD promoter activity by global regulatory proteins. Mol Microbiol. 2003;**49**:639-654.

[41] Römling U, Bian Z, Hammar M, Sierralta WD, Normark S. Curli fibers are highly conserved between Salmonella typhimurium and Escherichia coli with respect to operon structure and regulation. J Bacteriol. 1998;**180**:722-731.

[42] Gerstel U, Römling U. Oxygen tension and nutrient starvation are major signals that regulate agfD promoter activity and expression of the multicellular morphotype in Salmonella typhimurium. Environ Microbiol. 2001;**3**:638-648.

[43] Adams P, Fowler R, Kinsella N, Howell G, Farris M, Coote P, et al. Proteomic detection of PhoPQ-and acid-mediated repression of Salmonella motility. PROTEOMICS: International Edition. 2001;**1**:597-607.

[44] Robbe-Saule V, Lopes MD, Kolb A, Norel F. Physiological effects of Crl in Salmonella are modulated by σS level and promoter specificity. J Bacteriol. 2007;**189**:2976-2987.

[45] Sambanthamoorthy K, Luo C, Pattabiraman N, Feng X, Koestler B, Waters CM, et al. Identification of small molecules inhibiting diguanylate cyclases to control bacterial biofilm development. Biofouling. 2014;**30**:17-28.

[46] Liu Z, Niu H, Wu S, Huang R. CsgD regulatory network in a bacterial trait-altering biofilm formation. Emerging microbes & infections. 2014;**3**:1-5.

[47] Teplitski M, Al-Agely A, Ahmer BM. Contribution of the SirA regulon to

biofilm formation in Salmonella enterica serovar Typhimurium. Microbiology. 2006;**152**:3411-3424.

[48] Teplitski M, Goodier RI, Ahmer BM. Pathways leading from BarA/SirA to motility and virulence gene expression in Salmonella. J Bacteriol. 2003;**185**:7257-7265.

[49] Ng W, Bassler BL. Bacterial quorum-sensing network architectures. Annu Rev Genet. 2009;**43**:197-222.

[50] Atkinson S, Williams P. Quorum sensing and social networking in the microbial world. Journal of the Royal Society Interface. 2009;**6**:959-978.

[51] Smith D, Wang J, Swatton JE, Davenport P, Price B, Mikkelsen H, et al. Variations on a theme: diverse N-acyl homoserine lactone-mediated quorum sensing mechanisms in gram-negative bacteria. Sci Prog. 2006;**89**:167-211.

[52] Winzer K, HARDIE K WP. LuxS and autoinducer-2: their contribution to quorum. Adv Appl Microbiol. 2003;**53**:291.

[53] Dyszel JL, Smith JN, Lucas DE, Soares JA, Swearingen MC, Vross MA, et al. Salmonella enterica serovar Typhimurium can detect acyl homoserine lactone production by Yersinia enterocolitica in mice. J Bacteriol. 2010;**192**:29-37.

[54] Zhao T, Podtburg TC, Zhao P, Schmidt BE, Baker DA, Cords B, et al. Control of Listeria spp. by competitive-exclusion bacteria in floor drains of a poultry processing plant. Appl Environ Microbiol. 2006;**72**:3314-3320.

[55] Tang X, Flint SH, Bennett RJ, Brooks JD, Morton RH. Biofilm growth of individual and dual strains of Klebsiella oxytoca from the dairy industry on ultrafiltration membranes. Journal of industrial microbiology and Biotechnology. 2009;**36**:1491.

[56] Brooks JD, Flint SH. Biofilms in the food industry: problems and potential solutions. Int J Food Sci Tech. 2008;**43**:2163-2176.

[57] Shi X, Zhu X. Biofilm formation and food safety in food industries. Trends Food Sci Technol. 2009;**20**:407-413.

[58] Steenackers H, Hermans K, Vanderleyden J, De Keersmaecker SC. Salmonella biofilms: an overview on occurrence, structure, regulation and eradication. Food Res Int. 2012;**45**:502-531.

[59] De Oliveira, Débora Cristina Vidal, Fernandes Junior A, Kaneno R, Silva MG, Araujo Junior JP, Silva NCC, et al. Ability of Salmonella spp. to produce biofilm is dependent on temperature and surface material. Foodborne pathogens and disease. 2014;**11**:478-483.

[60] Lianou A, Koutsoumanis KP. Strain variability of the biofilm-forming ability of Salmonella enterica under various environmental conditions. Int J Food Microbiol. 2012;**160**:171-178.

[61] Chia T, Goulter RM, McMeekin T, Dykes GA, Fegan N. Attachment of different Salmonella serovars to materials commonly used in a poultry processing plant. Food Microbiol. 2009;**26**:853-859.

[62] Storgards E, Simola H, Sjöberg A, Wirtanen G. Hygiene of gasket materials used in food processing equipment part 1: new materials. Food Bioprod Process. 1999;**77**:137-145.

[63] Galinari É, Nóbrega JEd, Andrade NJd, Ferreira, Célia Lúcia de Luces Fortes. Microbiological aspects of the biofilm on wooden utensils used to make a Brazilian artisanal cheese. Brazilian J Microbiol. 2014;**45**:713-720.

[64] Bendinger B, Rijnaarts HH, Altendorf K, Zehnder AJ.

Physicochemical cell surface and adhesive properties of coryneform bacteria related to the presence and chain length of mycolic acids. Appl Environ Microbiol. 1993;**59**:3973-3977.

[65] Arnold JW, Boothe DH, Suzuki O, Bailey GW. Multiple imaging techniques demonstrate the manipulation of surfaces to reduce bacterial contamination and corrosion. J Microsc. 2004;**216**:215-221.

[66] Adetunji VO, Isola TO. Crystal violet binding assay for assessment of biofilm formation by Listeria monocytogenes and Listeria spp on wood, steel and glass surfaces. Global Vet. 2011;**6**:6-10.

[67] Guobjörnsdóttir B, Einarsson H, Thorkelsson G. Microbial adhesion to processing lines for fish fillets and cooked shrimp: Influence of stainless steel surface finish and presence of gram-negative bacteria on the attachment of listeria monocytogenes. Food Technol Biotechnol. 2005;**43** :55-61.

[68] Brown HL, Reuter M, Salt LJ, Cross KL, Betts RP, van Vliet, A. H. M. Chicken juice enhances surface attachment and biofilm formation of Campylobacter jejuni. Appl Environ Microbiol. 2014;**80**:7053-7060.

[69] Li J, Feng J, Ma L, de la Fuente Núñez, C., Gölz G, Lu X. Effects of meat juice on biofilm formation of Campylobacter and Salmonella. Int J Food Microbiol. 2017;**253**:20-28.

[70] Wong ACL. Biofilms in Food Processing Environments. J Dairy Sci. 1998;**81**:2765-2770.

[71] Speranza B, Corbo MR, Sinigaglia M. Effects of Nutritional and Environmental Conditions on Salmonella sp. Biofilm Formation. J Food Sci. 2011;**76**:M12-M16.

[72] Yang Y, Mikš-Krajnik M, Zheng Q, Lee S-, Lee S-, Yuk H-. Biofilm formation of Salmonella Enteritidis under food-related environmental stress conditions and its subsequent resistance to chlorine treatment. Food Microbiol. 2016;**54**:98-105.

[73] Lamas A, Fernandez-No IC, Miranda JM, Vázquez B, Cepeda A, Franco CM. Biofilm formation and morphotypes of Salmonella enterica subsp. arizonae differs from those of other Salmonella enterica subspecies in isolates from poultry houses. J Food Protection. 2016;**79**:1127-1134.

[74] Wang H, Wu N, Jiang Y, Ye K, Xu X, Zhou G. Response of long-term acid stress to biofilm formation of meat-related Salmonella Enteritidis. Food Control. 2016;**69**:214-220.

[75] O'Leary D, McCabe EM, McCusker MP, Martins M, Fanning S, Duffy G. Acid environments affect biofilm formation and gene expression in isolates of Salmonella enterica Typhimurium DT104. Int J Food Microbiol. 2015;**206**:7-16.

[76] Nguyen HDN, Yuk H-. Changes in resistance of Salmonella Typhimurium biofilms formed under various conditions to industrial sanitizers. Food Control. 2013;**29**:236-240.

[77] Tan CH, Lee KWK, Burmølle M, Kjelleberg S, Rice SA. All together now: experimental multispecies biofilm model systems. Environ Microbiol. 2017;**19**:42-53.

[78] Fagerlund A, Langsrud S, Møretrø T. Microbial diversity and ecology of biofilms in food industry environments associated with Listeria monocytogenes persistence. Curr Opin Food Sci. 2021;**37**:171-178.

[79] Chen D, Zhao T, Doyle MP. Single- and mixed-species biofilm formation by Escherichia coli O157: H7 and

Salmonella, and their sensitivity to levulinic acid plus sodium dodecyl sulfate. Food Control. 2015;**57**:48-53.

[80] Pang X, Yuk H-. Effect of Pseudomonas aeruginosa on the sanitizer sensitivity of Salmonella Enteritidis biofilm cells in chicken juice. Food Control. 2018;**86**:59-65.

[81] Jahid IK, Han N, Zhang C-, Ha S-. Mixed culture biofilms of salmonella typhimurium and cultivable indigenous microorganisms on lettuce show enhanced resistance of their sessile cells to cold oxygen plasma. Food Microbiol. 2015;**46**:383-394.

[82] Cotter PD, Hill C, Ross RP. Bacteriocins: developing innate immunity for food. Nature Reviews Microbiology. 2005;**3**:777-788.

[83] Riley MA, Wertz JE. Bacteriocins: evolution, ecology, and application. Annual Reviews in Microbiology. 2002;**56**:117-137.

[84] Yang S, Lin C, Sung CT, Fang J. Antibacterial activities of bacteriocins: application in foods and pharmaceuticals. Frontiers in microbiology. 2014;**5**:241.

[85] Jenssen H, Hamill P, Hancock RE. Peptide antimicrobial agents. Clin Microbiol Rev. 2006;**19**:491-511.

[86] Mills S, Ross RP, Hill C. Bacteriocins and bacteriophage; a narrow-minded approach to food and gut microbiology. FEMS Microbiol Rev. 2017;**41**:S129-S153.

[87] Umu ÖC, Bäuerl C, Oostindjer M, Pope PB, Hernandez PE, Perez-Martinez G, et al. The potential of class II bacteriocins to modify gut microbiota to improve host health. PLoS One. 2016;**11**.

[88] Grande Burgos MJ, Lucas López R, López Aguayo, M. D. C., Pérez Pulido R, Gálvez A. Inhibition of planktonic and sessile Salmonella enterica cells by combinations of enterocin AS-48, polymyxin B and biocides. Food Control. 2013;**30**:214-221.

[89] Bag A, Chattopadhyay RR. Synergistic antibacterial and antibiofilm efficacy of nisin in combination with p-coumaric acid against food-borne bacteria Bacillus cereus and Salmonella typhimurium. Lett Appl Microbiol. 2017;**65**:366-372.

[90] Kim N-, Kim WJ, Kang S-. Anti-biofilm effect of crude bacteriocin derived from Lactobacillus brevis DF01 on *Escherichia coli* and Salmonella Typhimurium. Food Control. 2019;**98**:274-80.

[91] Seo H-, Kang S-. Inhibitory effect of bacteriocin produced by Pediococcus acidilactici on the biofilm formation of Salmonella Typhimurium. Food Control. 2020;**117**.

[92] Woo J, Ahn J. Probiotic-mediated competition, exclusion and displacement in biofilm formation by food-borne pathogens. Lett Appl Microbiol. 2013;**56**:307-313.

[93] Das JK, Mishra D, Ray P, Tripathy P, Beuria TK, Singh N, et al. In vitro evaluation of anti-infective activity of a Lactobacillus plantarum strain against Salmonella enterica serovar Enteritidis. Gut Pathogens. 2013;**5**.

[94] Gómez NC, Ramiro JMP, Quecan BXV, de Melo Franco, B. D. G. Use of potential probiotic lactic acid bacteria (LAB) biofilms for the control of *Listeria monocytogenes*, Salmonella Typhimurium, and *Escherichia coli* O157: H7 biofilms formation. Front Microbiol. 2016;**7**.

[95] Cruciata M, Gaglio R, Scatassa ML, Sala G, Cardamone C, Palmeri M, et al. Formation and characterization of early bacterial biofilms on different wood

typologies applied in dairy production. Appl Environ Microbiol. 2018;**84**.

[96] Howard-Varona C, Hargreaves KR, Abedon ST, Sullivan MB. Lysogeny in nature: mechanisms, impact and ecology of temperate phages. The ISME journal. 2017;**11**:1511-1520.

[97] Endersen L, O'Mahony J, Hill C, Ross RP, McAuliffe O, Coffey A. Phage therapy in the food industry. Annual review of food science and technology. 2014;**5**:327-349.

[98] Ross A, Ward S, Hyman P. More is better: selecting for broad host range bacteriophages. Frontiers in microbiology. 2016;**7**:1352.

[99] Loessner MJ. Bacteriophage endolysinsâ€"current state of research and applications. Curr Opin Microbiol. 2005;**8**:480-487.

[100] Schmelcher M, Loessner MJ. Bacteriophage endolysins: applications for food safety. Curr Opin Biotechnol. 2016;**37**:76-87.

[101] Oliveira H, Melo LD, Santos SB, Nóbrega FL, Ferreira EC, Cerca N, et al. Molecular aspects and comparative genomics of bacteriophage endolysins. J Virol. 2013;**87**:4558-4570.

[102] Oliveira H, Azeredo J, Lavigne R, Kluskens LD. Bacteriophage endolysins as a response to emerging foodborne pathogens. Trends Food Sci Technol. 2012;**28**:103-115.

[103] Loessner MJ. Bacteriophage endolysinsâ€"current state of research and applications. Curr Opin Microbiol. 2005;**8**:480-487.

[104] Bakhshinejad B, Sadeghizadeh M. Bacteriophages as vehicles for gene delivery into mammalian cells: prospects and problems. Expert opinion on drug delivery. 2014;**11**:1561-1574.

[105] Tiwari BR, Kim S, Kim J. A virulent Salmonella enterica serovar Enteritidis phage SE2 with a strong bacteriolytic activity of planktonic and biofilmed cells. Journal of Bacteriology and Virology. 2013;**43**:186-194.

[106] Hosseinidoust Z, Tufenkji N, van de Ven, Theo GM. Formation of biofilms under phage predation: considerations concerning a biofilm increase. Biofouling. 2013;**29**:457-468.

[107] Karaca B, Akcelik N, Akcelik M. Effects of P22 bacteriophage on Salmonella enterica subsp. enterica serovar Typhimurium DMC4 strain biofilm formation and eradication. Archives of Biological Sciences. 2015;**67**:1361-1367.

[108] Yüksel FN, Buzrul S, Akçelik M, Akçelik N. Inhibition and eradication of Salmonella Typhimurium biofilm using P22 bacteriophage, EDTA and nisin. Biofouling. 2018;**34**:1046-1054.

[109] Garcia, Keila Carolina de Ornellas Dutka, de Oliveira Corrêa, Isadora Mainieri, Pereira LQ, Silva TM, Mioni, Mateus de Souza Ribeiro, de Moraes Izidoro, Ana Carolina, et al. Bacteriophage use to control Salmonella biofilm on surfaces present in chicken slaughterhouses. Poult Sci. 2017;**96**:3392-8.

[110] Gong C, Jiang X. Application of bacteriophages to reduce Salmonella attachment and biofilms on hard surfaces. Poult Sci. 2017;**96**:1838-1848.

[111] Milho C, Silva MD, Melo L, Santos S, Azeredo J, Sillankorva S. Control of Salmonella Enteritidis on food contact surfaces with bacteriophage PVP-SE2. Biofouling. 2018;**34**:753-768.

[112] Milho C, Silva MD, Alves D, Oliveira H, Sousa C, Pastrana LM, et al. Escherichia coli and Salmonella Enteritidis dual-species biofilms:

interspecies interactions and antibiofilm efficacy of phages. Scientific reports. 2019;**9**:1-15.

[113] Kosznik-Kwaśnicka K, Ciemińska K, Grabski M, Grabowski Ł, Górniak M, Jurczak-Kurek A, et al. Characteristics of a series of three bacteriophages infecting Salmonella enterica strains. International journal of molecular sciences. 2020;**21**:6152.

[114] Esmael A, Azab E, Gobouri AA, Nasr-Eldin MA, Moustafa M, Mohamed SA, et al. Isolation and characterization of two lytic bacteriophages infecting a multi-drug resistant Salmonella Typhimurium and their efficacy to combat salmonellosis in ready-to-use foods. Microorganisms. 2021;**9**:423.

[115] Sadekuzzaman M, Mizan MFR, Yang S, Kim H, Ha S. Application of bacteriophages for the inactivation of Salmonella spp. in biofilms. Food Sci Technol Int. 2018;**24**:424-433.

[116] Zhang Y, Huang H, Duc HM, Masuda Y, Honjoh K, Miyamoto T. Endolysin LysSTG2: Characterization and application to control Salmonella Typhimurium biofilm alone and in combination with slightly acidic hypochlorous water. Food Microbiol. 2021:103791.